NEW WORLDS

Discoveries from Our Solar System

NEW WORLDS

Discoveries from Our Solar System

Wernher von Braun and Frederick I. Ordway III

with editorial assistance and preface by Eric Burgess

Anchor Press/Doubleday
Garden City, New York   1979

Library of Congress Cataloging in Publication Data

Von Braun, Wernher, 1912–1977.
New Worlds.

1. Solar system.   I. Ordway, Frederick Ira,
1927–   joint author.   II. Title.
QB501.V66 1979   523.2
ISBN: 0-385-14065-7
Library of Congress Catalog Card Number 78–55854

CONTENTS

FOREWORD

No two people could be better qualified to survey what we have learned about our Solar System during the opening decades of the Space Age than Wernher von Braun and Frederick I. Ordway III. I knew and worked closely with them for at least twenty years and thus appreciate at first hand their intimate involvement in America's space endeavor from its genesis back in the 1950s to its flowering in the '60s and early '70s.

Wernher von Braun will almost certainly go down in history as the father of the incredibly complex and brilliantly successful Apollo/Saturn manned lunar landing program. Frederick I. Ordway III, a graduate of Harvard and author of some thirty books and hundreds of articles on rocketry and space travel, was a close and admired friend of Von Braun's and with him shared a deep interest and involvement in the space program for over twenty-five years. They have collaborated on many works in the past.

New Worlds represents the last major literary effort undertaken by Von Braun before his death over a year ago. In it, the authors set the stage by placing the Solar System within the cosmic whole and then proceed to examine how the Sun, the planets and satellites, comets, and meteors may have come into being. They then guide us throughout the Solar System, beginning with the Sun and the interplanetary environment and continuing outward to Mercury, Venus, the Earth/Moon pair, Mars, the asteroids, Jupiter and the other giant planets, and finally Pluto. Much of their research was based on the vast Wernher von Braun and Frederick I. Ordway collections

housed at the Alabama Space and Rocket Center in Huntsville, Alabama. Although earthbound, both men share with us in *New Worlds* their feelings, findings, and prospects for what's out there at and beyond the space frontier.

Von Braun was a great man of vision, a child of the future. Despite his fascination, so evident in this book, with the Moon and planets, he came more and more to realize the significance and the beauty of the world on which we live. He once wrote, "From outer space our home planet appears in a new perspective. Our firm and familiar Earth becomes a beautiful but exceedingly tenuous blue and white globe. National, racial, and linguistic boundaries fade away and what remains is a picture of an unpowered, unskippered spaceship which, equipped with an exceedingly vulnerable life support system, carries three and one half billion astronauts through a lonely, black, star-spangled infinity to an unknown destination."

While the destination of the Solar System remains unknown, man's conception of the universe has expanded incredibly since the opening years of the space age two short decades ago. *New Worlds* serves as a telescope that permits us to peek through and get a glimpse of what we have learned about our companions in space.

Edward O. Buckbee
Curator, Wernher von Braun Collection
Director, Alabama Space and Rocket Center
Huntsville, Alabama

August 1978

PREFACE

There are some people who can be categorized only as optimists, incurable optimists. The authors of this book fall into such category. It was because they do that I was extremely grateful to be invited to participate in helping this book along to publication. I believe it is the incurable optimists who fashion the future and that their ideas should be made available to the greatest possible readership.

I first met Fred Ordway in Paris in 1950, at a time when a mere handful of people were forming a now prestigious international organization of interplanetary enthusiasts to push for an exploration of the Solar System. In the ensuing twenty-eight years, mankind's understanding of the other worlds of our system has completely changed. And just as expansion from village to town and from country to continent broadened human horizons and reflected back into social and technological improvements for those who stayed at home, so the expansion into space is already providing us with a different view of ourselves and the planet on which we originated.

It was ten years after the Paris congress when I met Wernher von Braun. The place was Los Angeles, where I chaired a banquet of the American Astronautical Society and he was a speaker. I had followed his career, of course. Most of us interplanetarians knew about each other's work. Even during World War II the enthusiasts from the years before the war kept tabs as best we could on where everyone was and what they were doing. Naturally there was also much speculation on the progress being made behind the screens of military secrecy. I recollect also how I scanned intelligence reports when I was in the British Royal Air Force and marveled at the naïveté of some of the comments in them about German rocket development. They were so obviously written by people who had not been aware of or acquainted with the prewar activities of the rocket and interplanetary groups, or of their "crazy" schemes to place people in space in permanent space stations, land men on the Moon, and send unmanned spacecraft on missions to other planets.

Often, the optimists are thought of as skilled and enthusiastic engineers involved solely in the engineering challenge of making a big, new machine that works. Anyone who has witnessed the awesome spectacle of a Saturn 5 climbing into the night sky and has felt the Earth-shattering roar of its exhaust as it sends men to the Moon might well regard such an engineering feat as an end in itself, a demonstration of man's control over nature even if in a mechanistic sense.

The size of the Saturn 5 was almost unimaginable. At the National Air and Space Museum, in Washington, D.C., a visitor today can stand and look at the business end of this mighty rocket vehicle. But the real vehicle was too large to be accommodated easily within the museum, so the end of the rocket has to be shown through a system of mirrors to reflect the big rocket engines. Imagine the courage needed for an engineer to attempt the task of scaling up from a rocket motor the size of an egg that would function for only a few minutes before exploding in a shower of molten metal, to an engine bigger than a large automobile, and then coupling five of those engines to work smoothly together.

But beyond the engineering and the engines was always the dream of space exploration, of a permanent manned station in space, of people walking on the Moon and later on Mars—fascinating and stimulating projects that Wernher von Braun did much to popularize in the early 1950s.

The engineers who make such feats possible impressed me with their keen perception of much beyond mere rocket vehicles. I saw dreamers who lived their dreams, explorers who developed advanced tools to complete their explorations, visionaries who looked beyond the pain and disharmony of a world of conflicting ideologies toward a promising future of great achievements. I saw unique individuals who knew that the future of mankind holds infinite promise, that limitations are of purely human making, and that we can reach for and attain the stars if we wish. Wernher von Braun represented all these impressions. To him the rocket was only a tool and not an end in itself; it was a tool that would be replaced by other tools as they were invented and perfected by a new generation of engineers that would follow him. For Von Braun led the way and developed the first spaceship.

As a result, in a little more than twenty years we came to know more about Mars, Venus, and Mercury than we had known about the Moon before the space age. We had taken a close look at Jupiter, too, and found it to be so interesting that other spacecraft were sent to inspect the giant planet and its family of big satellites more closely. These spacecraft would also be sent to examine the ringed planet Saturn and its satellites, and possibly Uranus.

Exploring the planets by spacecraft brought important side effects. There was a surge of professional interest in these bodies, our neighbor worlds, where for many decades their investigation had been left to bands of enthusiastic amateurs. Not since the hundred years or so following the invention of the telescope had the Solar System and its worlds been studied with such enthusiasm. Earth-based observations produced many startling results, discoveries that in many respects rivaled those made by the spacecraft. Rings were

found around the planet Uranus. Additional satellites were discovered. A strange object orbiting the Sun beyond Saturn hinted at possibilities of another belt of asteroids in the outer Solar System. Clouds of sodium were discovered in space around Jupiter associated with one of its satellites, Io. Storms and magnetic fields of the Sun were investigated in unprecedented detail. New instruments were developed to link with telescopes, and new computing techniques were devised to interpret the wealth of new observations. A golden age in astronomy blossomed, and stimulated many new students and an upsurge of amateur interest that revitalized many of the amateur societies.

Even more important, when people generally started to look outward with renewed awe and wonder at the universe in which they live, they also looked more closely at their own planet and themselves and realized more fully the precarious balance between technology and the natural processes of planet Earth and its biosphere. A new awareness of Earth as a unique planet dawned, and we accepted a responsibility to our environment. This was the first time that mankind had accepted such a responsibility. In the past the story has always been one of exploitation without regard to the consequences.

We may thus be at a unique point of history, the point at which life on Earth can consciously control its future on a planetary scale, and having mastered the problems of living harmoniously with the environment of this planet, may transfer life to other worlds of the Solar System and beyond in a process of making the universe aware of itself. Thus we may be witnessing today the initial stages of a transformation from terrestrial mankind to interplanetary mankind, from adolescence to adulthood.

Space-age thinking has affected many of our activities. It has triggered enormous changes, many of them most subtle and not easily related directly to spaceflight. There has been a stirring of the human consciousness as people everywhere see technical miracles performed and the impossible achieved. If we can land on the Moon, why can't we . . . ? became an increasingly common question. And the question remains. Why can't we? Since Sputnik I girded the Earth in eighty minutes, beeping its radio message to amateurs around the world as it streaked through the familiar constellations, a new generation has been born. This new generation has the opportunity to supply the answer. Tradition, fashionable thinking, stereotyping—all have to be overcome to solve the social and supply problems of today.

The breakthroughs are invariably made by young minds. The aged savants are too steeped in habits of a lifetime, too active in safeguarding their image with their contemporaries, to make the quantum jumps in thinking that are needed to lead mankind into a new future of interplanetary travel following our interplanetary exploration. The hope is that this new breed of children of the space age will as fearlessly break with traditional thinking as did the Von Brauns of their age. They can then be the individuals who will drive the golden spikes into our trackless railroads of interplanetary trajectories. In the universities of today there are students who will surpass the deeds of those visionary optimists of the past. In our underprivileged youths there are those who will surmount difficulties and emerge as leaders despite their disadvantages. Such people will lead the future Lewis and Clark expeditions across the Mississippis of space.

One of the important lessons learned from Apollo—possibly the greatest technological endeavor in the history of civilization—has been pointed out by Fred Ordway elsewhere. It is that such major technological feats cannot be accomplished by any single group alone. The support of this type of human endeavor has to come from a public majority. John Q. Citizen has to support the effort with tax dollars if it is to succeed. And this requires an informed citizenry who understand what is being done, why, and how, and what may result. For many years Fred Ordway has stimulated public interest in and the management of new applications of science and technology at the frontiers of human knowledge. In this book he has teamed with Von Braun to help disseminate the important discoveries of the space age about our Solar System.

At present the data collected by satellites and aircraft about our own planet are growing at the rate of 100 billion bits of information every day. Much of this information is of direct use in investigating the weather, climate, land use, water, and other natural resources of our Earth, and it is available to anyone who wants to go to the trouble of seeking it out. Enormous amounts of information are also being amassed about the other planets for comparison with Earth. The impor-

tance of our expansion into space has not been solely that of greater understanding of the other planets, though that would have been laudable and worthwhile in itself. The exploration has also brought tremendous increase in our understanding of Earth. We now have a broader view, greater insight, greater appreciation of our own planet's uniqueness. Many scientists confidently predict that the wealth of new information being collected about the other worlds will enable us to understand more completely why our planet is unique, why it nurtured life, and why we are here to think and to write about it.

In researching and writing this book about our Solar System, Wernher von Braun and Frederick Ordway expressed their desire to show what twenty years of space exploration has meant to our increasing comprehension of the universe around us.

If one day we send probes to the remote planet Pluto, we will have looked closely at all the known planets of the Solar System, physically repeating the earlier work of astronomers who had to do their probing through telescopes based on Earth. In just over twelve human generations since Galileo first looked at the planets through the newly invented telescope, we have explored at first hand all the planets of which he and other people of his time were aware. But the endless quest for understanding has only just begun. Out beyond Pluto the diamond points of stars beckon from the blackness of interstellar space. From tiny rocket engines the size of hen's eggs tested at the Raketenflugplatz, in a suburb of Berlin, in the 1930s, through the V-2 and the Redstone and the Jupiter, to the mighty Saturn 5 that hurtled our astronauts to the Moon, we have seen the development of the enormous capability of men and women of Earth. They have been shown that they can successfully reach for the stars. Whether or not we continue this fantastic progress that has been wrought during our lifetimes cannot be foretold. In a small way, each of us has helped mankind reach out from Earth to a new dimension of space. Now we can see a future of electric rockets carried into space in the great new transportation system of the space shuttle; we can see ion drives, solar sails, and perhaps photon drives and interstellar ramjets.

The possibilities of expanding beyond our Solar System seem today almost as tantalizing to the human intellect as were the possibilities of reaching for the Moon when the authors began their careers. The question is not so much one of technology but of purpose. Where do we want to go? What is our purpose? Is it to remain on Earth and accept the limitations of a finite planet? Or is it to expand into the virtually unlimited commons of the Solar System and thence to other star systems?

Already we are thinking in terms of extraterrestrial intelligences, of a cosmic community. We are proposing listening stations consisting of huge systems of mighty antennas to try to amplify the whispers of our future from the stars themselves. If there are intelligent beings living on planets of nearby star systems we might thereby hear their communications and find out how they solved their problems. And if we search thoroughly the surfaces of the planets of our Solar System, there is always the remote possibility that we may discover evidence of unmanned probes sent to our system by alien life forms just as we have sent several of our probes out among the stars. That is why, among many other reasons, a manned mission to Mars, as envisioned by Von Braun so many years ago, could be very important.

The people who urged mankind into space through living their dreams are now reaching the stage in their lives when they have to pass the torch to others of their kind. The dreamers of today, children of the space age, are just beginning their professional careers. They, too, can live their dreams. And if they accomplish during their careers as much as the rocket and interplanetary enthusiasts of the 1930s did during theirs, despite a world war that intervened, then who can really predict where mankind might be in the twenty-first century?

The message from this book to those young professionals is that the future is unlimited. The opportunities need only be grasped. The Solar System has started to receive us, and a whole universe awaits us, ready to reveal its secrets if we have the will and the stamina to search them out. What could be more exciting? What could be more challenging? This really is the golden age of unprecedented opportunity.

Eric Burgess
Sebastopol, California
March 1978

AUTHOR'S NOTE

This book is dedicated to the memory of my friend of twenty-five years, mentor, collaborator, and coauthor, Wernher von Braun. When writing got under way, several years ago, we had hoped to complete the manuscript in time for publication on the twentieth anniversary of the orbiting of Explorer 1, the first U.S. artificial satellite. Our aim was to survey knowledge of the Solar System accumulated during the two first decades of the space age.

Obviously, we did not make the January 1978 target. In fact, we are more than a year behind our original schedule. When we completed some three quarters of the manuscript, Von Braun's health had deteriorated to the point that only with great difficulty could he continue his work. Nevertheless, he persisted almost to the end. Once, over the phone from the hospital, he told me he had just finished work on a chapter while lying flat on his back.

I went on to complete the first draft alone. Then, to accelerate the publication of the book after Von Braun's untimely death, in June 1977, our close friend Eric Burgess generously offered his expert editorial assistance. Skillfully and rapidly, he polished the manuscript and prepared the Foreword. The result is this final collaborative effort with Wernher von Braun, offered to those who share his fascination with the Solar System and his long-term conviction that the rocket-boosted spacecraft would play a crucial role in unraveling its mysteries.

Frederick I. Ordway
Washington, D.C., and Huntsville, Alabama

September 1978

NEW WORLDS

Discoveries from Our Solar System

1
THE SOLAR SYSTEM AND THE UNIVERSE

AND IN THE BEGINNING?

An immense cloud of dust and gas, dark, almost inert, stirred slightly under a gentle urging of magnetic and electric forces. From the distant stars, a faint light illuminated gently moving wraiths of cosmic dust and diffuse groups of leisurely cruising atoms riding the rivers of the electromagnetic fields. Across the billions of miles of this nebula, only a few of these atoms ever met with each other, for mostly they pursued lonely paths. Infrequently, they greeted each other with an exchange of electrons and became bound into simple molecules. And even less frequently, they bumped into grains of dust, sometimes to stick fast to them.

There was little order and no organized motion, only the aimless wanderings of atoms, molecules, and dust.

A casual observer might have imagined that here was tranquillity, here the ultimate region of peace, destined to remain so for eternity.

Then the light came, searing, intense, in unimaginable brightness. There was no warning before every atom, every speck of dust became incandescent in the reflected glory of a tidal wave of brilliance that surged across the cloud. Within fifteen hours of a time that had not yet been invented the wave of photons flashed across the nebula from end to end. Electrons jumped to new orbits in countless atoms. Molecules flew apart and new molecules fused together. Dust particles recoiled under the shock of the energetic photons

and trailed along behind the wave of brilliance like flecks of spray behind a breaker. And simultaneously, a dark wave of neutrons cut through the cosmic cloud like an army of surgeons operating on the nuclei of atoms. They chopped and changed and rebuilt the nuclei to produce mutant elements, atoms that presented the same face of chemistry to the world around them but were of different weights. And cosmic rays, nuclei of other atoms moving almost at the speed of light, arrived shortly afterward, weaving more new and unusual isotopes of the atoms within the gas and dust of this primordial nebula. Five billion years later, these isotopes would allow intelligences, on a planet yet to be born, to re-create creation in minds built from these clouds of gas and cosmic dust.

Somewhere not too far away from the cloud, two gyrating close stars had exchanged matter in a fantastic burst of energy. The resulting nova, or new star, had sprayed the surrounding light-years of space with immense surges of radiation—with light, X rays, gamma rays, neutrons, and the cosmic-ray bullets of speeding atomic nuclei. The gyrations of these two stars had conceived a solar system, our Solar System.

The act of creation sent that languid cloud of gas and cosmic dust along an evolutionary path that led to the spawning of consciousness within its organized matter, an intelligence that first questioned, then searched and began to find answers. We are that consciousness. Today we believe that we understand, at least in part, the answer to the question: And in the beginning?

For the energy imparted to that gas cloud some 4–5 billion years ago stimulated it into activity and started a gestation that ultimately led to the birth of a nuclear-powered Sun and a retinue of planets. On one of those planets conditions favored a continuing chemical evolution: complex molecules of carbon, hydrogen, and nitrogen developed miraculous qualities of being able to produce copies of themselves and, even more miraculously, to produce improved models of greater and greater complexity until they developed the capability of being aware of their own existence and to question where they came from and where they might be going.

From the quietness of that incredibly ancient nebula to the appreciation of the wondrous activity of today's Solar System took 4.5 billion years. If one year is represented by the thickness of a page of this book, 4.5 billion years would have to be represented by a stack of these pages 250 miles (400 kilometers) high. Now all is motion. Yet this motion depends very much upon viewpoint—your viewpoint.

If you go outdoors and look at the stars on a clear, moonless night, away from the glare of city lamps, the crisp, diamond points strike you as being of the very essence of stability—fixed, unmoving. But continue to watch the night sky and you might see the quick, streaking flash of a meteor breaking the serenity of the stars. And if you keep on watching you will notice that the stars themselves seem to move as a body around the sky, circling the Earth once each day.

After weeks of watching, you will see that some "stars" move among the others. These are the (planetary) "wanderers," the worlds that resulted from the agitation of the primordial nebula. Much longer observations will reveal that the stars themselves move relative to each other.

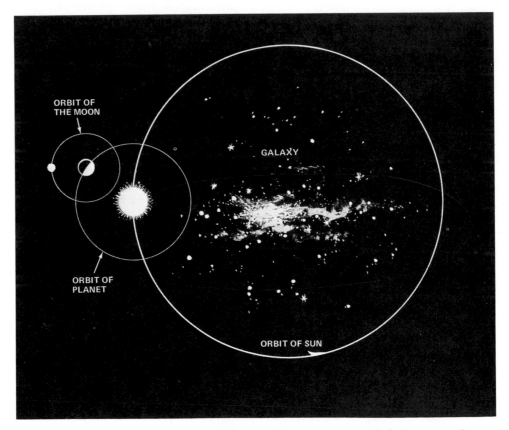

In the universe, everything appears to be in orbit—moons around their parent planets; planets around their parent suns, suns around the centers of their respective galaxies, and galaxies around the center of attraction of the galactic family to which they belong. In our Solar System, asteroids, comets, meteoroids, and micrometeoroids also travel in orbits around our Sun. (Arfor Picture Archives)

In fact, all is motion. The universe is not still but is dynamically active. Much of this action we cannot see in the brief time we are permitted to observe. But through our minds we can speed up the action. We can plunge back to the beginning and forge ahead into the future.

When we do, we see ordered motion all around us. The Earth spins on its wobbly axis and revolves in orbit around the Sun. The Moon also spins, with one face tied to the Earth as it weaves in and out of Earth's orbit. The Sun itself rotates on its axis and follows a path around the great conglomeration of stars and dust, plasmas and gases known as the Milky Way Galaxy. Even galaxies gather in clusters, as numerous as the stars within each of them. It is a system of space carousels to which we ourselves have added satellites and spacecraft that orbit our Sun and some of its planets. We have even dispatched a couple of interstellar wanderers, and two more spacecraft will probably follow them out of the Solar System.

Whether we see the motion of the heavenly bodies or not depends entirely upon our viewpoint: whether we take a snapshot with our everyday senses or make a time-lapse motion picture with our minds, compressing billions of years into microseconds of thought.

While astronomers have probed to the far reaches of the known universe with observations and theory, we really know very little concrete beyond our Solar System—that small family of planets and satellites associated with the Sun. This system also includes debris from its formation—comets, meteoroids, and dust—and is filled with complex electromagnetic fields and with speeding subatomic particles sprayed out by the Sun and some of the big planets.

Our knowledge of this Solar System has been revolutionized more than any other branch of astronomy over the past two decades. Never in previous human history did the frontiers of knowledge about our environment expand so quickly. This mental breakthrough from abysmal ignorance to expanded awareness of the local system of which our planet Earth forms a part, resulted

The space age moved mankind from observational to experimental astronomy; we walked on and touched the stuff of the Moon. (NASA)

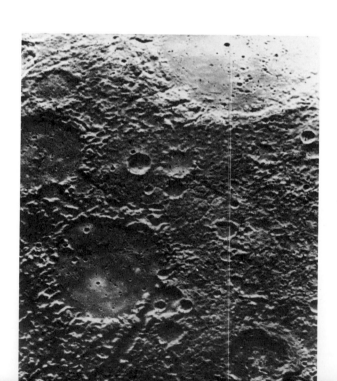

By the wonders of microminiaturized electronics, we have been able to stand on other planets and look around as though actually there. Through Viking we have visited the icily cold, dusty plains of Mars and watched a Martian sunset in the eternally pink sky. (NASA)

Our spacecraft have plunged in toward the Sun to give us our first look at the surface of Mercury, showing details that eluded astronomers over generations of telescopic observations from the Earth. (NASA)

And spacecraft have climbed out of the gravitational pit of the Sun to the first of the giant outer planets, Jupiter, to look down on its poles and obtain views impossible from Earth. (NASA)

from the development of big rockets and their application as launch vehicles for complex electronic machines that acted as probing extensions of human senses at the new frontiers of space.

In a few hours, as these spacecraft flew by distant planets, they provided us with more hard facts than generations of astronomers had been able to do despite centuries of observations through Earthbound telescopes. We actually stood and looked around and touched the stuff of the Moon. We effectively stood and looked around on Mars and on Venus; we peered closely at the twilight planet, elusive Mercury, and we looked down on the bizarre poles of mighty Jupiter.

We have moved from interplanetary ignorance to interplanetary consciousness. In a short time, we have progressed from a parochial society wondering what might be beyond the hills surrounding our village to explorations of the worlds beyond the gravitational pit of Earth. And this book recounts the story of this unique period in the history of mankind, a story rivaled only if a primeval amphibian were to have chronicled the emergence of our planet's life from the sea onto the land.

THE SYSTEM OF SOL

The Sun, sometimes called the star Sol, is the central body of the Solar System. It is a bright yellow star believed to be about 5 billion years old and about halfway through its expected life. Nine known planets orbit this star: Mercury, Venus, Earth, Mars, Jupiter, Saturn, Uranus, Neptune, and Pluto. Their orbits are close to being in one plane, and they all go the same way around. The orbits are not circles but ellipses. The closest planet to the Sun, Mercury, approaches to within 28.6 million miles (46 million kilometers) at perihelion, which is the closest point to the Sun on the planet's elliptical orbit. The most distant planet, Pluto, reaches 4.7 billion miles (7.5 billion kilometers) at its most distant point from the Sun, called aphelion.

Mercury, Venus, Mars, and Pluto are smaller than Earth. Jupiter, Saturn, Uranus, and Neptune are much larger. The inner planets, Mercury, Venus, Earth, and Mars are rocky planets. The large outer planets, Jupiter, Saturn, Uranus, and Neptune, are planets of liquefied lightweight gases. The composition of Pluto is still unknown.

Most of the planets have satellites revolving around them. Some of these satellites are quite small, others are bigger than our Moon—planetary size. Of the Solar System's thirty-four known satellites, twenty-two are at least 90 miles (150 kilometers) in diameter and one day might be visited by our spacecraft.

Between Mars and Jupiter there is a zone of asteroids, in which there are small planetary bodies ranging from the size of large rocks to the minor planet Ceres, some 635 miles (1,022 kilometers) in diameter. A few of the asteroids stray inside the orbit of Mars; others go out beyond Jupiter. A very recent discovery is that of an asteroid-type body far out in the Solar System. Named Chiron by its discoverer, Charles Kowal, this 200-mile (320-kilometer) -diameter world swings inside the orbit of Saturn at perihelion and almost to the orbit of Uranus at aphelion. It may be the first of another system of asteroids, in the

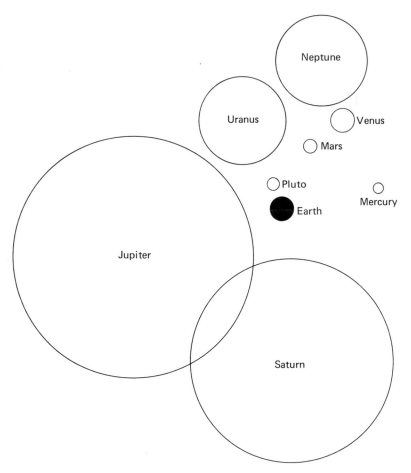

The planets of the Solar System vary greatly in size. Shown here are the two groups— the small, terrestrial planets and the large, outer ones.

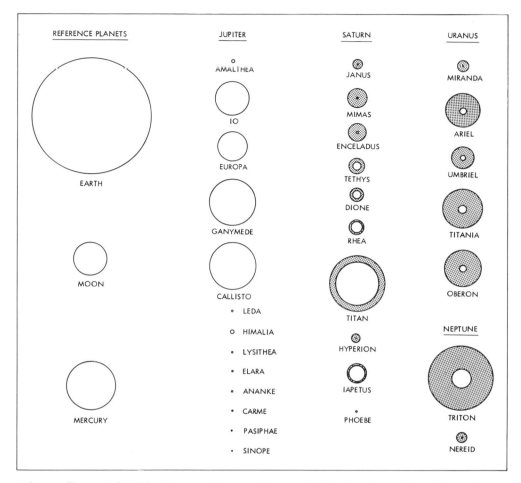

The satellites of the Solar System also vary immensely. Notice that at least 6 are about the same size as or larger than our 2,160-mile (3,476-kilometer) -diameter Moon. Jupiter's Europa has a diameter of 1,818 miles (2,926 kilometers); Io, 2,075 miles (3,340 kilometers); Callisto, 2,938 miles (4,727 kilometers); and Ganymede, 3,170 miles (5,100 kilometers). The smallest Jovian satellite shown is Amalthea, which has a diameter of only 87 miles (140 kilometers). Aside from 3,032-mile (4,880-kilometer) -diameter Titan and 870-mile (1,400-kilometer) -diameter Rhea, the Saturnian satellites are less than 600 miles (1,000 kilometers) across. Uranus' five companions range from 125 to 600 miles (200 to 1,000 kilometers) in diameter, while Neptune's Triton measures at least 2,500 miles (4,000 kilometers). All these moons are potential targets for space probes of the future. All are unusual by terrestrial standards, and all are extremely interesting worlds for future exploration. (Arfor Picture Archives)

outer Solar System. Chiron is a million times fainter than the faintest star you can see on a clear night. Such bodies in the outer Solar System are extremely difficult to detect.

Asteroids may be very important to the future of mankind because of their potential for supplying almost inexhaustible raw materials for manufacturing in space during a new industrial revolution.

There are other lesser members of the Sun's family, including comets and meteoroids. Comets are very mysterious objects about which we still know very little other than the fact that they provide spectacular displays in the night sky and have sometimes been bright enough to be seen in daylight. Astronomers have likened them to dirty snowballs that originate in the outer reaches of the Solar System, possibly in interstellar space.

An exciting future mission in space might be to mine an asteroid for its raw materials and use those materials at industrial facilities established in space. This artist's concept shows such a mission as it might be undertaken early in the next century. Veteran space artist Chesley Bonestell, whose prophetic pictures showed the beginning of the space age, here gives us a view of the next major step of our expansion into space. (NASA)

As these bodies approach the Sun, along highly elongated orbits, solar heat vaporizes their ices so that gases are pushed away from the cometary nucleus by radiation pressure—the pressure of light from the Sun—to form a spectacular tail pointing away from the Sun. American scientists have made plans to send spacecraft to look closely at these mysterious objects, but such missions have not been approved so far. A unique opportunity to visit Halley's comet, a spectacular comet that returns to Earth's vicinity only once every seventy-five years, was missed: the comet's next visit is in 1985/1986; a project to send a spacecraft to it needed a start about ten years ahead. The nation missed its chance.

Meteoroids are small bodies moving in space between the planets. Sometimes they encounter the Earth and appear as "shooting stars," or meteors. Chemical analysis of meteorites (fallen meteoroids) provides important clues to the origin of the Solar System.

At one time, scientists believed that interplanetary space was virtually empty except for these bodies, that it was a desolate void through which the planets, moons, comets, and meteoroids followed their eternal orbits around the Sun. But today, as a result of research with spacecraft, we see an entirely different picture. Energetic particles, that is, parts of atoms moving at high speeds, crisscross what was once considered "void." These emanate from the Sun and the big planet Jupiter, possibly also from Saturn, and probably from other star systems. There are also fields of forces: gravity and electromagnetic forces. There are micrometeoroids, dust, and gases. Some of these are harmless to people, instruments, and spacecraft. Others are potential hazards.

The size of the Solar System is difficult to define exactly. We believe that it certainly extends to the orbit of Pluto. But even twice this extent, a distance of 30 billion miles (48 billion kilometers) across, with the Sun at the center, may not represent the total size. The Sun has control over vast numbers of comets, and many of these travel out to a distance of one tenth light-year. Since light travels at approximately 186,000 miles per second (299,000 kilometers per second), a light-year, which is the distance it travels in one year, represents a distance of 5.9 trillion miles (9.5 trillion kilometers). The extent of the Solar System may even be to one light-year, which would make the diameter of the system 12 trillion miles (19 trillion kilometers).

How big is this? If a period in this text represents the size of the Earth on which we live (8,000 miles, or 12,000 kilometers, in diameter), the Solar System would have to be represented by a ball some 465 miles (750 kilometers) across.

OUR GALAXY: THE MILKY WAY

The Sun is one of 100–50 billion stars that make up the vast assemblage known as the Milky Way Galaxy. Spiral, like a huge pinwheel, the Milky Way measures some eighty thousand light-years across. Though it is huge, the disparity in size between the Solar System and the Milky Way Galaxy is nowhere near as large as between the Earth and the Solar System. If we let a period in this text represent the size of the Solar System, the Milky Way Galaxy has to be forty-four yards (forty meters) in diameter.

The Solar System is on one of the spiral arms of the Milky Way Galaxy, nearly thirty thousand light-years from the center. The Sun and all its planets are moving around the galaxy at a velocity of approximately 137 miles per second (220 kilometers per second). Even at this speed it takes more than 250 million years for the Solar System to make a complete circuit, a period that astronomers refer to as a cosmic year.

STAR TYPES

Many types of stars are present in the Milky Way Galaxy. Some are quite similar to the Sun in terms of age, size, temperature, and color. Some are quite different. At one end of the scale of size are giant and supergiant stars, including ones that are over 600 million miles (1 billion kilometers) in diameter, compared with the Sun's diameter of 865,000 miles (1,392,000 kilometers). At the other end there are dwarf stars that are thought to be smaller than the Earth yet thousands of times denser than the Sun. One of the smallest dwarfs is only one eighth the diameter of the Earth. Some stars are so dull they can hardly be detected in our best telescopes even when comparatively close to the Earth, while others are thousands of times brighter than our Sun. Some stars change in size and brightness, following reg-

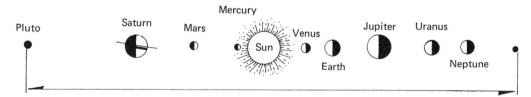

The relative distances of the planets from the Sun—not shown to scale in this drawing— give an idea of the extent of our system in terms of Earth's distance, which is 92.96 million miles (149.57 million kilometers), a distance referred to as one astronomical unit. The diameter of the system is often considered as bound by the aphelion (greatest distance from the Sun) of Pluto, which is 4.58 billion miles (7.38 billion kilometers), or 49.3 astronomical units. Many comets, however, travel far beyond Pluto, perhaps to a third of the distance to the next star. Such comets are still bound to the Sun and are thus part of the Solar System. (Arfor Picture Archives)

ular cycles; others are sporadic in their changes.

Our Sun is an average star and is expected to last as it is for a total of 10 billion years. Already 5 billion years old, it is approximately middle-aged. More-massive stars have shorter lives; they burn up matter more quickly. For example, a star with ten times the mass of our Sun burns up its nuclear fuel a thousand times faster and lasts for only 100 million years. Conversely, a star of only one tenth the mass of the Sun may last an incredible trillion years.

Stars are powered by the burning of the hydrogen within them. But it is nuclear rather than chemical burning. Four atoms of hydrogen combining to make one atom of helium lose mass in the process. This mass is converted into energy within the core of the star. Exhaustion of the hydrogen of a star signals its approaching end as a radiant body. To compensate for the diminishing supply of hydrogen, the star begins to expand its nuclear activity from its core and starts to consume hydrogen in its outer layers. The result is that the star swells as it produces enormous quantities of energy in those layers. It behaves like a slow-acting fusion bomb. Such swollen stars are red. They often expand to one hundred times their earlier size, engulfing any nearby planets they might have. This phase of their life as a giant star may last up to a billion years, depending upon the initial mass of the star. But the end is inevitable.

Once the hydrogen in the layers surrounding the core has been transformed into helium, the star can no longer generate sufficient internal energy to balance the weight of its own mass. It starts to contract under its own weight. And once this process of contraction starts, it proceeds rapidly. The star implodes. As it does so, tremendous internal pressures are generated. Just as air compressed in a bicycle pump heats up under pressure, the interior of the star heats up. Extremely high temperatures are generated, sufficient indeed to start a new fusion reaction within the core, the burning of helium into carbon. New energy surges from the core and swells the star into another giant stage.

If the star is not more than four times the mass of our Sun, the sequence after it has consumed its helium is for another contraction to occur. This time, however, it does not have sufficient mass to raise internal temperature high enough to fuse carbon into heavier elements. So now it cools off,

like a hot cinder radiating its heat into space. Its color changes to yellow, then to red, and finally it becomes black. Possibly the internal pressures have converted its carbon into diamond. The star may be truly a diamond in the sky, but it has lost its brilliance; its internal fires have at last blinked out in a stellar death. We believe billions of these dead stars litter our galaxy, like hidden treasure chests of cosmic jewels.

But the story is different for stars that are more than four times as massive as our Sun. The end of such stars is seen ten or twenty times each year in our Milky Way Galaxy. They suddenly increase in brightness some hundreds of thousands of times. Then, in less than thirty years, they fade to their original brightness. This violent outpouring of energy from stars of between four and eight solar masses occurs when nuclear fusion involves oxygen. This reaction can become explosive and form novas, which pour out a tremendous amount of energy for a relatively short period.

There are other ways in which stars can suddenly produce a great outburst of energy. If a white dwarf is one member of a binary system and the other star reaches its expansion stage, the outer hydrogen shell of the expanding giant can approach close enough to the dwarf to be pulled in by its gravity. The hydrogen shell formed around the white dwarf becomes compressed and produces temperatures high enough for a hydrogen fusion reaction to start and spread all around the white dwarf. Since there is nothing to contain the hydrogen shell, it explodes, bursting away from the white dwarf, accompanied by a surge of intense radiation. The star suddenly increases in brightness hundreds of thousands of times before fading back to its original dullness.

Such an outburst on a nearby white dwarf circling a giant companion might have been the trigger that caused the Sun and the Solar System to form from the primordial nebula. Clues to such a process have recently been obtained from inspecting the composition of meteorites; scientists found that these rocks from the sky, presumed to be relics of the early Solar System, contain isotopes of elements that were most likely formed by an explosive wave of neutrons passing through a primordial nebula containing a normal mixture of elements. Material from the Allende meteorite, which fell in northern Mexico in 1969, contains isotopes of calcium, barium, and neodymium that are not in the same proportions as the average of these materials in rocks of the Earth, Moon, and other meteorites. The analysis suggests that the primordial nebula was not a uniform mixture but contained injected or converted material from another source. The isotopes of oxygen and magnesium in this meteorite were also found to be different from normal. Many of those unusual isotopes probably were derived from a neutron bombardment of the primordial nebula or actually within an exploding star. One theory is that as much as 0.1–1 per cent of the material of the Solar System may have been produced in an exploding star that sprayed the Solar System with these materials. The theory also suggests that this same system formed within 200 million years after the explosion. The event might possibly have been a supernova rather than a simple nova explosion; supernovas are even more spectacular cosmic explosions.

Such stars, up to twenty times the mass of the Sun, go through several fusion cycles beyond the fusion of hydrogen into helium. They fuse carbon and helium into oxygen; carbon into helium, neon, and manganese; and oxygen into helium, silicon, and sulfur. Some of the biggest stars even attain a silicon-into-nickel fusion process with a subsequent decay of the nickel into iron. In the cores of stars of twenty solar masses, silicon burning occurs, and then, when it exhausts its fuel, the strong gravitational collapse pushes electrons and protons together to form neutrons. The core of neutrons collapses very quickly and then bounces back and sends a shock wave through the outer shells of oxygen and other materials. The shock wave gives rise to quick increases of temperature that are sufficient to trigger an explosive oxygen fusion reaction that literally blows off all the outer layers of the star. So incredibly energetic is the explosion that a supernova outburst temporarily develops enough energy to shine brighter than all the other stars combined in a

Sometimes astronomers see stars flare up and become as bright as a whole galaxy. This pair of pictures is believed to represent such a supernova explosion, shown by the bright star above the galaxy in the right picture (taken in May 1972), where there is no star in the left picture (taken in 1959). (Mount Wilson and Palomar Observatories)

galaxy. Such supernova explosions are observed to occur at the rate of about one per galaxy every thirty years. All the heavy elements in the universe today are thought to have been spread through space by such explosions.

The collapsed remnants of at least some of these supernovae form incredibly dense spheres consisting almost entirely of neutrons. Astronomers call them neutron stars. In them, a piece of material the size of a cube of sugar has a mass of 10 million tons, compared with only one ton in a typical collapsed white dwarf star, and five and a half grams in material of the Earth.

Rapidly spinning neutron stars with powerful gravitational fields are observed to pulsate and so are known as pulsars. By 1978, fourteen had been discovered. The most famous of these objects is connected with the collapsed remains of a Milky Way supernova that was observed and recorded by the Chinese in A.D. 1054. The cloud of gases surrounding this pulsating neutron star is a diffuse area of nebulosity named the Crab nebula. It is in the constellation of Taurus, the Bull. This nebula is also the strongest source of radio waves from outside the Solar System. We think that pulsars are produced by X-ray beams radiating like beacons into the universe.

The fate of bigger neutron stars is even more bizarre. Effectively, they drop from the material universe, becoming what scientists have called black holes. Sometimes called collapsars, black holes occur when the gravitational forces of a neutron star become greater than the nuclear forces that hold the neutrons apart from each other in the core of a neutron star. As a result, the star continues to collapse, all its material being concentrated toward a mathematical point, which has position but no dimensions. The gravity of the point becomes so concentrated that no object and no radiation can escape from it. It cannot be observed from our universe except by the effects of its tremendous gravitational field. If there is a nearby star, some of its material can be sucked into the black hole. As this material falls into the immense gravitational pit produced by the black hole, it is pulled apart and stretched to infinite length. The violence of this process acting on the infalling material produces intense radiation, including X rays.

Our first indication of the presence of a black hole came in 1972. A satellite called Uhuru discovered that a star in the constellation of Cygnus (the Swan) emits X rays. Astronomers interpret this observation as meaning that the star, named Cygnus X-1, is a binary system consisting of a visible star and an invisible star. The invisible star is believed to be a black hole sucking material from the visible star. The observed X rays are being generated as the matter plunges at ever-increasing velocity out of the universe into the black hole.

STAR GROUPS AND GALAXIES

The various kinds of stars exist alone and in groups. When two stars revolve around each other on a common center of gravity, they are referred to as a binary system. Such stars probably form when a single star rotates fast enough to form a disc but does not have magnetic fields to form planets as explained in a later chapter. The disc of gas condenses into a companion star. Three or more stars, which may arise from subsequent disc formation, are known as multiple-star systems.

When tens or hundreds of stars group loosely together, astronomers call them galactic or open clusters. More-compact groupings containing many more stars, often numbered in the thousands, are referred to as globular clusters. These clusters are distributed like a halo in a sphere around a galaxy. In our galaxy some are close enough to be seen as faint fuzzy spots to the unaided eye. M-13, the most famous, is in the constellation Hercules and contains about one hundred thousand individual stars. M-13 refers to a catalogue of fuzzy-looking objects compiled in 1782 by the French astronomer Charles Messier. He wanted to make a list of nebulous objects that might be mistaken for comets.

Beyond the Milky Way Galaxy are other Messier objects, quite different from M-13. One of these is M-31, the Great Nebula in Andromeda. It is another galaxy similar to our Milky Way Galaxy, one of countless galaxies, many at the limit of observation by our telescopes. Some astronomers have estimated that there may be as many of these other galaxies as there are stars in our own galaxy. They differ in both size and shape. Some are spiral, others elliptical. Some are quite irregular in shape. Our Milky Way Galaxy is a spiral, characterized by coiled arms. About 75 per cent of known galaxies are of this shape,

Some three hundred of these open clusters of stars have been catalogued within our Milky Way Galaxy, each containing up to several thousand individual stars. These clusters are found either in or close to the galaxy. This one is in the constellation Cancer (the Crab). It is listed in the New General Catalogue of stars as NGC 2682. The photograph was obtained with the 200-inch (5-meter) telescope on Palomar Mountain. (Mount Wilson and Palomar Observatories)

There are much more compact clusters of stars, called globular clusters, of which some one hundred are distributed around the Milky Way Galaxy like a halo. Ranging from 22,000 to 225,000 light-years from our Solar System, they are each spheroidal in shape and contain thousands of stars. This cluster is also from the New General Catalogue and is listed as NGC 5272. It appears in the constellation Canes Venatici (the Hunting Dogs). (Mount Wilson and Palomar Observatories)

20 per cent are elliptical, and the remaining 5 per cent are irregular. The two galaxies closest to the Milky Way, the Magellanic Clouds, are irregular and relatively small but are part of a local cluster to which our galaxy belongs. The large cloud is 145,000 light-years away; the small cloud is 160,000 light-years away. In the large cloud the unique star S Doradus is the most brilliant star known. It shines constantly more than a million times as brightly as our Sun.

Just as planets occur in families called solar systems, and just as stars are found in groups, clusters, and galaxies, the galaxies also form asso-ciations. Our local group has eighteen known members, ranging in diameter from two thousand light-years for a galaxy in the constellation Draco (the Dragon) to one hundred thousand light-years for the giant galaxy M-31 in Andromeda. This latter is the most distant object visible to the unaided human eye, being 1.5 million light-years from Earth. The closest galaxy in our local cluster is the Large Magellanic Cloud; the most distant known is the Wolf-Lundmark galaxy, at 1.6 million light-years. But beyond the local family of galaxies there are many other clusters, such as those in Virgo, Leo, and Coma. The cluster in

The great Andromeda galaxy, NGC 224, is sometimes referred to as M-31, the number it had in Messier's first catalogue of nebulous objects in the sky. It is one of billions of galaxies stretching to the end of the visible universe. It is the closest of the spiral galaxies and part of a local cluster to which the Milky Way Galaxy belongs. It is the most distant object we can see with the unaided eye, being about 140,000 light-years in diameter and 1,500,000 light-years distant. Nearby are two other galaxies, NGC 205 and NGC 221. (Mount Wilson and Palomar Observatories)

Coma has at least a thousand members. Beyond galactic clusters are even superclusters, in which several clusters of galaxies are grouped together, bound by hot and tenuous gas.

As nearly as astronomers can determine, clusters of galaxies are the largest groupings of matter in the universe. And if our interpretations of observations are correct, some clusters are receding from us at velocities as high as 78,000 miles per second (125,000 kilometers per second), which is more than 40 per cent of the velocity of light. Such galaxies are believed to be 4 billion light-years from Earth. The light we are seeing now, set out from those galaxies when our Solar System had only just formed from its primordial nebula.

Even farther away are the mysterious objects called quasars, which were discovered in 1960 by radio astronomers. All previously observed radio sources in the heavens had been diffuse. In 1960, with newly developed equipment, radio astronomers for the first time found radio emitters that were starlike: points in the sky, rather than diffuse clouds. They called them quasi-stellar objects, and the name was later contracted to quasars. Much later, these quasars were identified on photographic plates as very faint sources of light, and astronomers were surprised to find that the light from them was shifted to the red end of the spectrum more than from any other object observed in the sky.

This shifting of light toward the red means that the source of the light, in this case the quasar, is traveling away from us. Vesto Slipher discovered the red shifts of galaxies in the early 1900s. Just as sound from a source such as a train whistle or a siren moving away from the listener decreases in pitch and sound approaching the listener increases in pitch, light from stars and galaxies shifts toward the red and the blue for movements away from or toward the observer, respectively. All galaxies have shifts toward the red, referred to by astronomers as the red shift. They are all moving away from the Earth. Astronomers were surprised to find that the light from the quasars is displaced strongly toward the red end of the spectrum, more strongly than any of the galaxies, even the most distant. If this can be interpreted as meaning a velocity away from the Earth following the same pattern as the galaxies, namely the greater the distance the greater the speed of recession, the quasars are the most distant objects we have observed. They would also have to be the most energetic objects, to be observable at such enormous distances.

The closest quasar is receding from us at about 15 per cent of the speed of light; the most distant is traveling at 90 per cent of the speed of light. The most distant is thus close to the edge of the observable universe, which is about ten billion light-years away. This means that the radiation started from it toward Earth billions of years before our Sun was born from its primordial nebula.

As we look toward the edge of our observable universe (for when the velocity of recession reaches 100 per cent of the velocity of light we can see no farther), we are also looking back into time. The most distant objects we see may be the birth of galaxies. There has been speculation that the quasars are, indeed, the beginning of galaxies. But whatever they are, they appear to be radiating into space 12 billion times as much energy as our Sun.

No one knows if quasars and galaxies stretch out to infinity, or if there are limits beyond which they end. Astronomers suspect that we are approaching the "edge" of our observable universe and that this may be the edge of an expanding universe that originated some 13 billion years ago in a cataclysmic explosion called the "big bang." But into what is the universe expanding?

If such an edge in space and time is one day found by improving our instruments, it still leaves us with the philosophical and perhaps unanswerable question of what might be beyond. Other universes? Nothing?

The Sun, Earth, and the other members of the Solar System somehow evolved from the agitated region of gas and dust that we term the primordial solar nebula. It in turn probably originated from a large interstellar gas cloud that also spawned hundreds of thousands of similar suns and planets. And all these formed only a small part of an enormous galactic cloud that became the Milky Way Galaxy, while that galaxy itself, with its hundreds of millions of stars and vast clouds of dust and gas and intertwining electromagnetic fields and unimaginably vast plasmas, is but one of a myriad formed at one time or another from collapsing clouds of intergalactic gas and dust. The late Russian-American cosmologist George Gamow referred to the original fireball that spewed forth all this gas and dust into the universe as the big bang. That expression stuck. If the matter that subsequently collapsed into

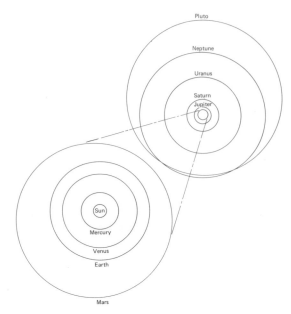

The inner Solar System (lower left) in relationship to the outer Solar System (upper right). Although the inner planets received most of our attention during the opening years of the space age, by the 1970s Pioneer 10 and 11 probes had brought the nearer of the outer planets within the range of our astronomical endeavors. The scale of the Solar System can be judged by the length of time it takes light to reach each planet from the Sun. The exploration of the outer planets, which has just started, will take many years to complete.

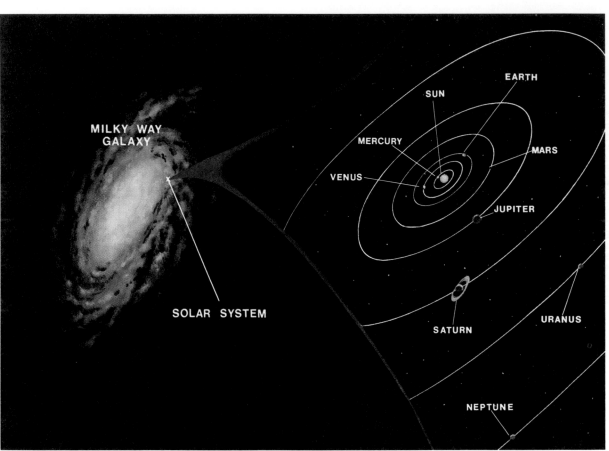

In a universe of galaxies, globular clusters, quasars, pulsars, novae, neutron stars, black holes, and other puzzling phenomena, the Sun and its family of nine planets seem prosaic. Yet myriad other such solar systems, some harboring intelligent life, may stretch from one end to the other of our Milky Way Galaxy—and throughout other galaxies beyond.

galaxies, stars, planets, and satellites did indeed originate in an incredibly hot, dense fireball, the question remains as to what existed before, and what caused the explosion.

Could our expanding universe have been preceded by Herbert Friedman's collapsing universe? Did heat released by the abrupt annihilation of kinetic energy of an imploding universe set off the initial thermonuclear explosion that spawned the present universe? In his unique mixture of wit and awe, Gamow had a term for the time preceding the big bang. He called it St. Augustine's era, since it was that worthy saint who first asked what God was doing before he made heaven and Earth.

In this book we are not concerned with the universe at large except to establish the backdrop against which the local scene is directed. We are not concerned either with the Milky Way Galaxy or the local star clusters. Rather, we examine one star of that Milky Way Galaxy and its family of planets, moons, and other bodies. This is the realm of nature that mankind has been able to explore during the first two decades of the space age, two decades that were made possible by the development of large rocket vehicles, starting with the V-2 and culminating in the Saturn 5. And except for the fact that the star is our Sun, it would otherwise pale into insignificance amid tens of billions of others.

At this beginning of the third decade of space exploration, we have to confine detailed discussion to the bodies of the inner Solar System, because the expeditions to the outer giants have only just begun. It will not be much before the end of this century that the outer planets will be able to be explored as thoroughly as the inner planets. Relative to terrestrial standards, the distances are so vast that missions to the outer planets will consume large parts of a professional space scientist's career.

This book shows how we groped for generations for understanding, through conventional astronomy making observations from Earth, and how the search for knowledge accelerated enormously when it became possible to send missions to other worlds and enter the age of experimental rather than purely observational astronomy.

When the authors first became interested in spaceflight, several decades ago, the idea of landing on the Moon and visiting other planets was generally thought of by most astronomers and scientists as hopelessly farfetched. But in recent years the inner Solar System has seemed to shrink in size and open up in detail, much as the Earth-Moon system changed with the Apollo lunar missions of the late 1960s and early 1970s. The landing of astronauts on Mars, the mining of the asteroids, the exploration of the satellite worlds of the outer planets, and the dispatch of unmanned probes to explore distant star systems seem no more fantastic today than did the conquest of the Moon a generation ago. The limiting factor in planetary astronomy and space exploration is the will of mankind, not the science and technology we have at our disposal.

2
FROM NEBULA TO PLANETS

LET THERE BE LIGHT

The dark nebula had stirred, agitated by the breaking wave of energy and radiation from the exploding star, maverick neighbor in the Milky Way Galaxy. Magnetic fields coiled and uncoiled. Grains of dust coalesced into fluffy lumps like fairy castles in the vacuum of space. Gradually, in one part of the incredibly ancient nebula, energy and matter focused into an irregular spheroid many millions of miles long. Gravity began to tug lightly, urging the particles of dust and gas to come together.

Obeying the command of the weakest of the universal forces, the gas and dust congregated toward a common center, pressing and pushing against each other in a developing pressure. Gradually, temperature increased as the gas compressed under its own weight, which developed as its infall under gravity was resisted by material that had raced it to the common center. Within 100 million years the temperature produced by overcrowding reached millions of degrees. Within the clouds of hydrogen gas there was a small proportion of atoms that had a neutron as well as a proton within the nucleus. These heavy-isotope atoms of hydrogen, called deuterium, reached a temperature at which they fused together when they collided with each other. Their nuclei merged to form nuclei of helium atoms and in so doing lost some matter in the form of energy.

The dull red glow of the sphere of gas, which is referred to as the proto-Sun, became a bril-liant incandescence as the nuclear fire was ignited within it. The bright surface rapidly expanded to a diameter of about 7 million miles (11 million kilometers). Gases and dust from the ancient nebula continued to fall into the newborn star that was our Sun, and it spun rapidly on its axis at the rate of one revolution in about twenty days. Its magnetic field swept through the space around it, hurtling infalling debris along wide arcs which collided in the equatorial plane to form rings of matter. The first materials to be thrust into a ring were heavy silicates and metals. But they were in such small amounts that their presence scarcely affected the huge Sun. Within another hundred million years or so, lighter materials condensed from the nebula and poured into the Sun, to be deflected again into rings farther out than the rings of rocky and metallic materials. The biggest of the infalls took place about 80 million years after the Sun started to shine, and the mass of materials produced the giant planet Jupiter. This planet was so massive it stole rotational energy from the Sun, slowing the Sun's rotation to a rate of once every twelve years.

During the same period, the streams of particles and gases ringing the Sun coalesced into the planets and other bodies. Mercury, Venus, and Earth may have arisen from one set of rings, Mars and the Moon from another, and Saturn, Uranus, and Neptune from a third. Pluto and Neptune's moon Triton probably originated from heavy materials in a remote ring around the Sun.

It was about this time that the Sun used up all

its deuterium. Without the internal pressure of nuclear energy, it began to collapse. As its diameter shrunk its spin increased, like a skater pulling in his arms to spin more rapidly. With a diameter of less than 1 million miles (1.6 million kilometers), the developing Sun attained its present rotation rate—about twenty-five days. And at that time, too, internal heat generated by the contraction started a new nuclear reaction. Now ordinary hydrogen atoms, which require higher temperatures than deuterium, were able to fuse into helium atoms. And since the Sun consists mainly of hydrogen, it has continued burning until today and has enough hydrogen to continue doing so for another 5 billion years or more.

Meanwhile the streams of particles around the Sun had condensed into planets, which in turn had developed magnetic fields able to form secondary rings from material falling into them from the remains of the nebula. These secondary rings ultimately accreted into satellites.

The Solar System as we know it today had been born. And during the 4.5 billion years that followed, it changed very little, though the individual planets progressed through various stages of evolution. On some, the action halted billions of years ago. On others, it continues to this day.

This is a modern view of how our planetary system came to be. It incorporates many of the discoveries made during the two decades of space exploration. But the enigma of the Solar System's genesis is by no means completely resolved. There are still many unanswered questions. We have progressed so far in our understanding because of the information gathered by spacecraft not only about the Sun and the planets but also about the space environment between them, a strange realm of electrical and magnetic fields and of high-speed subatomic particles. But during this period of discovery the viewpoint of mankind changed drastically.

EARLY BELIEFS

Throughout most of recorded history, man accepted religious explanations, of faith rather than fact, to account for the origin of the Solar System. He attributed it to a creative event initiated by a manlike being, which was the commonly accepted view of God. People generally explained the origin of everything in terms of the most advanced thought of the day; and they still do. Early matriarchal societies postulated the birth of Earth from a female deity. The hatching of a cosmic egg satisfied the understanding of other early civilizations. Today we are steeped in the awesome power of nuclear explosions, and perhaps our explanation of a big-bang origin of the universe stems from this nuclear mesmerism. How much closer we are to the truth than those Babylonians who believed in a beginning from the lifeless body of a dragon slain by a god is debatable.

But, beginning in the eighteenth century, natural scientists began to try to substitute observations for faith in explaining creation. From such observations of natural processes, they developed laws of physics and attempted to explain how the Solar System originated in terms of those laws. Reading old astronomical textbooks soon makes one painfully aware that often people forget that

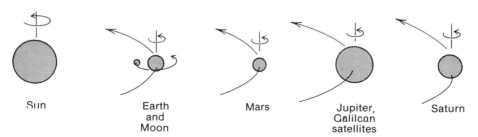

Sun   Earth and Moon   Mars   Jupiter, Galilean satellites   Saturn

The Solar System today shows ordered motion, with all the planets and their satellites generally following a common traffic pattern. Theories of the origin of the system have to account for some members of the system not following this pattern.

the laws of physics are only best-for-the-time-being explanations of what we have observed. No matter how much faith these laws may inspire, they must not be regarded as ultimate truths. Even today people still argue that some sort of original start-up substance, some equivalent of the cosmic egg, must have existed at the time of creation. And no matter how awesome and how organized the physical universe might be, we have always had trouble accepting the notion of something coming from nothing.

In the age of natural science following the discovery of new worlds on Earth, efforts to apply scientific reasoning to creation and subsequent evolution were first focused on the Solar System. This was understandable, since the stars are so remote. Astronomers armed with the newly discovered telescope were excited about the details they saw on the Moon and some of the planets but disappointed to find that their instruments revealed very little more about the stars than they had gleaned with the unaided eye. All the telescope did was to show them that there were many more, fainter stars.

It eventually became clear to these early scientists that the laws of nature they had observed in our Solar System were most likely repeated among the stars. Slowly they accumulated evidence for a number of basic facts. First, astronomers learned from the movements of the planets among the "fixed" stars that the Solar System is essentially a disc; all the planets move around the Sun in almost the same plane. And all the planets move around the Sun in the same direction, the one in which the Sun rotates on its axis and most of the planets rotate on theirs. Moreover, most of the satellites go the same way around, too. Only six of the thirty-three known satellites disobey this one-way traffic pattern.

While the motion suggested a single process responsible for the origin of the Sun, the planets, and their satellites, such a process had also to account for the exceptions: the few planets and satellites that disobeyed the rules.

CHANCE ENCOUNTERS

The first scientific theory for the origin of the Solar System appeared, in 1739, in the book *Histoire Naturelle,* by the Frenchman Georges Louis Leclerc, Comte de Buffon. He speculated that a comet had approached the Sun and torn from it the material that subsequently became the planets.

Later investigators thought that the planets had somehow been captured by the Sun as it journeyed through space among the stars. But neither of these theories explained how the planets traveled around the Sun in such relatively regular orbits. Objections to theories of chance encounters for the origin of the planetary system are mainly based on celestial mechanics, the ways in which the bodies of the Solar System move under the gravitational influence of other bodies and of each other. A body like a planet intruding into the Solar System from the vast spaces between the stars would be moving so quickly that it would enter the Solar System, speed around the Sun, and escape out again unless some braking force slowed its passage so as to change its orbit to an ellipse around the Sun.

Such a force might be created by a cloud of dust in the Solar System. Or it might be a large planet already in the system that did the braking, through its gravity. However, chance encounters for the origin of planets require so many remote possibilities that astronomers quickly sought other ways to explain the complex system of planets, in a more logical manner.

BORN IN RINGS OF NEBULOSITY

The most widely accepted theory came from another Frenchman, the philosopher René Descartes, in 1644. He said that the planets originated from the same nebula that spawned the Sun. This idea was expanded and refined by the German philosopher Immanuel Kant and, a year later, in 1756, by the French astronomer-mathematician Pierre Simon Marquis de Laplace. In the final chapter of his popular book *Système du Monde* he stated that he believed the Solar System originated from a great nebula which, in condensing, detached various rings that later condensed into the planets and their satellites. The process started with a cloud of gas and dust that rotated slowly, but gradually contracted under its own gravity. The contraction caused the cloud to spin more rapidly, so that it bulged at its equator. As the spin continued to increase, a ring of matter separated from the central body. This process repeated as the gas continued to contract, until

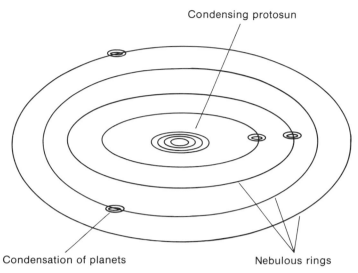

Condensing protosun

Condensation of planets

Nebulous rings

One of the early theories for the origin of the Solar System was the nebular hypothesis, of the French astronomer Laplace. This theory, which held sway for three quarters of a century, said that a gradually contracting Sun spun more and more rapidly as it contracted and successively flung out rings of material that condensed into the planets and their satellites.

there remained the central Sun surrounded by rings of matter. These rings coalesced into the planets, and as they spun in contraction they, too, formed rings, which became their satellites. About this time, the central Sun, said Laplace, heated up under pressure of its own collapse and started to shine as a star.

The Laplace theory received considerable support from Herschel's discovery of other nebulae, but this nebular hypothesis had no way of explaining how the Sun could continue to shine as a star for billions of years. Yet it seemed to be supported by observations of the rings of Saturn, which were explained as inner rings thrown off by that planet but not yet condensed into satellites.

INSTABILITIES AND THE MYSTERY OF THE MOMENTA

The nebular hypothesis of Laplace remained in vogue for eighty years. Its supporters found ways to explain obvious problems such as the unusual motions of the satellites of Uranus and Neptune and the rapid orbital motions of the tiny Martian satellites, one of which buzzes around the planet faster than Mars rotates on its axis. Then the followers of Laplace received a shock from which they never really recovered.

The time was 1857 and the man the brilliant Scottish physicist James Clerk Maxwell. At the age of fifteen he had presented a paper to the prestigious Edinburgh Royal Society, and in 1854 capped his second wrangler graduation from Cambridge with a masterful mathematical analysis of the rings of Saturn. It earned him the Adams Prize, together with a great deal of antagonism from those who believed strongly in the nebular hypothesis. Maxwell showed mathematically that a solar disc containing particles of dust would be unstable and could not coalesce into planets unless its total mass was some hundreds of times greater than that of all the planets combined. In other words it was more likely to form a second Sun than a system of planets. Maxwell also broached the subject of the problem of the distribution of spin among the bodies of the Solar System, especially the missing momentum from the Sun itself. This mystery of the momenta has not been fully solved even today.

It goes like this.

A planet moving around the Sun can be likened to a rock on the end of a piece of string being swung around in a circle. The rock corresponds to the planet, the string to solar gravity keeping the planet in its orbit. The planet, and the stone, have a quality physicists refer to as momentum; it is the mass of the body multiplied

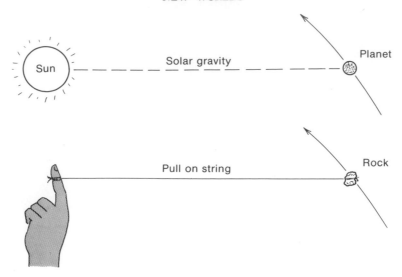

A rock being swung around on the end of a length of string is analogous to a planet in orbit about the Sun. If the string is shortened by allowing it to wind around the finger, the stone travels faster so as to maintain its angular momentum. Theories about the origin of the Solar System have to account for the unusual distribution of angular momentum, most of which is in the planets and not in the Sun.

by its velocity. For a body whose path is a closed curve, angular momentum is defined as the product of not only the mass and the velocity but also the distance from the center. In the case of the planet, this distance is its distance from the Sun, and in the case of the stone, it is the length of the string.

An important property of matter and motion is that momentum is conserved. Thus if you are spinning the stone on the end of a string attached to your finger and allow the string to wind up around your finger, you will see that the speed of the stone increases as the length of free string decreases. The angular momentum thereby remains the same as the stone moves faster around a smaller orbit. You see this law being put to practical effect on ice rinks. A skater wishing to spin faster pulls in his arms so that their mass is closer to the center of spin. The spin of the whole body becomes faster to conserve the angular momentum. And to stop spinning, the skater suddenly extends his arms, slowing down rapidly to a rate at which an extended skate tip can provide final braking to a stop.

Now, how does this apply to the nebular hypothesis and the origin of the Solar System? Astronomers found from observations that angular momentum is strangely distributed among the

bodies of this system. The spin of the Sun about its axis accounts for only about 2 per cent of all the angular momentum. Here angular momentum is summed for each particle of the Sun moving around its center at its own particular distance from the center. The nebular hypothesis supposed that the primordial nebula collapsed into the star we now call the Sun and so its rate of spin increased like that of a skater. Yet today the Sun spins relatively slowly. Scientists were at a loss to explain how the Sun could have extended arms of matter to slow it down like a skater at the end of a rapid spin.

By contrast, the planets in their motions around the Sun possess 98 per cent of the momentum of the Solar System. Somehow, the early central condensation called the proto-Sun had to transfer most of its angular momentum to the planets. The nebular hypothesis would have it do just the opposite. As a consequence, the hypothesis had to be abandoned.

CATACLYSMIC ENCOUNTERS

Between 1919 and 1928, Sir James Jeans in England and Forest R. Moulton and Thomas C. Chamberlain in the United States revived the old idea of a chance encounter forming the Solar System. Whereas Buffon had suggested a comet, the new approach was more cataclysmic. It assumed that a star comparable in size to the Sun approached close enough to raise mighty tidal waves on the solar surface. Great tongues of matter siphoned from the top of these monster waves into space. Hurtling toward the invading star, they were dragged into motion around the Sun as the star swept past it at high speed. In this manner the distribution of angular momentum might be explained together with the fact that all the planets orbit in the same plane, for these planets condensed out of the material pulled from the Sun into space.

Although the concept sounded plausible, it led to the depressing picture that solar systems and hence life are extremely rare in the universe, because the probability of such close encounters was so low, only one in several billions. Also, some doubted whether the tidal filaments drawn out from the Sun would condense into planets. The Princeton University astronomer Lyman Spitzer demonstrated that the torn-off matter would be more likely to dissipate into space in tenuous gas clouds than to condense into planets. Moreover, the theory did not fully account for the distribution of angular momentum. Several variants were produced in an attempt to do this. One of them, by Lyttleton (1936), assumed that the Sun originally formed one of a pair of stars; the tidal filaments from which the planets formed were raised from the Sun's companion, which was afterward whirled from the

Solar System by the gravity of the passing intruder.

Other cataclysmic theories suggested the origin as an ejection of Sun and planets from a special class of star known as a Cepheid variable (Banerji, 1942), as part of a supernova explosion of a companion to the Sun (Hoyle, 1945), and as the fission of a rapidly rotating star during encounter between two stars (Gunn, 1950).

A THEORY OF NEBULAR CLOUDS

Toward the end of World War II, Carl Friedrich von Weizsächer showed that the massive disc of the solar nebula required by Maxwell to generate the Solar System might indeed have existed. Moreover, if it was made up of about 99 per cent gas and 1 per cent dust it would be viscous enough to rotate as a solid body. In such a nebula the high-velocity gas molecules in the outer regions would gradually dissipate into interstellar space. At the same time, molecules with lower velocities would spiral inward, toward the central Sun. In this way the amount of gas in the disc would be reduced by about one half every 5 million years, thus getting rid of much of the matter in a way required to make the nebular hypothesis feasible. So Laplace's ideas were again revived but in a more advanced and modified form.

While the gases of the nebula were being thinned out, dust particles would have acted differently. This was because gas molecules are

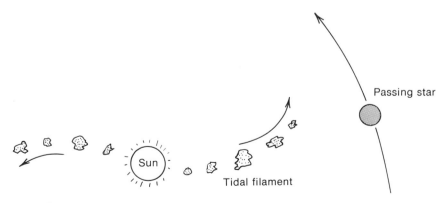

As a possible way in which the Solar System might have formed and its angular momentum been distributed as it is, some astronomers speculated that the planetary material originated from tongues of matter pulled out from the Sun by the gravity of a passing star.

elastic; they bounce off each other like billiard balls without losing energy. But when dust particles collide with each other in space they tend to adhere together, like pieces of damp clay, and form lumps of loosely connected particles. Such groups of dust grains would gradually grow in size depending upon the total amount of material in the region at which the clumps formed. Close to the Sun, where a path around that luminary does not cover as much space as paths farther out, the aggregations of matter would be relatively small. In this region, the planets Mercury, Venus, Earth, and Mars formed. Farther out from the Sun, planets were larger; and still farther out, where the matter itself thinned out, the planets were smaller.

This theory for the origin of the Sun's planetary system was refined by its originator and by other astronomers in subsequent years. Even more recently, Hannes Alfvén and Gustaf Arrhenius developed an ingenious electromagnetic version of the nebular hypothesis which accounts for nearly all the observed facts about the Solar System, including anomalies. But first, how did nebulas themselves, such as that which we believe spawned the Solar System, originate? This is the realm of cosmology, of the origin of the universe itself.

COSMOLOGY OF BIG BANGS AND UNIVERSAL EXPLOSIONS

In the early 1920s a young Russian mathematician, Alexander Friedman, constructed a model of the whole universe based on Albert Einstein's then new theory of general relativity. Relying upon observations and still following the Copernican creed, he assumed a universe filled with galaxies but without any preferred center. He concluded that under the effect of an all-pervading gravity such a universe would be unstable. It would inevitably collapse upon itself unless it was already expanding at a rate sufficient to overcome the trend to collapse. Such an expansion, however, seemed to require that the universe must have originated in an explosive event of enormous energy in which the matter that formed the galaxies had been hurled apart at tremendous velocities.

At about the same time, the American astronomer V. M. Slipher made the important discovery, mentioned in the previous chapter, that meant that the galaxies were moving away from the Solar System. Meanwhile, Edwin P. Hubble had been investigating the distances of near galaxies by observing a special class of star, of changing brightness, called Cepheid variables. The period in which the brightness changed depended upon the star's intrinsic, or true, brightness. By comparing actual brightness with apparent brightness, the distance of the star could be estimated. By 1953 Hubble had amassed data on the velocities and distances of galaxies showing that all galaxies are moving away from us. Even more startling, he found that the more distant a galaxy is the faster it is receding. Actually Hubble had made his discovery in 1927 but downplayed it in a paper in which he attempted to show that the red shifts and distances of the galaxies could be used to measure the motion of our Sun around the Milky Way Galaxy.

This discovery of the recession of the galaxies had tremendous implications. Our galaxy was

Astronomers found that the spectral lines of galaxies are shifted toward the red end of the spectrum, thereby indicating that they are moving away from the Solar System. They also found that the more distant a galaxy is the faster it is moving away. (Mount Wilson and Palomar Observatories)

shown not to be like one of many polka dots on the surface of an expanding balloon; if it were so all the dots would move apart from each other at the same rate. Instead, the implication of the Hubble discovery was that all the galaxies are explosively scattering from some initial center. And if we trace their motions back in time for some 12–15 billion years we find that all the material of all these galaxies, that is, all the material in the universe, must have been together at one time. Some cataclysmic event of creation must have flung the material outward to form the universe as we now see it. This event has been called the big bang.

During the first few seconds after this beginning the explosive fireball was so hot that some of the initial atoms of hydrogen fused into helium. And as the fireball expanded and cooled, blobs of gases were drawn together to form protogalaxies —the galaxies in their embryonic form. One of these was the proto-Milky Way.

From conglomerations of gas, each protogalaxy gave birth to stars, possibly from eddies generated at the beginning. Astronomers believe that the early stars were arranged symmetrically around the center of the then spherical Milky Way Galaxy. But the immense system of stars and gas and dust rotated and stretched itself into its present disc shape.

The shrinking gas clouds of the stars reached high internal temperatures due to gravitational compression in which the weight of the outer layers pressed on the inner layers. Temperatures at the cores of stars became sufficiently high to start deuterium fusion reactions and, later, ordinary hydrogen fusion reactions, each producing helium. Massive stars use hydrogen more quickly than their less massive fellows. They then start to fuse heavier and heavier elements. Some of them reach a stage of explosive nuclear reactions and blow apart, scattering their heavy elements into new clouds of dust and gas. Astronomers suspect

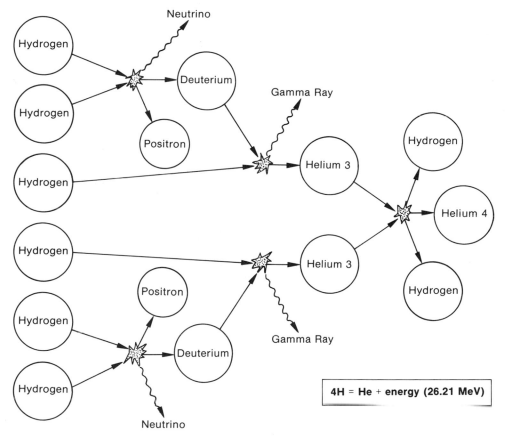

4H = He + energy (26.21 MeV)

Stars are powered by nuclear reactions. Stars such as the Sun fuse hydrogen into helium in the way shown here. Bigger stars use carbon in another set of reactions.

that all the heavy elements in the universe today were "transmuted" in the cores of giant stars early during the history of the galaxies. This includes the carbon, oxygen, and nitrogen and other molecules making up our bodies, and the metals we use to operate a technological civilization. In truth we can call ourselves children of the stars.

Fortunately our Sun and many similar stars are not so massive. They burn hydrogen more slowly. The Sun is, indeed, a relatively middle-aged star. It is also a common star, which leads us to suspect that it was created by the same processes that gave rise to countless other stars in our galaxy and other galaxies. Even today, astronomers observe stars being born in the spiral arms of galaxies much like the Milky Way. For example, Dr. Donald Hall, of the Kitt Peak National Observatory, in Tucson, believes that he has discovered a star in the Orion nebula born only one thousand to two thousand years before the light detected began its journey across the universe. A team of scientists from the Seward Observatory of the University of Arizona in Tucson and NASA's Ames Research Center under Dr. Roger Thompson have found in the constellation Cygnus a highly luminous disc-shaped object they believe to be a star no more than a thousand years old. Referred to as MWC 349, it is surrounded by a disc of glowing gas with a diameter some twenty times that of the central star. Planets may now be forming in the disc. Such observations strengthen the belief that there is nothing unusual or special about our Sun. And if this is so there may be nothing special about its Solar System and the fact that one of the planets developed a form of life. The galaxy, and other galaxies, may harbor hundreds of thousands of similar planets.

MODERN COSMOGONY

While cosmology attempts to describe the origin of the universe, cosmogony attempts to describe how our Solar System came into being. Based on recent discoveries by spacecraft and increasingly sophisticated techniques for observing,

Sagittarius Star Cloud. (U. S. Naval Observatory)

scientists have a good picture of how the Solar System formed and subsequently evolved. It is a complex one, far removed from earlier, simplistic theories. It is a picture painted in the many colors of classical astronomy, astronautics, gas dynamics, magnetohydrodynamics, microscopic examination of meteorites and moondust, and study of the plasmas and electromagnetic fields of the Solar System examined by spacecraft.

This most recent theory visualizes the cloud of gas and dust being subjected to two opposing forces: gravitation pulling it together and internal pressure holding it apart. Under normal conditions, internal pressure would prevent the gas cloud from collapsing. However, at some point the cloud did collapse, possibly under the shock of a nova or a supernova explosion, an abrupt change in the interstellar magnetic fields, or some other catastrophic event in our galaxy.

After this collapse had started, variations in density coupled with swirling volumes of turbulence would have produced centers of attraction, some of which coalesced into large masses which became protosuns. One of these became our Sun. It was born as a vast mass of relatively cool gas and dust some 30 billion miles (48 billion kilometers) across and probably twice the mass of the Sun today. This nebula continued its collapse until its density reached the point at which too much heat was generated to be radiated into space from the central mass. As a result, inner regions heated up to 5,900° F (3,000° K, where K stands for Kelvin, which is an absolute unit of temperature). Although very hot by terrestrial standards, the temperature was still far below that needed to start a nuclear fusion reaction.

According to current theories the change from proto-Sun to the Sun with its system of planets and satellites probably occurred fairly rapidly. We know from radio observations of the interstellar clouds that there is deuterium in space; probably this heavy isotope formed part of the initial gas cloud from which the Sun formed. Probably the deuterium nuclear fire ignited within the center of the Sun when it had a radius of about 6 million miles (10 million kilometers) and was rotating in a period of about twenty days.

During this early phase of the Sun, which lasted about 100 million years, the planetary system was formed. Helium fell in from the nebula, accelerated by the gravitational attraction of the Sun. It reached sufficient speed to become ionized —stripped of electrons. Then the magnetic field of the rotating Sun was able to affect the infalling atoms more than gravity could. They were flung out again but this time along the lines of magnetic force so that they followed curved paths that deposited them in a ring of gas surrounding the Sun in the magnetic equatorial plane. This formed a first planetary cloud, analogous to one of the rings of matter in the nebular hypothesis of Laplace.

Later infalls of heavier materials also ended up as rings around the Sun, and there was some filtering of the infalling material because the first clouds stopped the motion inward of most of the lighter elements. Only heavy metals and silicates could penetrate inward. The result was a selective process of ringed cloud formation around the developing Sun. Jupiter formed while the Sun was burning its deuterium. Its mass acted tidally upon the Sun to slow the solar rotation to the same period as Jupiter in orbit, namely twelve of our years. Later, when the Sun had used up its deuterium, it contracted again and increased its rate of spin to its present twenty-five days as it achieved its present diameter of 865,000 miles (1,392,000 kilometers) and started to burn its hydrogen into helium.

The other planets were never massive enough to affect the rotation of the Sun. We think they formed from different clouds of infall. Thus, the Moon and Mars formed from an early cloud of fairly lightweight substances. Mercury, Venus, and Earth formed from a cloud rich in metals and rocky materials. Jupiter, Saturn, Uranus, and Neptune formed from another cloud, and Pluto and Triton from yet another.

The planets formed because particles moving in orbit around the Sun would tend to clump into larger particles. The magnetic field of the Sun would urge them into streams, and the tendency of streams of bodies would be to coalesce into larger bodies, which ultimately merged into the planets we know today.

Once planets had condensed and started spinning on their axes and generating magnetic fields, they could emulate the Sun and create their own rings of infalling material, which, in turn, became their satellites.

Since hydrogen and helium were not able to condense in the inner Solar System, they did not provide clumps of matter to form planets. The material for the planets came from impurities in the hydrogen and helium clouds. It was only when Jupiter and Saturn became large enough that they were able to attract gravitationally the hydrogen and helium from the plasma clouds in the region where they formed. The hydrogen and helium in the inner Solar System either became part of the Sun or diffused out to Jupiter. There is also the possibility that the Sun went through what is termed a t-Tauri phase (named after the star in which the phenomenon was first identified), during which intense radiation during the Sun's early history developed sufficient radiation pressure to blow all remaining light gases from the Solar System.

While this theory explains the differences in mass among the planets, the asteroids, and the satellite systems, and the unusual distribution of angular momentum, other theories have been put forward in recent years to account for the nature of the individual planets. Observed differences might be explained, in part, at least, by a difference in temperature within the regions of the solar nebula at the time of planetary formation. It is speculated that near the center of the nebula temperatures would have been high enough to prevent easily volatized materials, for example light gases, from condensing. This could account for the inner planets consisting of rocky materials rich in metals. In the outer system, however, water could condense around rocky cores of the planets, so that these outer planets would be larger, even large enough to gather in hydrogen and helium also as gases. The difference in temperature also seems to explain differences among the inner planets such as why Mercury is richer in heavy elements than Mars. But bodies such as the Moon and the asteroids seem oddly out of place.

MISFITS

There are other objects in the Solar System besides the planets and their satellites that have to be accounted for in any theory of cosmogony. The asteroids present a puzzling problem. Some astronomers have contended that these Lilliputian worlds represent the fractured remains of one or more planets that broke apart. Others contend that the great mass of Jupiter prevented a planet from forming in the region of the as-

teroids, between the orbits of Mars and Jupiter. Yet another possibility is that the infall of matter was not able to make a band of very great density at that distance from the Sun, so there has never been enough material to form a planet there. Indeed asteroids may then represent material that is still trying to come together to form a planet.

Pluto, too, is somewhat of a mystery. Dense and small like the terrestrial planets, it may be an escaped satellite of Neptune, or perhaps one of two original outer planets (the other being Triton, which at some point was captured by Neptune). According to the latter theory, Mars and the Moon were two planets of a kind formed in a belt that could produce only lightweight planets. The Moon would then have been captured by the Earth, during which process it destroyed the Earth's original natural satellites, the impacts of which produced the immense lunar basins. The capture of Triton by Neptune also destroyed all the satellites of that planet except for Nereid, which was flung into its unusual orbit, —the satellite moves in the opposite direction to the general traffic pattern of the Solar System.

The comets also have to be explained in any theory of the origin and evolution of the Solar System. Comets seem to consist of clumps of fine-grained rocky material and ice. In fact, they have been likened to dirty snowballs. When they approach the Sun, toward the perihelion of their very elongated elliptical orbits, their materials evaporate and form tails pushed away from the Sun by radiation pressure. Comets also veer from the orbital pattern of the planets by zooming into the Solar System from all directions; they are not confined to the orbital plane of the planets.

Back in 1963, Jan Oort, of the Leiden Observatory, Holland, proposed that a cometary reservoir exists about one tenth of a light-year out from the Sun. That is about one fortieth the distance to the next-nearest star. He speculated that comets residing in this reservoir form an immense halo around the Sun that inevitably became known as Oort's cloud. Typically one of these comets would spend nearly a million years in the cloud and only a few years inside the orbit of Jupiter. The comets, said Oort, originated from the same solar nebula as the planets and the Sun. They were created in the vicinity of Jupiter but were hurled out into the halo by the gravity of that mighty planet. Other astronomers disputed this view, contending that comets originated far-

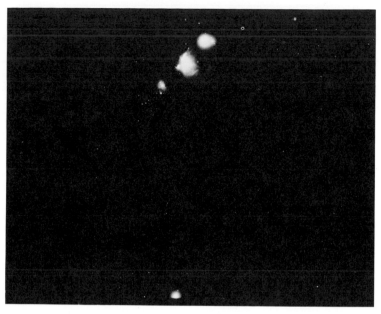

Astronomers believe that stars form within vast clouds of gas and dust as instabilities develop within these clouds. Here is shown a region of gas and dust in which stars are forming. (Mount Wilson and Palomar Observatories)

ther out in the Solar System, perhaps at the orbits of Uranus and Neptune, while Hannes Alfvén suggested that they originated in the spaces between the stars, where they congregate today in the Oort cloud or beyond.

While some astronomers believe that streams of meteoroids originate from the breakdown of comets, others think that comets are formed from meteoroid streams. While there is evidence either way, if the latter is true we may be witnessing today the first step in the process from which planets are made, the concentration of individual grains, condensed from gas clouds, into larger bodies.

Generally astronomers agree that the planetary system has remained relatively unchanged since it was first formed. This seems to be borne out by exploration of the Moon, which revealed that the lunar rocks have remained relatively undisturbed for billions of years. So a big question that is often asked is: When did the Earth, Moon, and other worlds form?

Studies of meteorites here on Earth and of samples brought back by the Apollo astronauts reveal that rock solidification took place between 3.7 and 4.5 billion years ago. Allowing several hundreds of thousands of years for the planets to have formed in the first place, the rocks to have cooled, and the meteoroid materials to have crystallized from the original nebula, one can place the age of our planetary system at somewhat less than 5 billion years.

Science can thus give us what seems to be a reasonable picture of how and when the Solar System was formed. How realistic our theories are is difficult to say. Every space mission brings new evidence, which often overturns earlier theories. How much closer we are to the truth than the biblical statement about "In the beginning . . ." only time and many more space missions will tell.

3
THE SUN—INSIDE AND OUT

BLEMISHED PERFECTION

At the beginning of the seventeenth century the Sun was regarded by laymen and ecclesiastics alike as the purest symbol of celestial incorruptibility. Even though spots had been sometimes seen on the face of the Sun, these had been attributed to the passage across it of planets such as Mercury and Venus. Early records, for example, mention a large spot marring the face of the Sun for eight days during the reign of Charlemagne.

In 1610 a new and exciting era of astronomical discoveries started with the invention of the telescope. Scientists turned the new instrument toward the familiar objects in the sky including the Sun, and they were surprised and mystified by what they saw. One of the first to look at the Sun through a telescope was a Jesuit priest, Christoph Scheiner, of Ingelstadt, Germany. He was astonished to see spots blemishing the immaculate face of the Sun, and as a result faced a terrible dilemma. To admit the existence of blemishes on the face of the Sun was tantamount to sacrilege. Yet as he continued to watch day after day the spots did not disappear but only moved gradually across the face of the Sun. Undoubtedly the face of the Sun was marred, thought the worthy father and he could no longer keep his discovery to himself. So he went to report it to the local superior of his order to seek advice on what he should do.

The advice was typically simplistic. Said his superior: "I have read my Aristotle several times, and I can assure you that he mentions nothing about spots on the Sun. So, my son, calm down and rest assured that what you take to be sunspots are defects in your lenses or your eyes."

But the truth could not be suppressed for long. Elsewhere Galileo Galilei saw a spotted Sun. And so did J. Fabricius and many others. A year after Father Scheiner's discovery, the existence of spots on the Sun had to be admitted. Reluctantly the savants and theologians of the day had to accept the fact that the solar perfection was, indeed, blemished.

But what were these spots? Early astronomers, including Herschel, thought that the Sun was a cool body surrounded by a luminous ocean. The great French astronomer-mathematician Lalande speculated that sunspots were mountain peaks revealed by the ebb and flow of this luminous ocean.

The English astronomer Walter Maunder described the naked-eye appearance of sunspots as being like the heads of nails driven into the Sun. Through a telescope the spots have a delicate structure of streaked "penumbra" surrounding a very dark "umbra," a darkness that is actually the result of contrast with the surrounding brilliance of the glowing visible surface of the Sun called the photosphere. Today we know that sunspots are cooler regions of gas within the photosphere.

A major discovery about sunspots came from the work of an amateur astronomer, H. S. Schwabe, a druggist who was pursuing his hobby

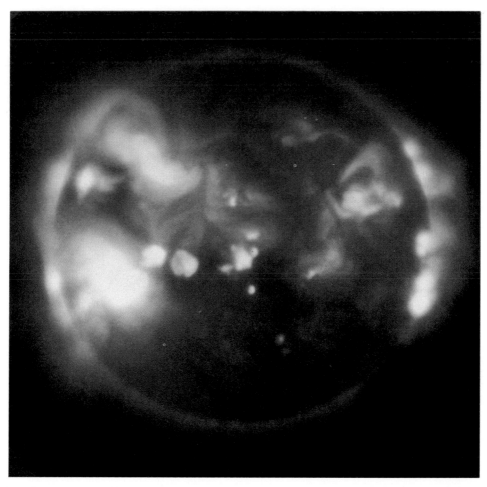

Our Sun—an incandescent body of gas—is by far the largest and most massive, as well as the central, object of our Solar System. Here is a "false color" reproduction of an X-ray photograph of the Sun obtained on eclipse day, March 7, 1970, by a sounding-rocket (NASA Aerobee) experiment of American Science and Engineering, Inc. Note especially the bright points in otherwise quiet regions, the active loop, and the extended coronal emission near active regions. The quiet "polar caps" are also evident. The picture was taken near "contact" between solar limb and Moon. The outline of the Moon can be seen. (NASA)

of looking for the mythical planet Vulcan, thought to orbit the Sun within the orbit of Mercury. Because he was employed at night, the method he picked for his search was to look for the planet crossing the face of the Sun, a task that he could attempt during his off-duty daylight hours. Beginning in 1826 he systematically observed the Sun for forty-three years, hoping to glimpse the elusive planet. During all these years he diligently recorded the position of the sunspots on the solar disc so that he would not mistake them for Vulcan. But while he never found

the mythical innermost planet, he produced a unique, forty-three-year record of sunspots. And this record showed that the number of spots not only varies from day to day but also from year to year in an eleven-year cycle.

Astronomers searched earlier records and found that this cycle extended back for at least three quarters of a century. So not only had astronomers discovered that the Sun was blemished, but also they had found it variable. In subsequent years the nature of this variability was more clearly defined as affecting not only the spots but also the magnetic fields of the Sun and the whole of interplanetary space around it, including the Earth itself.

The movement of the sunspots across the face of the Sun allowed astronomers to make another important discovery: that the Sun rotates on its axis. Finally, because spots at various latitudes went around the Sun in different times, they found that the Sun was not a solid body.

With these discoveries scientists were freed of much restrictive thinking, so that they were able to allow their minds to probe into the Sun's nature. As years went by and observational details mounted, they proved that the Sun is a variable star, a vast globe of incandescent gas, and of very great interest to the whole of science as well as to astronomers.

Except for nuclear fission and fusion, the Sun is our only source of energy. While large-scale direct solar power awaits development on a commercial scale, indirect solar power provides our mechanical and biological energy resources. The photosynthesis of complex carbon-based molecules by plants starts the food chain. The structure of plants provides all our carbon fuels for combustion: wood and the fossil fuels of coal, petroleum, and natural gas. And in the future the Sun may offer a key to the understanding of other processes whereby we will be able to meet all our energy needs.

THE SUN AS A STAR

Astronomically the Sun is the nearest star. It is only 92.9 million miles (149.5 million kilometers) from Earth, whereas the next-nearest star is some 21 trillion miles (34 trillion kilometers) distant. So in studying the Sun we are able to study the stars.

Our Sun is an average yellowish star that astronomers classify as class G. This means that its surface temperature is in the range of 9,800–11,300° F (5,200–6,000° K) and the distinguishing features of its spectrum are lines of strongly ionized calcium and neutral metals such as iron. It is, however, composed mainly of hydrogen with some helium and traces of all the other known elements. It is about average as stars go, being only 865,000 miles (1,392,000 kilometers) in diameter. Some stars are fifty times as big, others fifty times as small. It is not too

bright, either, compared with many other stars, but it is bright enough to cause blindness if looked at with unprotected eyes for more than a few seconds. Sunglasses are not suitable protection, and using a telescope or field glasses to look directly at the Sun is extremely dangerous. Most astronomers observe the Sun by projecting its image through a telescope onto a screen. Early astronomers did not appreciate the danger. Galileo's blindness probably resulted from his observing the Sun without adequate safeguards.

The most important thing we have learned about the Sun is that although it is an ordinary star it is still unbelievably complex, much more so than pictures in white light, showing a few sunspots, would have us think. An astute inspection shows that the brilliance of the Sun is not uniform across its whole disc. It fades toward the edge, which astronomers refer to as the limb of the Sun. And in the darker areas close scrutiny reveals irregular patches of brightness that have been called flocculi, or loose, fluffy markings. An even closer look at the bright surface reveals that it consists of granules like grains of rice. This apparent surface of the Sun is called the photosphere, or sphere of light. It is where the atoms of the solar material are far enough apart for photons, which are discrete amounts of radiation emerging from the Sun, to escape into space. As a consequence, we see light from its surface. Even though there is no abrupt change in the density of the gas from below to above the photosphere, this region, which is about 220 miles (350 kilometers) thick, provides a surface to the Sun and a sharp edge to the illuminated disc as seen from space.

Darkening toward the limb is caused by murkiness in the atmosphere of the Sun. Light received here on Earth from the center of the solar disc passes straight up through the solar atmosphere, whereas light from the edges of the solar disc comes to us from a higher region of the photosphere. Because increasing height in the solar atmosphere leads to lower temperature, the limb regions appear less bright than the central regions. The bright flocculi seen near the edges are thought to be hot gases rising high in the atmosphere.

The rice-grained look of the photosphere is believed to be caused by hot gases rising from the interior into the equivalent of cloud tops. These grains, now referred to as granules, change rap-

This picture of the Sun shows groups of sunspots and the fine details of structure on the bright surface known as the photosphere. (Mount Wilson and Palomar Observatories)

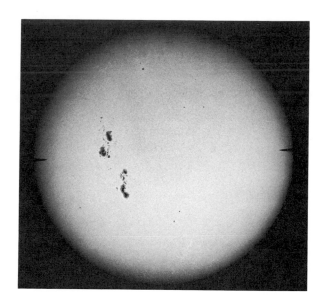

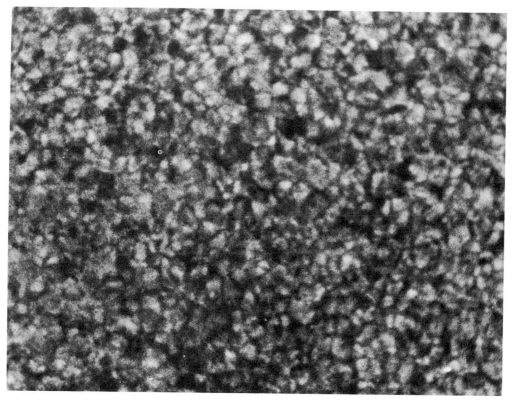

In close-up this fine structure has the appearance of rice grains; they are actually columns of photospheric gases rising like the tops of cumulus clouds. (Mount Wilson and Palomar Observatories)

idly, and we believe this shows that they are rising columns of gas that spread out at their tops and in cooling disappear from view. The effect is like liquid in a saucepan coming to the boil, with small volumes of water rising toward the surface so that the whole pan of water is continuously churned by this convective movement of heat from below toward the surface.

The Sun is thus a turbulent star, even though it looks placid when shining brightly in our terrestrial skies.

The chemical composition of the Sun became well known with the invention of the spectroscope, which allows scientists to identify elements by the lines they produce across a spectrum. Over sixty different elements were quickly identified in the solar atmosphere. However, hydrogen, the lightest element, accounts for over 73 per cent of the mass of the Sun. About 25 per cent comes from the next element, helium. And the remainder consists of the other elements. Some estimates place the percentage of hydrogen much higher: at 95 per cent of the Sun's mass.

Like the Earth, Jupiter, and Mercury, the Sun has a general magnetic field the intensity of which varies from zero to 20 gauss. In comparison, the Earth's field at the surface is about 0.5 gauss at the magnetic poles, and that of Jupiter at its polar cloud tops, about 12.5 gauss. On the Sun, local magnetic fields are immensely stronger than the general field and often reach thousands of gauss in the vicinity of sunspots.

SUNSPOTS

The sunspots are fascinating features of the Sun. Forever changing, they have lifetimes of anywhere from one day to several months. Historical reports tell of spots of enormous size and visible to the unaided eye. Those spots that have a short life span develop first as mere pores in the

photosphere, small black dots about 1,250 miles (2,000 kilometers) in diameter. If they are destined to become big spots they start to develop a lighter region around their dark center. Originally astronomers likened spots to shadows cast upon the Sun, so they referred to the parts of the spots with terms used in optics. The black central part was called the umbra and the lighter, surrounding region the penumbra. And these names stuck. Some spots develop several umbrae within a single penumbra. Others become huge, irregular features in a region of great complexity. Chains of spots also form, strung out in a line across the face of the Sun.

It may take as long as a week for a large spot to develop fully. A complex group may last for over two months and be seen through several rotations around the Sun's axis. Seventy or eighty spots are sometimes formed as a group. Often associated with the spots are bright streaks and spots called faculae. As these faculae rapidly

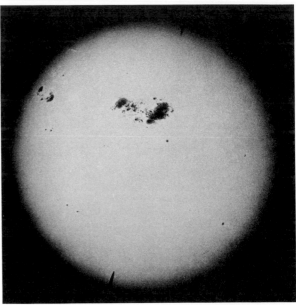

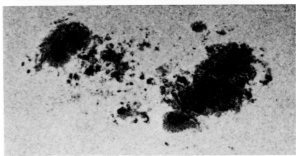

Sunspots are fascinating features of the visible surface of the Sun and have been recorded with the naked eye back to the time of the ancient Chinese. We now know they are magnetic maelstroms of photospheric gases that appear dark only because they are cooler than the surrounding gases. This huge group of spots was photographed on April 7, 1947. (Mount Wilson and Palomar Observatories)

grow, they are sometimes accompanied by a simultaneous increase in the number of sunspots in their vicinity.

Individual spots and groups of spots wax and wane in cycles that for the past 250 years have averaged eleven years. Rudolf Wolf, of Berne, refined Heinrich Schwabe's earlier cycle of about ten years to the 11.1 years now accepted. However, the Sun does not follow the cycle consistently. There was a period, from 1645 to 1715, when there were hardly any spots, while observations before then indicated a possible cycle of fifteen years. As a rule, a sunspot cycle gets under way with the first few spots appearing on the solar disc at mid northern and southern latitudes. Within a year or so the character of the cycle becomes apparent. If the maximum number of spots during the cycle is going to be high the number of spots increases rapidly and the maximum is reached within about four years. But if the maximum number is going to be low the increase is more gradual and the peak occurs within five or six years. Because of the variability in the time from minimum to maximum, astronomers measure the cycles of eleven years from the minimum point.

While the sunspot cycle is taken as being the count of all the spots appearing on the Sun, a closer look shows that it is not that simple. The spots appear at various latitudes, which change as the cycle progresses. Early in each cycle the spots are about 30° north and south of the solar equator. As the cycle draws to its close, the spots have moved toward the equator. Around the time of the minimum there are frequently spots of the old cycle along the equator and spots of the new cycle at high northern and southern latitudes.

MAGNETIC MAELSTROMS

Sunspots are attributed to an uneven rotation of the Sun which produces eddies in the hot plasma of a zone close to the photosphere in which heat is transferred upward by convection. This plasma, which is a mixture of electrons and atoms stripped of one or more electrons, contains sufficient numbers of free electrons to create magnetic fields as they are whirled around by the convective turbulence. A flow of electrons is a current, and a current generates a magnetic field. In turn, a magnetic field can give rise to magnetic

flux tubes along which plasma flows to produce the dark spots in the photosphere. Magnetic walls prevent the brighter, hotter material from engulfing the cooler, dark spots.

We know of the magnetic nature of sunspots from observing their spectra. In 1896, the Dutch physicist Pieter Zeeman was examining the spec-

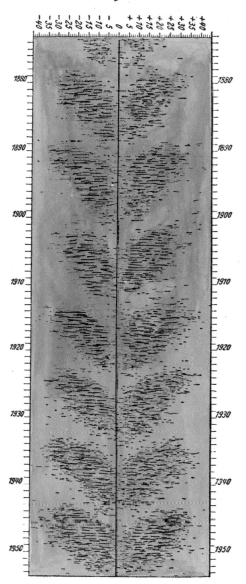

LATITUDE *of* SUNSPOTS

When the positions of sunspots are plotted against time, the resulting diagram has the butterfly shape shown here. This diagram, after Karl Kiepeheuer, in The Sun, *is based on spot distribution during the period from 1874 to 1932.*

tra of gases in his laboratory. He made a gas sample luminous and passed the light from it through a spectroscope. He was excited to find that when he placed the luminous gas within a magnetic field, each of the single lines that the gas produced in the spectrum became split into two lines close together. He withdrew the field and the lines merged back into one. He had discovered a way to detect the presence of a magnetic field at a distance, beyond the range of the effects of that field. Later, in 1908, the famous American astronomer George Ellery Hale used this effect, now called the Zeeman effect, to detect the presence of magnetic fields in sunspots. Later astronomers discovered that the field of a sunspot can often reach an intensity of ten thousand times that of the Earth's magnetic field. Astronomers found that sunspots were veritable magnetic maelstroms.

A simple pair of sunspots usually has a straightforward magnetic structure. If the leading spot has a north polarity the spot following it around the Sun's axis due to solar rotation has a south polarity. In the opposite hemisphere, however, the reverse is true; the leader has south polarity and the follower north polarity. Even more strangely, Hale discovered that this polarity reverses from one sunspot cycle to the next. Things become quite confusing at times of sunspot minimum, for the spots of the new cycle, in mid north and south latitudes, have opposite polarities from the spots of the old cycle, near the equator. Thus, from the point of view of the magnetic nature of sunspots, the complete solar cycle is twenty-two years rather than eleven.

Solar physicists generally agree that the driving force behind the sequence of sunspot activity is that the Sun does not rotate as a whole, i.e., like a solid body. Spots in equatorial regions travel once around the Sun in close to twenty-five days. But spots at 35° latitude take twenty-seven days. Other markings on the Sun, observed at 80° latitude suggest that the rotation there is about thirty days. The resultant twisting of the material of the photosphere probably extends deep below the surface. Most likely, this creates currents and

eddies that, coupled with the convection heat cells, create the magnetic fields that, in turn, give rise to the sunspots and other, related magnetic phenomena.

Many people have questioned whether the solar cycle observed today is a permanent characteristic of the Sun. As mentioned earlier, doubts were cast by Walter Maunder, and the spot-free period he alluded to was also recorded by Gustav Spörer, of Germany, about the same time. These observations were forgotten until recently, when J. A. Eddy, of the National Center for Atmospheric Research, again examined the old records and correlated them with other events that are thought to be associated with solar activities, particularly the aurorae. He found that there were no records of aurorae during this same sunspot-free period, from 1645 to 1715. These glows, high in Earth's atmosphere surrounding each pole, are believed to be associated with solar activities transmitted across space to the Earth. When aurorae reappeared after two generations without them, Scandinavians crowded into the streets, wondering what the strange apparitions foretold.

There is considerable scientific argument as to whether or not solar activities affect weather conditions on Earth. In the 1920s, A. E. Douglass suggested that the growth rings of trees followed an eleven-year cycle, like that of sunspots. Moreover, he found that there were no traces of this cycle during the seventy years of inactivity reported by Maunder and Spörer. During 1977, scientists at the Kitt Peak National Observatory, in Tucson, Arizona, found that the temperature of the surface of the Sun dropped by about a tenth of 1 per cent. Should the drop continue, it would have severe effects on the world's climate. Dr. William C. Livingston, of Kitt Peak, estimates that a 2 per cent decline in solar temperature over a fifty-year period could trigger the formation of glaciers over much if not most of our planet. But, he stated, "We presume the changes we see are cyclic and that the temperatures will stop falling sometime in the near future. I can't imagine anything else happening." While this matter has certainly not yet been resolved, it is

When the Sun is photographed by radiation from a very small part of the complete spectrum, astronomers produce a spectroheliogram, as shown in this picture. These fascinating views of our Sun show how it looks in red light emitted by certain elements such as hydrogen. Thus they show the detailed structure of hydrogen clouds on the Sun. (Mount Wilson and Palomar Observatories)

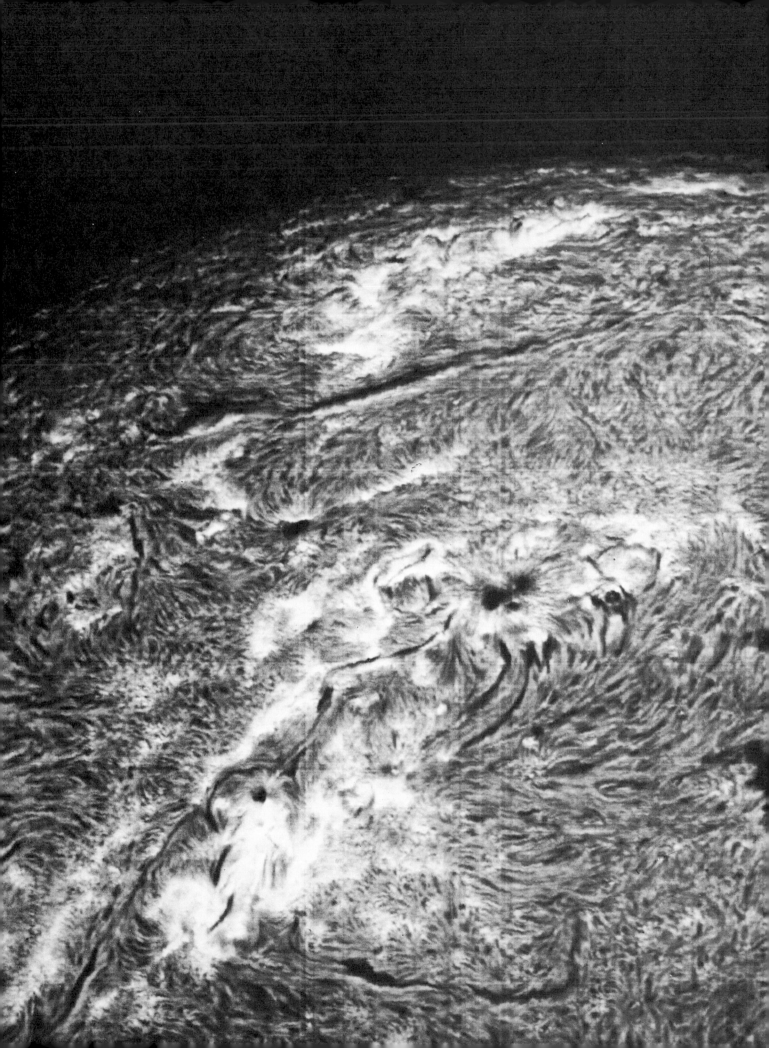

a very important one, because we are entirely dependent upon plant growth for our food chain irrespective of the more physical effects of floods and droughts.

One of the modern tools for astronomers to find out more about the Sun is that of observing it in light from only a very small region of the spectrum, like tuning in to an individual radio station instead of listening to many stations (white light) at once. This tuning is achieved by spreading the light from the Sun through a spectroscope into a long band of colors and looking at only one very narrow region of this band, say the region occupied by the emissions from a single gas in the photosphere of the Sun. For example, an astronomer might choose to look at the Sun by the light emitted from hydrogen atoms. In this way he sees just the hydrogen gas of the photosphere. Or he might decide to look at the Sun in the violet region of the spectrum, where there is a strong line of calcium. Then he sees the calcium in the photosphere of the Sun. This method of observing features on the Sun was developed initially by Johann Zöllner, of the University of Leipzig, in the mid-nineteenth century to study prominences at times other than during a solar eclipse. But it was George Hale, of Mount Wilson Observatory, who invented the apparatus known as the spectroheliograph, in 1908, to photograph the whole Sun in light emitted by hydrogen or calcium. His pictures showed hydrogen and calcium sweeping around sunspots as though part of gigantic whirlpools.

Astronomers now use this method of solar observation extensively. The view of the Sun in hydrogen light shows a solar surface that has been described as stringy, whereas the view in light from calcium atoms shows a mottled solar surface.

In addition to the differing structures within the photosphere seen in light of various wavelengths, there are also common areas of brightness that appear at many wavelengths. Called plages, they are areas of the photosphere where atoms are being emitted with sufficiently intense radiation to engulf individual sunspots and even groups of spots. They are also sites of strong magnetic fields and appear to be regions where spots later develop.

From years of observation of the Sun, astronomers have concluded that the sequence of sunspot creation is as follows: Spots begin with the generation of a strong magnetic field. This leads to the development of the hot spot of glowing atoms that we call a plage, or a solar activity center. Soon a dark pore appears like the top view of the funnel of a magnetic tornado, and grows into a spot, to be followed, shortly afterward, by another spot, of opposite polarity. As the spots enlarge, more spots may begin to form and also grow. Ultimately, a well-developed plage-sunspot group includes many individual spots, some of which may be large and irregularly shaped. This complex may last for several solar rotations until first the spots and then the plage itself merge back into a normal, rice-grained photosphere. Finally the magnetic field fades away too, removing all traces of the fantastic disturbance upon the face of the Sun.

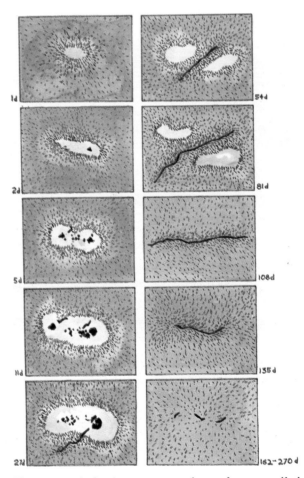

Hot spots of glowing atoms produce what are called activity centers, or plages, on the face of the Sun. The development of one of these plages is shown here. (After Kiepeheur, The Sun)

FLARES OF INTENSE ENERGY

Several times during the lifetime of such a spot complex, a small part of the plage will suddenly brighten enormously. Within twenty minutes or so it reaches an unusual brilliance, only to disappear a couple of hours later. This brightening is associated with the release of a prodigious amount of energy in what astronomers call a solar flare.

A flare emits intense ultraviolet radiation and X rays, together with high-velocity electrons and protons that spray like machine-gun fire across the Solar System. When the ultraviolet and X radiation reaches the planets, it affects their atmospheres. Here on Earth nitrogen and oxygen atoms are stripped of electrons (become ionized), producing a magnetic storm which interferes with radio communications. The protons and electrons enhance the solar wind into a sub-

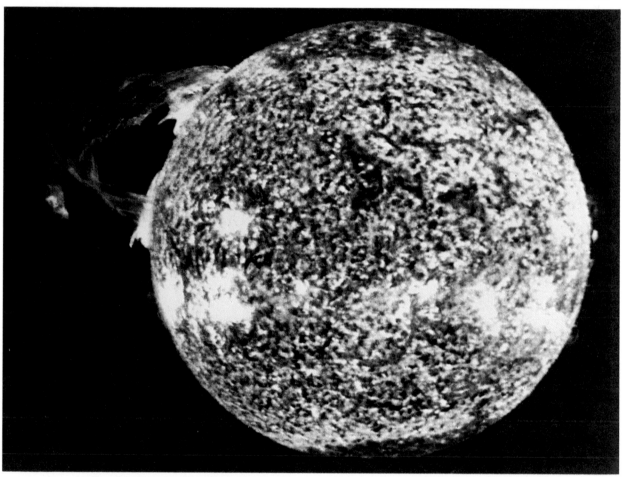

This photograph of the Sun, taken December 1973 by NASA's Skylab 4, shows one of the most spectacular solar flares (upper left) ever recorded, spanning more than 367,000 miles (588,000 kilometers) across the solar surface. The lost previous picture, taken some seventeen hours earlier, showed this feature as a large quiescent prominence on the eastern side of the Sun. The flare gives the distinct impression of a twisted sheet of gas in the process of unwinding itself. Skylab photographs such as these may provide clues to the mechanism by which such quiescent features erupt from the Sun.

In this photograph, the solar poles are distinguished by a relative absence of super granulation network and a much darker tone than the central portions of the disc. Several active regions are seen on the eastern side of the disc.

The photograph was taken in the light of ionized helium by the extreme-ultraviolet spectroheliograph instrument of the U. S. Naval Research Laboratory. (NASA)

atomic blizzard that also affects the mag-
netospheres of the planets and produces aurorae
at the Earth. At Jupiter the increased solar wind
squeezes the giant planet's magnetosphere and
squirts cosmic rays across the whole Solar System.

BLOOD-RED SPRAY FROM THE PHOTOSPHERE

When we look at the Sun at single spectral
lines of specific elements we see dark, narrow,
and irregular filaments that obscure the light from
the brighter photosphere beneath them. Often
these filaments become extremely active at the
time a solar flare bursts out of a plage. Other
times they are quiescent. Sometimes they disappear from view and then equally as suddenly
reappear. Seen on the edges of the sun, they rise
high above the photosphere and appear as bright,
blood-red clouds standing out clearly against the
darkness of space beyond. Then they are called
prominences. They were first seen as such by astronomers when the Sun was totally eclipsed by
the Moon. At first they were thought to belong to
the Moon, but this notion was dispelled by the
eclipse expeditions of 1851. And at the time of
the 1860 eclipse, Angelo Secchi and Warren de la
Rue both secured photographs of these flamelike
tongues extending from the limb of the Sun.

Nowadays they are photographed without waiting for a total solar eclipse, by using the spectroheliograph to look at the Sun at the wavelength of a single line in the solar spectrum, or by
using an instrument called a coronograph, which
produces (within the instrument) an artificial
eclipse of the Sun. Time-lapse pictures of prominences have been obtained that produce a breath-
taking series of pictures in which the motion of
the prominences is speeded up. The spectacular
sequence of plasmas and incandescent gases
streaming out from and into the Sun in arches,
fountains, curtains, and geysers is unforgettable
when viewed on a large screen. Sometimes these
prominences appear as though matter is shooting

*Against the bright photosphere there appear snake-
like darker filaments, which, seen on the edge of the
Sun, appear as blood-red prominences, huge flames
reaching into space from the photosphere. (Big Bear
Solar Observatory)*

away from the photosphere into space. Other
times, it seems that the flow is from space into the
Sun. Some prominences slide horizontally across
the surface, while others lie quietly for hours before suddenly exploding into space in celestial
fireworks that reach hundreds of thousands of
miles above the solar limb.

Prominences and their alter egos, filaments, are
phenomena of the chromosphere, a turbulent
layer above the photosphere. Its name was first
suggested by eclipse watchers who observed that
a ring of brilliant red surrounded the eclipsed Sun
when the Moon just covered the solar disc. There
is a relatively cool layer between the photosphere
and the chromosphere, but then temperatures rise
rapidly to about 25,000° F (15,000° K). The
chromosphere consists of myriad spicules like a
stubble of short hairs rising from the photosphere. These thin columns of gas last for only ten
or fifteen minutes. Each is about 450 miles (725
kilometers) in diameter but thousands of miles
high. They are believed to transfer heat from the
photosphere into the chromosphere, which has
been described as the spray from the turbulent
photosphere below it.

PEARLY RADIANCE OF MILLIONS OF DEGREES

Surrounding the chromosphere is another important region of the solar atmosphere, which is

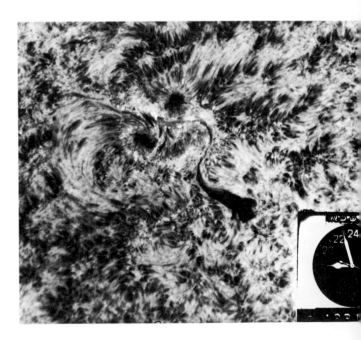

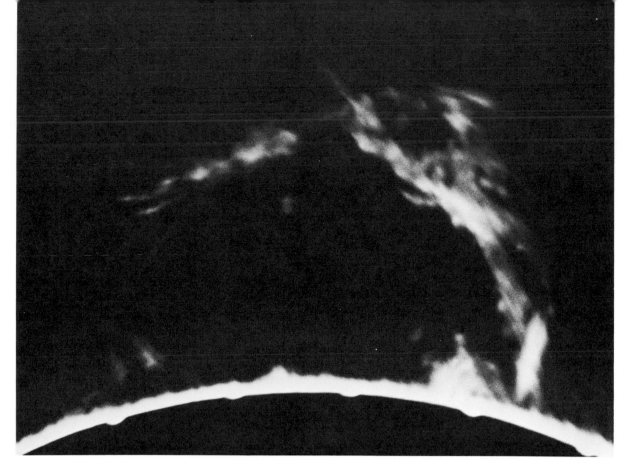

This prominence erupted to a height of 205,000 miles (330,000 kilometers) above the photosphere on July 2, 1957. (Mount Wilson and Palomar Observatories)

Prominences take diverse shapes: arches, fountains, and curtains. Here is a giant solar prominence reaching upward from the edge of the Sun. (Sacramento Peak Observatory, Air Force Cambridge Research Laboratories)

called the corona. Here temperatures are extremely high. This thin, tenuous outer atmosphere of plasma appears as a bright halo to the Sun during a total eclipse. Its beautiful pearly radiance assumes differing forms at differing stages of the sunspot cycle. At the minimum of activity, the corona stretches out from the solar equator as wide streamers, while elsewhere around the disc it is closer to the limb. At the poles, short "brush strokes" of light push toward the blackness of space.

At sunspot maximum, the corona appears more uniform around the Sun, but it has a few faint streamers at high latitudes that extend to several solar diameters, millions of miles, into space.

The view of the corona during a total eclipse is unforgettable. As the last few brilliant diamonds of the solar photosphere peeking between lunar mountains are finally snuffed out, the dark orb of the Moon is abruptly surrounded by a pearly iridescence that is never captured in the whiteness of a photographic print or slide. Though this

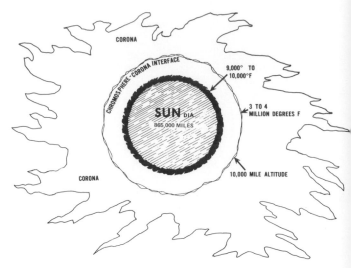

Above the photosphere is a region called the chromosphere, and farther out still is the corona. This diagram shows the interface between these two regions of the atmosphere of the Sun that we can see only on special occasions or with special instruments. (NASA)

At times of total eclipse of the Sun, the pearly corona becomes a spectacular, even awe-inspiring sight around the jet-black orb of the Moon. This eclipse photograph was obtained in Wyoming, at Green River, during the total eclipse of June 8, 1918. (Mount Wilson and Palomar Observatories)

corona extends over a much greater area of sky than the photosphere that we normally see as the Sun, the corona gives off only about as much light as that of two full moons. It vanishes immediately when the photosphere emerges from behind the Moon, at the end of the eclipse.

Until 1930, when the French astronomer Bernard Lyot invented the coronograph, astronomers could observe this fascinating region of the Sun only during the brief minutes of infrequent total solar eclipse. Despite many expeditions around the world, often at great physical hardship to the astronomers involved, only about one hundred minutes of total observation time had been accumulated since the invention of the telescope-camera-spectroscope combination. The coronograph considerably extended this period of observation, but it was limited to use on high mountaintops, where the atmosphere was relatively free of dust particles that would scatter the sunlight and obscure the faint corona. The really big breakthrough in observing the corona came with the space age, when big rockets carried instruments into space free from all the glare of Earth's atmosphere. Out in space the corona can be observed at any time with a very simple coronograph that actually gives better viewing than a total solar eclipse, because even during an eclipse the Earth's atmosphere scatters light.

SUBSTANCE OF THE NEAREST STAR

The spectroscope has played an important part in solar studies almost since the time when Joseph von Fraunhofer, the self-taught Bavarian master of optics, discovered that the spectrum of the Sun was barred by hundreds of dark lines. His untimely death, in 1826, prevented him from interpreting the meaning of these lines. It was another thirty-three years before Gustav Kirchhoff and Robert Bunsen, of the University of Heidelberg, demonstrated that the lines were the signatures of elements of the substance of the Sun.

Dark lines are produced when an element is present as a gas between a light source and the observer. The gas absorbs light at discrete frequencies and produces dark lines. If the element is incandescent, it emits energy at these same wavelengths and produces bright lines.

The light of the photosphere spread into a spectrum is striped with vertical dark absorption lines caused by the gases above the bright layers. Just after the photosphere is covered, during a total eclipse, there are a few seconds in which the solar spectrum reverses and all the dark lines become bright lines. This flash spectrum arises from hot photospheric gases above the bright limb that has been obscured by the disc of the Moon. It lasts for only a short while, because the emitting layer is not very deep, and rapidly the flash spectrum changes into the red-dominated spectrum of the chromosphere.

Toward the end of the nineteenth century, scientists found that some of the Fraunhofer lines in the solar spectrum could not be identified: they did not correspond to known elements whose lines had been catalogued by experiments in terrestrial laboratories. An important discovery was that the unidentified lines of this material on the Sun might be from a gas that would fit into the periodic table of the elements between hydrogen and lithium. If so it would be a noble* gas, similar to neon, argon, and krypton; in fact, it would be the lightest of the noble gases.

The new element was called helium, after Helios, the Sun. Shortly afterward, it was found here on Earth. Helium is now known to be an important element with a cosmic abundance second only to that of hydrogen, but its presence had escaped chemists on Earth until it was discovered on the Sun.

Spectroscopic examinations of the solar corona also revealed strange spectral lines that astronomers couldn't identify. Fresh from the discovery of helium, they suggested that a new element, coronium, had been found in the solar corona. But they were overenthusiastic. In the 1930s, B. Edlen, of Sweden, identified the coronal lines as belonging to normal elements such as iron and calcium under extreme conditions of low pressure and high temperature. Under these conditions atoms lose electrons and become ionized, or electrically charged. The amount of ionization depends upon the temperature, because higher temperatures provide energy to strip more and more electrons from their shells around the atomic nucleus. The importance of the discovery was far-

---

* These are called "noble" because they do not combine chemically with other elements to form molecules of other substances as, say, hydrogen combines with oxygen to form water.

reaching: By identifying the ionization states of iron and calcium needed to produce the lines of "coronium," scientists had developed a way to measure the temperatures present in the corona and elsewhere on the Sun. They were astounded to find that the gases of the Sun's outer atmosphere were unimaginably hotter than the surface of the photosphere itself, much hotter even than the outer chromosphere. The material of the corona has temperatures approaching 5.5 million degrees F (3 million degrees K).

We do not yet understand how such extremely high temperatures are produced. Certainly, the corona is by no means a quiet region. There are abrupt and extensive changes in its form. One theory of the high temperature is that shock waves are produced in the photosphere by the energetic processes occurring there, and these are transmitted upward through the rarefied gases of the outer solar atmosphere. Analogous to a supersonic airplane producing noise by concentrating energy in a sonic boom that can shatter windows miles beneath it, the solar shock waves concentrate energy to produce high temperatures far above their origin in the photosphere. Another theory is that electrons and ions in the coronal plasma interact with the changing magnetic field of the Sun. On Earth we use changing magnetic fields within transformers to induce electric currents in secondary circuits. If you touch such a transformer, even though it is a very efficient device, you will be aware that some of the electrical energy is appearing as heat. In fact, big transformers have to be cooled to operate. The changing fields of the Sun may be generating heat in a similar way in the plasmas of the corona.

Both processes may be operating together, and there may be others, of which we are not yet aware. We still have a lot to learn about plasmas and particles and fields.

WHAT MAKES THE SUN TICK?

While many people are still queasy about the safety of nuclear power stations, the Sun has been steadily generating nuclear power for some 5 billion years. The question of what keeps the Sun shining had intrigued astronomers and physicists for many generations. Often their explanations seem naïve and strange in view of what we have learned today. But we ourselves are still not sure that we know all the answers, because the Sun refuses to perform as our theories predict. However, for a long time the source of solar heat was a major mystery and a continuing challenge to the human intellect.

Each square meter of the Earth on which the Sun's rays fall directly, receives approximately 1.36 kilowatts of radiated energy. Thus, if we think of a spherical collector completely surrounding the Sun at the distance of Earth's orbit, so that it collects all the radiation pouring from the Sun, it would intercept some 4×10^{23} kilowatts. This is sufficient to supply 100 trillion worlds with Earth's population at the per capita electrical consumption of affluent Americans.

During the coal-burning age on Earth this type of energy abundance was an impossible challenge to scientific explanation. Even if the Sun had consisted of the best-quality anthracite coal it would have burned all of its mass in less than six thousand years. Yet scientists of that time already knew of fossils of life that had lived on Earth millions of years ago, so the Sun must have been radiating its energy for at least that long.

Toward the end of the nineteenth century the great Scottish physicist Lord Kelvin and the equally eminent German physician-philosopher-physicist Hermann Ludwig von Helmholtz attempted another explanation. They pointed out that the gravitational field of the Sun must store a vast amount of potential energy—energy of position. All the material high in the Sun's outer layers could be visualized as innumerable weights supported by the internal solar pressure but poised, ready to drop toward the center of the Sun, if the pressure were ever relieved. If a weight is supported by a cord that is then cut, the weight drops and converts some of its energy into heat when it hits the floor. Similarly, argued Helmholtz and Kelvin, the outer layers of the Sun, dropping just a short way toward the center before being stopped by internal pressure, would change gravitational energy into heat.

They suggested that as the solar interior lost heat by radiation into space, the Sun cooled slightly and contracted. As a result, infalling material converted some of its energy into heat so as to maintain a high temperature and keep the Sun shining. However, they were dismayed when they found that such a source of energy would suffice to let the Sun shine for a mere 24 million years.

Other astronomers speculated that meteors might be plunging into the Sun to raise its temperature. Again, calculations showed that this was unlikely. A glimpse of the truth we accept today came when the English astrophysicist Sir Arthur Eddington published his book *The Internal Constitution of the Stars,* in 1926. In the introductory chapter he wrote that the inside of a star is a hurly-burly of atoms, electrons, and other waves; it is therefore necessary to assume that subatomic energy of some kind is liberated within the star so as to replenish the store of radiant energy.

The subatomic processes had been hinted at even earlier. An encyclopedia published at the beginning of the century suggested: "It [the Sun's] extinction may be indefinitely postponed by unknown or barely suspected modes of action, such as the disintegration within its substance of elements akin to radium."

In 1928 two young scientists, Robert Atkinson and Fritz Houtermans, working in Germany, suggested that at the tens of millions of degrees temperature at the center of the Sun the kinetic energy of thermal motion would be so great that violent collisions among regularly moving atoms would be just as destructive to atomic nuclei as the then newly fashionable bombardment by subatomic projectiles that was being explored in many physics laboratories. They suggested that at such a high temperature, gases would become mixtures of atomic nuclei stripped of their electrons and the free electrons moving independently of the nuclei, what physicists today refer to as a plasma.

About this time too, Eddington proposed that, rather than relying upon the disintegration of heavier elements such as uranium and thorium, of which supplies within the Sun might allow only a lifetime measured in tens of millions of years, we should concentrate on the possibilities of hydrogen being converted into helium in accordance with the mass-to-energy conversion that Einstein had formulated in 1905, when he developed his special theory of relativity. How four atoms of hydrogen might be converted into one atom of helium was obscure in the 1920s, but it was a possible source of energy for the Sun. George Gamow also calculated what he referred to as the "disintegration ability" of atomic nuclei whirling around in a plasma at a temperature of millions of degrees. He concluded that while two nuclei of

atoms were more likely to collide at high temperatures, heavy nuclei would be repelled from each other more strongly by electrical forces between them, because they possessed greater electrical charges than did the lighter nuclei. Thus, he singled out hydrogen as being a more likely candidate for nuclear collisions than heavier elements such as uranium as a source of energy within the Sun.

The 1930s witnessed fervent activity worldwide in research into the nature of the atomic nucleus and how nuclei behave. By 1938, an answer to the riddle of the enormous energy of the Sun was found. Charles Critchfield proposed a reaction of four atoms of hydrogen fusing into helium and the conversion of about 0.8 per cent of the mass into energy. For one gram of hydrogen converted into helium, this energy amounts to about 600 megawatts. By deriving its energy from such a proton-proton reaction, the Sun could maintain its energy flow for 10 billion years.

There are several forms of the hydrogen-into-helium reaction known today. In one of them, two atoms of hydrogen fuse to form an atom of deuterium, with the release of an electron and a peculiar particle called a neutrino. This strange subatomic particle was suggested first by Wolfgang Pauli in the early 1930s, even though the particle was not positively identified in experiments until twenty years later. This particle has no electric charge or any appreciable mass unless it is moving, but it was required to account for part of the energy loss during the nuclear reaction. Next the deuterium combines with another hydrogen nucleus to form an isotope of helium, helium 3. Finally two nuclei of helium 3 combine to form the common helium 4 and two protons (hydrogen nuclei). There are two other possible paths from hydrogen to helium, which rely upon beryllium, boron, and lithium as intermediate stages from helium 3 to helium 4.

A year later, Hans Bethe in the United States and Carl Friedrich von Weizsäcker in Germany independently proposed another sequence of nuclear reactions leading from hydrogen to helium. In this series, carbon, oxygen, and nitrogen are involved; it is often referred to as the carbon cycle, because carbon is used at the beginning and released at the end to start another conversion. First, carbon and hydrogen fuse to produce nitrogen, which then decays into another isotope of carbon, releasing an electron and a neutrino.

Then the new carbon fuses with another hydrogen nucleus to produce another isotope of nitrogen. In turn, this nitrogen isotope fuses with a hydrogen nucleus to produce an oxygen atom that is unstable and decays to yet another nitrogen isotope by releasing an electron and a neutrino. Finally, this third nitrogen atom combines with yet another hydrogen nucleus and immediately breaks down into a carbon nucleus and a nucleus of helium 4. The carbon atom is thus made available to start a second cycle.

It is now believed that the proton-proton reaction is the energy-producing source within such stars as the Sun, while the carbon cycle acts in larger stars. Both processes consume four hydrogen nuclei to produce a nucleus of helium 4. Both release energy and neutrinos. But they require different conditions, resulting from the difference between the electric charge on the nucleus of the hydrogen atom compared with that on the carbon nucleus. The former has a charge of one unit, the latter one of six units. It is more difficult to bring charges of six and one together than two charges each of one unit. The proton-proton reaction requires particles that are less energetic than the carbon cycle and can accordingly begin at a lower temperature.

At the center of the Sun, suitably elevated temperature and the right pressure and density of the plasma are attained for the proton-proton reaction. The density is about fifty times that of water, the pressure nearly a billion times Earth's sea-level atmospheric pressure, and the temperature is 29 million degrees F (16 million degrees K). Since conditions at the center of a star depend on its mass, small stars such as the Sun can manage only to enter the proton-proton reaction. But larger stars, in which the central temperature can climb to 54 million degrees F (30 million degrees K), initiate the carbon cycle, which becomes their dominant means of producing energy.

The difference in temperature and nuclear power generation also leads to a major difference in the ways stars pass their internal heat to their surfaces. In a massive star the central core passes its heat toward the surface by convective processes similar to boiling, while the outer regions transfer the heat by radiation. In stars such as the Sun the opposite is true. Heat is transported through the core by radiation, but in the outer layers it is carried toward the surface by

convection. This boiling of the outer layers produces the turbulent photosphere so characteristic of our Sun.

The interior of the Sun thus consists of a core that has about 20 per cent of the mass of the Sun, in which hydrogen fuses into helium. Radiation from the core starts out as gamma rays released in several stages of the proton-proton reaction. The high-energy photons (discrete amounts of electromagnetic radiation) commence their long journey to the surface of the photosphere. On their way they interact with atomic nuclei and electrons and lose energy, so that they become less energetic; their wavelengths become longer.

At about 70 per cent of the distance from the core to the photosphere, the temperature of the solar material has dropped to about 3.6 million degrees F (2 million degrees K), and the density is about one hundredth that of water. These conditions allow a significant change to occur in the structure of the plasma. Some of the protons and electrons combine into hydrogen atoms, a process that releases more energy, which adds to the upward flux from the core. The additional heating and the changed nature of the plasma generate strong convective forces that bubble solar material upward toward the visible surface.

The energy carried upward by the buoyant motion of heated solar material is loosely similar to that on Earth which carries cumulus clouds as bubbling masses into the colder upper atmosphere when the air is warmed by the hot surface beneath. As the heated solar material reaches the outermost layers of the photosphere, it sheds its energy into space by radiation and, cooled thereby, then descends back into the lower regions to repeat the circulation pattern.

The granules seen in the photosphere are loosely analogous to the tops of a dense collection of cumulus clouds. But there the comparison ends. The solar granules are hundreds of miles long and move with speeds approaching 1,800 miles per hour (3,000 kilometers per hour). The upthrusting solar material also consists of plasma —completely stripped atomic nuclei at the bottom and partially stripped nuclei at the top of the grains. The electrons that have been removed from the atoms of the photospheric gases are free to be whipped about by the magnetic fields. Some of them are recaptured by atoms as they reach the surface of the photosphere. This capture releases more energy, which raises the tempera-

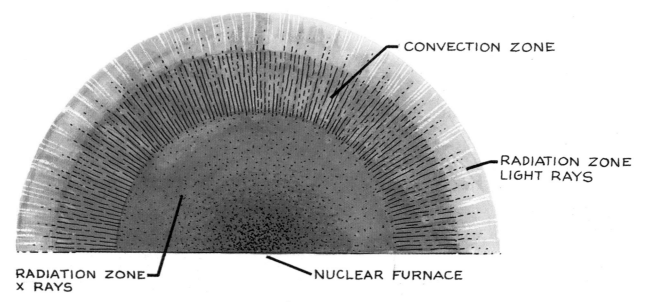

CONVECTION ZONE

RADIATION ZONE
LIGHT RAYS

RADIATION ZONE
X RAYS

NUCLEAR FURNACE

Energy generated within the Sun escapes into space at the same rate as it is generated. In the deep interior, X rays compete with convection, the flow of incredibly heated matter, in order to carry the energy of fusion away from the nuclear furnace within the Sun's core. After traveling some 300,000 miles (480,000 kilometers) toward the surface, lower density and temperature are reached and the energy produces convective swirls of plasma that transport the heat more rapidly. At more than 400,000 miles (640,000 kilometers) from the center, density and temperatures have dropped so much that radiation dominates. In this region, the photosphere, energy is emitted into space as visible light, as well as radiation at other wavelengths seen only with special instruments. (Arfor Picture Archives)

ture and forces the granules still higher. This, again, is somewhat analogous to condensation of water droplets from cloud vapors to release heat, causing the clouds to rise higher in the terrestrial atmosphere.

The rising granules in the solar photosphere are hotter than their surroundings and so appear brighter. This is what produces their rice-grained look.

MYSTERY OF THE MISSING NEUTRINOS

When Pauli developed his idea of a small particle with no mass and no electric charge, he called the particle a neutron. But before his particle was discovered, James Chadwick found a neutral heavy particle, formed from a proton and an electron, that he, too, named a neutron. Enrico Fermi, it is recounted, at a meeting in Italy of nuclear physicists, consoled Pauli by suggesting that his particle should be called a little neutron, *neu-*

trino in Italian. The name was accepted.

F. Reines and C. Cowan found evidence of neutrinos in 1953 with a cloud chamber—a device that reveals the tracks of subatomic particles —placed close to a nuclear reactor. The neutrino is an extremely penetrating particle and passes through matter like light through a window pane.

In 1968 a surprise discovery questioned a coherent and generally accepted scientific concept. The occasion was an attempt to detect neutrinos from the Sun by their absorption in great vats of perchloroethylene deep in a salt mine. The tanks were surrounded by jackets of water, and the experiment looked for the action of neutrinos to produce argon from chlorine. By placing the apparatus deep below ground (which hardly absorbs neutrinos but stops most other subatomic particles that could produce argon), the experimenter, Raymond R. Davis, of Brookhaven National Laboratories, avoided the formation of argon except by neutrinos.

The big surprise was that he did not find any

argon; the experiment failed to detect the expected flux of neutrinos from the Sun. This flux from the proton-proton reaction was five times the limit of the detection capabilities of Davis' experiment. Everyone was mystified. What could have happened to the solar neutrinos? Either the Critchfield-Bethe-Weizsäcker process did not apply to the Sun as everyone thought, or some unknown neutrino absorber was stopping these particles from leaving the Sun or reaching the Earth.

The discrepancy still remains unexplained. And it is very disturbing, for the Sun seems to be such a normal star and we would expect it to behave in what our theory says is a normal manner. But we could be wrong. Some physicists have speculated that perhaps the Sun is a variable star and its nuclear fires have died down. We have not witnessed the effect in the light and heat from the photosphere because this energy started from the core many years ago. The neutrinos, by contrast, penetrate quickly through all the matter of the Sun and they may be telling us something very important about the Sun's future. If the nuclear fire has died down, then things could be very bleak: our Sun will not be providing us with heat and light steadily for another 6 billion years as we now confidently expect.

Other scientists have speculated that the neutrinos may be absorbed between the Sun and the Earth, or that the nuclear reactions we have so painstakingly defined may be in error. This major mystery dramatizes how much we still have to learn about a relatively simple star such as our Sun, on which our very lives depend.

NEW TOOLS OF THE SPACE AGE

Since the advent of the space age, in 1957, powerful new tools for observation and actual experiment have been brought to bear to further our understanding of the Sun. As is usual in science, the new tools have not only confirmed and refined many earlier findings but have led to several new and deeper questions.

One of these concerns solar flares. Prior to the development of space-borne astronomy, these flares could be studied only by radiation that was capable of passing through the Earth's atmosphere. And even though telescopes were placed on very high mountains, this was not enough to avoid the screening effect of the terrestrial atmosphere, particularly when we wanted to find out how the Sun emits its energy at wavelengths other than those of visible light.

The first new information came from sensors mounted on Earth satellites to look at the Sun in X rays, gamma rays, and ultraviolet light. These sensors showed that the Sun observed at wavelengths beyond the visible spectrum has very different characteristics. Astronomers found out that there are hundreds of small regions that emit X rays and show up as bright spots on the X-ray photographs obtained beyond the Earth's atmosphere. These spots live for only six to eight hours before they disappear. These energetic eruptions on the Sun are much more frequent than solar flares and are believed to contribute to a major part of the solar wind, which sweeps outward from the Sun through all of interplanetary space surrounding it.

Later, the Skylab manned observatory in orbit around the Earth carried a battery of instruments designed to observe the Sun at wavelengths that do not penetrate to the Earth's surface. From this, we discovered that visual flares are preceded by powerful activities in the solar corona that were unexpected from the Earth-based observations. And astronomers were able to see pictures of the corona evolving and changing. It proved to be much more actively in motion than they had ever suspected. It pours outward from the Sun and is regarded as being the main flow of the particles that ultimately form the solar wind.

In August 1972 an instrument carried by an Orbiting Astronomical Observatory—an unmanned spacecraft placed into orbit around the Earth and carrying experiments to study the ultraviolet, X-ray, and gamma-ray regions of the electromagnetic spectrum—recorded gamma-ray emissions from the Sun that indicated violent nuclear reactions on the surface of the photosphere. Shortly afterward, this area developed a flare that was the most intense ever observed. Astronomers think that such gamma-ray emissions might provide advance warning of such intense flares that can cause disturbances such as magnetic storms on the Earth. One of the spectral lines observed during this event was of tremendous importance to physics, because it suggested that matter was actually being annihilated on the surface of the Sun by the collision of electrons and positrons (positive electrons) and the resultant cancellation

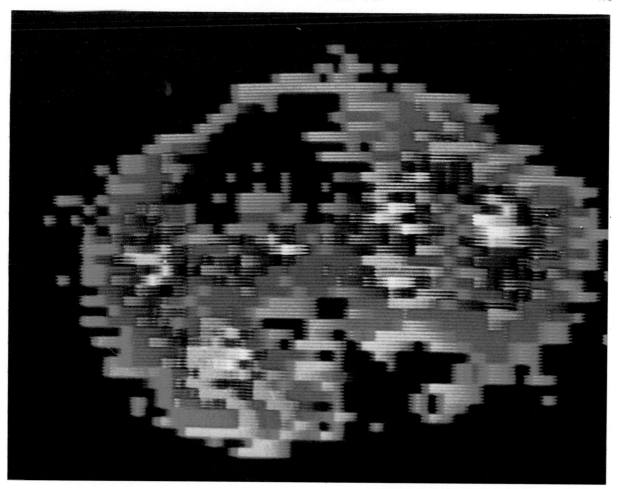

This is a map of the solar corona obtained by Orbiting Solar Observatory-7 (OSO-7) and presented on a computerized color display. The observations show the "polar caps" (dark areas upper left and lower right) in the photograph, which may be as much as 900,000 degrees Fahrenheit (½ million degrees Kelvin) cooler than the surrounding 3.6-million-degree-Fahrenheit (2-million-degree-Kelvin) corona. The north pole is the dark area upper left and the south pole is the dark area lower right. Embedded in the Sun's equatorial corona are hot spots where temperatures may rise as high as 72 million degrees Fahrenheit (40 million degrees Kelvin) during solar flares. The Sun is currently nearing the end of its eleven-year cycle, previously thought to be a relatively quiet period. (NASA)
(Note: The brighter the color, the greater the quantity of ionized gas. The very hottest areas appear white, while the cooler places appear turquoise and black.)

of the two oppositely charged particles in a blaze of energy. This was the first time that direct evidence had been obtained (outside of laboratory experiments) that such violent releases of energy occur in nature. The event on the Sun was on a much larger scale than that of the laboratory experiments.

Solar X rays have also revealed other strange characteristics of the Sun. Some regions do not emit this radiation; they appear as holes of darkness in X-ray pictures of the Sun. Astronomers interpret these dark regions as coronal holes where the coronal plasma is cooler than normal and possibly also of lower density. Some of these coronal holes extend for 500,000 miles (800,000 kilometers) or more. They were subsequently identified as being regions from which the blizzard of electrons and protons of the solar wind flows from the Sun into interplanetary space. Studies of the relationship between the dark regions and the shape of the corona led to the belief that hot material of the corona is contained within closed magnetic-field lines, i.e., lines that start at the solar surface and curve back to it elsewhere. However, when the field lines are open and extend into interplanetary space, the solar wind is emitted into space too. Loops in the coronal region have been revealed by extreme-ultraviolet photographs made from Skylab. Some of these extend hundreds of thousands of miles from the Sun and appear to originate in active regions of the photosphere.

Prior to the space age, astronomers thought that the solar wind blew through space in a constant stream from the Sun and that it was emitted from all parts of the photosphere. Now we know this is wrong. The wind blows in gusts, it consists of both slow and fast streams, and it seems to be coming from specific areas of the Sun. An Orbiting Solar Observatory—an unmanned spacecraft carrying instruments specifically designed to observe the Sun, as opposed to those in the Orbiting Astronomical Observatories, which looked at other celestial objects as well—discovered that the massive outbursts of the solar wind can have much stronger effects on Earth than even the most intense solar flares. Fortunately they affect us much less frequently, because they do not often score direct hits. Perhaps once in every ten thousand years one of these energetic streams achieves a bull's-eye on our planet. And when it does the result may be very abnormal weather conditions.

We also discovered that the Sun spews forth enormous quantities of dust; its pollution of the interplanetary environment makes our industrial pollution of Earth's atmosphere relatively insignificant. This dust is believed to cause noctilucent clouds that appear after sunset at very high altitudes in the skies of Earth. Rocket probes sent into these clouds to collect particles from them have shown that they contain dust of heavy metals including hafnium, nickel, osmium, lead, tantalum, and ytterbium. Scientists also found traces of these metals in the dust gathered by Apollo astronauts from the plains of the Moon. The dust particles of heavy metals may condense from the material of the solar photosphere in the cooler regions of sunspots. It is then sprayed into space by the violent activity and magnetic disturbances associated with the spots.

While the dust grains are extremely small, they are just about the right size to be pushed through the Solar System by the radiation pressure of ultraviolet light of extremely short wavelength. This action is similar to that which pushes out materials of comets to form their tails pointing away from the Sun. The pressure of radiation, long theorized by astronomers, was shown to be a real force to contend with in space when it was found to disturb the orbits of artificial satellites. In fact, the pressure was used to save a space mission from disaster. Engineers used the radiation pressure to control the orientation of a Mercury-bound spacecraft so that precious maneuvering gas could be conserved and three visits made to the innermost planet.

Skylab, which was placed in orbit in mid-May 1973, was our first manned space station. During subsequent months, crews of astronauts logged a total of 171 days operating in orbit. During this period they performed an incredible number of unprecedented scientific experiments, many of which were directed toward the Sun's mysteries. The great advantage of observations from orbit is to be above the absorbing layers of the Earth's atmosphere. This way we can record radiation from the Sun at many wavelengths inaccessible from Earth. Whereas terrestrial observations are limited to a band of wavelengths from about one thousandth to a little more than one ten-thousandth of a millimeter, the Skylab experiments pushed the range of wavelengths down to nearly one ten-millionth of a millimeter. This is down to the region of so-called hard X rays. With eight special telescopes observing the Sun over various spectral bands, Skylab provided the first opportunity to perform long-duration, highly detailed analysis of pictures of the Sun obtained simultaneously in visible, ultraviolet, extreme-ultraviolet, and X-ray wavelengths. The Skylab also carried a coronograph that made an artificial eclipse of the Sun free from the scattering effects of the terrestrial atmosphere. This instrument gave unprecedented views of the corona and its activities for several complete rotations of the Sun.

Although it will take years to assess the new data fully, most of which will be published in obscure scientific works and rarely noticed by the general public, solar astronomers have already reported on some of the months-long observations made by the Skylab astronauts. The corona, as

This photograph of the solar corona was made on June 5, 1973, by the High Altitude Observatory White Light Coronograph Experiment on Skylab's Apollo Telescope Mount. (NASA)

mentioned, changes much more than was expected. There are dramatic large-scale events occurring in a bewildering array of form and structure that may vary in a matter of hours. But despite the dynamic nature of the pearly outer atmosphere of our Sun, it is nowhere near as active as regions lower down. One of the coronal features observed consisted of huge expanding loops of material that have been called solar transients. They had actually been observed during eclipses, but astronomers thought they were rare. During the Skylab mission such solar transients were recorded forty times. So they are undoubtedly much more common than we thought. Skylab saw them as great loops of highly tenuous matter expanding outward from the Sun at speeds exceeding 1 million miles per hour (1,600,000 kilometers per hour).

X-ray investigations produced exciting information about solar flares and lesser but equally dramatic X-ray activities elsewhere on the Sun. The ultraviolet experiments produced images of the Sun far exceeding in crispness of detail those obtained earlier by unmanned satellites. On these new pictures many new features were revealed that are helping astronomers trace the complex story of solar activity. Other ultraviolet experiments provided pictures of the Sun similar to the spectroheliograms obtained in visible-light regions of the spectrum. They showed for the first time that helium erupts from the surface of the photosphere in enormous clouds that sometimes remain suspended 500,000 miles (800,000 kilometers) high. Series of pictures revealed that the helium cloud shot upward and was abruptly stopped as though its passage had been blocked. Each cloud may contain as much as fifty thousand tons of helium and hydrogen. A mysterious

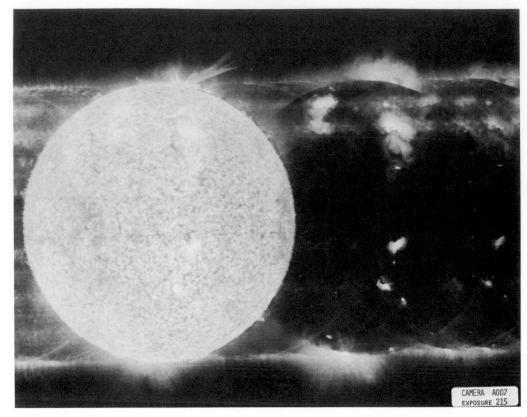

CAMERA A007
EXPOSURE 215

Contrasting images of the Sun in action. An eruption on the limb (at the top) is in progress, shooting out one-hundred-thousand-mile-long jets of helium. (Skylab 2) (NASA)

feature associated with the clouds was a rain of matter in fine threads back toward the photosphere. While astronomers can explain the cloud motions in terms of magnetic and gravitational fields, the rain cannot be accounted for in this way. Some other, unknown force appears to be responsible.

The Skylab mission was an excellent example of the scientific benefits of manned space flight as compared to reliance on automatic instruments or instruments controlled in part from Earth. The Skylab astronauts were able to overcome problems as they arose and, even more important, they were able to tailor the mode of operation of instruments to gather data in the best possible way. When something exciting began to take place, they could concentrate on preparing instruments to observe the new event.

A particularly good example of man's adaptability in space occurred toward the end of the Skylab mission. Controllers at NASA's Johnson Space Center, in Houston, Texas, heard the voice of one of the astronauts, Owen K. Garriott, coming over the communications channel from space. Excitedly, he told them of a huge magnetic loop forming on the Sun. He had been watching the TV-type display screen that displayed a visual-light conversion of ultraviolet images of the Sun being sent over the data links to Earth or recorded on cameras within the spacecraft. Quickly alerted to this unusually big solar event, astronaut Garriott was able to drop his normal schedule, abandon his sleep period, and train a battery of instruments to get as much information as possible on a fascinating solar feature that could not be recorded from the ground.

A solar occurrence that had been suspected by astronomers was confirmed by the Skylab observations. A flare occurring on one spot on the photosphere is followed a short while later by a flare elsewhere—a sympathetic flare. The observations from Skylab showed one flare popping up like an exploding flashbulb shortly after another, even sometimes simultaneously, but far away on the photosphere.

On September 7, 1973, the Sun put on an unusual display for the Skylab astronauts. Alan Bean, in orbit at that time, called the display the "big daddy" of solar flares. Lasting for about fifteen minutes, at its peak intensity the flare was ten times the diameter of the Earth and generated energy equivalent to more than 100 million one-megaton fusion bombs. Such enormous quantities of energy almost reach the limits of human comprehension.

As our awareness of physical processes in the universe is expanded through studies of our nearest star, we expect to find new means of tapping energy sources that will cause current energy limitations to seem ridiculous. Probing at the frontiers of scientific knowledge invariably brings to light discoveries that enterprising individuals rapidly apply to improving living standards in our everyday world. As we clear up the mysteries of how the Sun spews forth enormous amounts of energy across the Solar System, how it develops flares, and what has happened to the missing neutrinos, we will undoubtedly arrive at clearer understandings of how processes other than the burning of substances in the molecular reactions of fossil fuels can be used to power our technological civilization. Even now, NASA and the European Space Agency are developing some thirty experiments for a two-spacecraft team that will climb out of the ecliptic plane and examine the Sun from over its north and south poles.

The Sun may, indeed, be the key to our energy future but not necessarily in the form of the relatively simple gathering of its heat energy as currently used for solar energy applications. Rather, it could be the understanding of the physical processes involved in generating heat—the interactions of plasmas and subatomic particles, of magnetic and electric fields—that will enable us to develop advanced sources of energy that we scarcely can dream of today. For in our thrust for knowledge about the Sun we are indeed probing into one of the enormous powerhouses of our galaxy.

4
BETWEEN THE PLANETS

NATURE ABHORS A VACUUM

Look at any astronomical book published before the space age and you will find very little mention, if any, of the space between the planets, what we call the interplanetary environment. Some books may refer to the zodiacal light and *Gegenschein,* and they will certainly discuss comets and meteoroids, but until recently the great void of space was regarded as precisely that: a great void.

Gazing out from Earth at the other planets, it certainly does seem that space is quite empty. But we have discovered that this is not so. The new understanding has come from the exploration of space between the planets by many kinds of spacecraft. One has penetrated to within 27 million miles (43.4 million kilometers) of the Sun, and at the time of completing this book another had traveled far beyond Jupiter and crossed Saturn's orbit. Soon spacecraft will have explored the interplanetary environment, the weather of space, out to the orbits of Uranus and Neptune.

While the amount of matter in space between the Sun and Earth's orbit is less than one quintillionth that of Earth's atmosphere at sea level, a fantastically good vacuum by terrestrial standards, it is by no means a perfect vacuum, and the tenuous matter in space is by no means inert. The space between the planets is active and changing, swept by protons and electrons from the Sun, shot through by atomic nuclei from beyond the Solar System, sprayed with high-speed electrons from Jupiter, and twisted and snarled by electromagnetic fields.

BLIZZARD FROM THE SUN

The most important feature of the interplanetary environment is the blizzard of electrons and protons streaming outward from the Sun. By the mid-1940s scientists were generally agreed that there was some sort of particle radiation from the Sun associated with sunspots and solar flares. The theory that supported this also explained the formation of the aurorae, those high-altitude glowing banners and curtains that so spectacularly decorate Arctic and Antarctic skies. In addition, a few days after any great solar activity there is a disturbance of radio communications that rely upon reflection from ionized layers in the Earth's upper atmosphere. This, too, was interpreted as meaning that corpuscles had traveled from the Sun to the Earth. But scientists still wondered whether charged or neutral particles were involved, as well as electrons that spiraled in along lines of magnetic force.

The Sun radiates streams of electrically charged particles in the form of plasma. These streams squirt across the Solar System in bursts, sometimes moving very quickly, at other times more slowly, but spreading out from the Sun in all directions. Like rivers, electrons and protons swirl and eddy as fast currents catch up with slow currents that left the Sun earlier. Labeled the "solar wind" in 1958 by the American astrophysicist E. N. Parker, they reflect solar conditions and carry the effects of varying solar activities across the interplanetary reaches to the planets. Since the solar wind conducts electricity, it carries along with it a trapped magnetic field.

The first definite measurements of the solar wind in space were made by the Soviet spacecraft Luna 3 in 1959. Two years later the American Explorer 1 satellite made similar measurements and confirmed that streams of particles emanate from the Sun. Measurements have continued ever since with a variety of spacecraft including Earth satellites, so-called interplanetary monitoring platforms, Pioneers engaged on deep-space missions to Jupiter and other outer planets, and Mariners on missions to Mercury, Venus, and Mars. In 1962, while conducting extensive studies deep in space during its flight from Earth to Venus, America's Mariner 2 discovered that the solar

wind blows at variable speeds; prior to that, scientists had thought that the speed might be constant.

From the moment the plasma particles start on their journey from the Sun, they face a long, uphill fight against the powerful solar gravity. However, they are boosted on their journey by the pressure of the Sun's radiation. They achieve their initial velocity from the violent events taking place at and above the photosphere. By the time the streams reach the orbit of the Earth, the particles are traveling at an average speed of 250 miles per second (400 kilometers per second). But parts of the wind gust as high as 450 miles per second (725 kilometers per second), while slow streams may only reach 150 miles per second (240 kilometers per second).

Spacecraft measurements have revealed that at Earth's distance from the Sun the solar-wind plasma is amazingly tenuous, consisting of only

America's first satellite, Explorer 1, made a major discovery about the environment of space. Its instruments showed that Earth is surrounded by belts of charged particles —electrons and protons—that are called radiation belts. Later they were named after their discoverer, James A. Van Allen, and became known as the Van Allen belts. On the right is the late Wernher von Braun, leader of the Army's Redstone Arsenal team that built the first-stage Redstone rocket that launched Explorer 1; at left is William Pickering, head of the Jet Propulsion Laboratory that developed the satellite. (NASA)

about ten particles per cubic centimeter. But these particles are moving extremely fast away from the Sun and have a high temperature. This motion is able to propel the solar wind deep into the outer Solar System. Just as the individual motions of particles of a gas define its temperature, so the solar wind's particles define its temperature, which turns out to be approximately 280,000 degrees F (160,000 degrees K). Because of the small number of the particles and their high speed, each travels on the average a distance of 60,000 miles (100,000 kilometers) before colliding with another. Referred to by physicists as a "collisionless plasma," it has very high electrical conductivity.

However, the situation changes drastically when the fast-moving plasma stream encounters a magnetic field such as that of a planet. The highly conductive plasma cannot penetrate the magnetic field, with the result that the stream of plasma is "shocked" into a deflection around the planet's field like a stream of water from a hose hitting a curved surface. By measuring how the solar wind is deflected around planets, scientists have been able to map the shape of magnetic fields in space and outline their characteristics.

The solar wind reacts with magnetic fields even before it leaves the Sun. In fact, during violent activities within the solar corona local magnetic fields are torn from the Sun and pulled by the solar wind across the Solar System. Some astrophysicists refer to the cloud rushing from the Sun with its entrained magnetic field as a plasma "tongue." The field of the solar-wind stream also affects the charged particles within the stream. They are kept together as if they were in a magnetic bottle hurled across the Solar System.

If you could poise yourself high above the Solar System to look down on it, the streams of the solar wind (if made visible to you) would not emerge from the Sun like light rays. Rather, they would appear to be curved around the Sun in spirals. This is because their outward flow is affected by the spinning of the Sun on its axis. The effect is similar to what you would see by looking down from above on a rotating lawn sprinkler with a series of jets. Imagine an emitting point for the solar wind near the equator of the Sun. It spurts out the plasma particles vertically, and they stream off radially into space. But at the time they leave the solar surface they also have a horizontal velocity resulting from the solar rotation, just as

the inhabitants of Earth are spinning around with the planet. While people living near the Earth's equator have a horizontal speed of about 1,000 miles per hour (1,600 kilometers per hour), the solar particles begin their interplanetary journey with a horizontal speed of 4,300 miles per hour (7,000 kilometers per hour). They are traveling once around the Sun in twenty-five days. When they have reached a distance of one solar radius from the Sun's surface, they have only the same horizontal velocity but a much greater distance to travel, so they have lagged behind points on the surface. And as they rise higher and higher they lag farther and farther behind. So the emitting point seems to send out a curved stream of particles.

When this stream reaches the orbit of a planet, it does not move directly across the orbit but, rather, at an angle that depends on the planet's distance from the Sun. At Earth's orbit, for example, the solar-wind streams cross at an angle of about 45°. At the distance of Jupiter the spiral is more tightly wound and the streams cross Jupiter's orbit at an angle of only 10°.

Another important aspect of these streams is that fast streams catch up with slow ones. They act like billiard balls and rebound from each other when they collide. At the time of the collision there are big differences in the magnetic fields carried by the streams and these give rise to a magnetic gradient that acts as a barrier to cosmic rays coming from outside the Solar System. These high-velocity nuclei of atoms stripped of their electrons are scattered by the solar-stream collision areas and are prevented from plunging deep into the Solar System. Because of this scattering effect, no low-energy cosmic rays penetrate into the inner Solar System. One of the major discoveries from the first spacecraft to travel to and beyond Jupiter was that these scattering centers generated by the solar-wind streams extend more than five times Earth's distance from the Sun. How far out they reach at solar maximum to protect the system from cosmic rays is not yet known. It is anticipated that before the end of the century spacecraft now speeding into the outer Solar System will answer this question.

WHEN THE WIND BLOWS . . .

As the Sun rose over the Lunar Apennine and flooded with light the Hadley Delta site on which

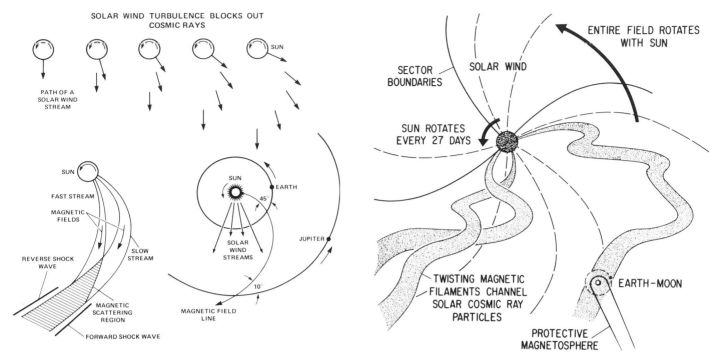

Rotation of the Sun causes the solar wind to follow a curved path through space. The drawing to the left shows how fast streams catch up with slow streams and produce scattering regions that prevent low-energy cosmic rays from penetrating into the inner Solar System. Turbulent fields flip the particles back before they can penetrate far. The drawing to the right shows a "weather map" of the solar-influenced interplanetary environment, looking down on the top of the Sun, i.e., from above Earth's orbit. The average angle of magnetic-field lines to the Earth-Sun line is about 45° and results from solar rotation. Once a magnetic filament rotates past the Earth, the high-energy particles in the solar-wind stream miss our planet. (NASA drawings)

the landing module of Apollo 15 nestled, astronauts Irwin and Scott prepared for another day of expeditions across the hostile lunar surface. When they emerged from the gaunt structure on its spindly legs and started their work on the lunar regolith of meteor-shattered debris, a howling wind was blowing around them. But they could not hear its cadence. Nor could they see any evidence of the wind from the Sun as its protons and electrons peppered microscopic holes into the surface of the rocks that they were gathering for scientists on Earth. Later those scientists were able to see the tracks in these rocks and deduce that the wind had not varied greatly over millions of years. The lunar rocks carried such

tracks as a record of the wind for all that time.

Particles of the solar wind hit unprotected worlds like the Moon as so many microscopic high-velocity bullets. But it is a different story when the solar-wind particles are deflected around the planet. Or, rather, most are, for some become trapped in the magnetic-field lines and oscillate backward and forward from pole to pole across the planet's equator.

The possibility that our Earth might be surrounded by a region of particles from the solar plasma trapped in Earth's magnetic field had been speculated about even before the first spacecraft flew into orbit about our planet. Scientists trying to find ways to release controlled nuclear

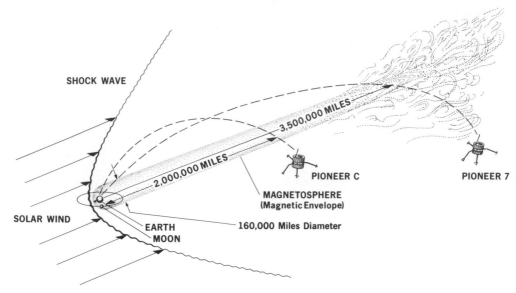

Measurements by spacecraft have shown us the extent and shape of the Earth's magnetosphere, which provides a protective magnetic shield around the Earth, holding off the solar wind. (NASA)

fusion had developed magnetic bottles to contain hot plasma. A Livermore Radiation Laboratory physicist, N. Christofilos, showed in the mid-1950s that a scaled-up version of such a magnetic bottle might correspond to the Earth's magnetic field and be able to hold particles of plasma. And he suggested that a nuclear explosion at high altitude could fill such a planet-sized magnetic bottle with plasma. Plans to conduct such an experiment were made before James A. Van Allen discovered the natural radiation belts of Earth, and indeed relevant atomic tests were made by the United States and the Soviet Union during the early 1960s.

The discovery of the natural radiation belts of Earth resulted from instruments carried aboard America's first satellite, Explorer 1, launched at the end of January 1958, almost four months after the first Soviet Sputnik ushered in the space age. For many years Van Allen had been investigating high-energy particles and cosmic rays with instruments carried to high altitudes by balloons and sounding rockets. When he analyzed the data returned from his instrument aboard Explorer 1, he concluded that the expected belts of energetic particles—electrons and protons—trapped in the

magnetic field of the Earth did indeed exist. These belts were later named in his honor.

Magnetic lines of the Earth's field arch far out over the equatorial bulge of our planet, then bend back toward the opposite pole, where they converge again. The trapped particles of the solar wind perform an endless dance spiraling along the magnetic-field lines. As electrons spiral around a field line in a clockwise direction (protons spiral the opposite way, they move along it toward higher latitudes. This is where the field lines curve toward the Earth's surface and get closer together. The magnetic field grows stronger. This has four effects upon the electron. First, it tightens the coils of its path by reducing the size of its sweeps around the field line. Next, it makes the coils of its path come closer together like a stretched spring being allowed to relax. Its orbital velocity increases and it moves along the field line more slowly. At one point, all these changes reach a critical situation and the electron cannot continue along the field line toward the Earth's surface. Effectively the magnetic field is too strong. The field actually reflects the electron, and it starts retracing its gyrations back toward the equator, reversing all the changes it made as it went toward the pole. It zooms across the equator toward the opposite pole and again is braked to a stop and then sent hurtling back again. Thus the electrons gyrate backward and forward in a seemingly endless dance along the field line from pole to pole. Protons behave similarly.

The particles also drift around the Earth (electrons drift east and protons west) and, because of their looping trajectories, go from high-intensity magnetic fields to low-intensity fields. The result is that particles are arranged in doughnut-shaped rings around the planet. Protons concentrate in an inner ring centered about 2,500 miles (4,000 kilometers) above the equator, while electrons form an outer ring about 10,000 miles (16,000 kilometers) above the equator.

At first the belts were envisioned as more or less symmetrically placed about the Earth, but detailed exploration by many spacecraft has revealed their structure to be more complex.

First, only part of the solar wind is absorbed by the belts; the rest is deflected around the Earth by the terrestrial magnetic field. As a result, much of the plasma flows around the Earth like the wake around a barely immersed rock in a swiftly flowing stream. This deflection of the solar wind produces a cavity in it, which is referred to as the Earth's magnetosphere. About 125,000 miles (200,000 kilometers) wide, it has several clearly defined features. There is a relatively abrupt boundary, between the magnetosphere and the solar wind, known as the magnetopause. In this region there is an electric current of several million amperes, a thin sheet of flow that results from the solar wind streaming around the magnetic field. Inside the current sheet is the inner magnetosphere, which includes the region of the Van Allen belts, where the trapped electrons and protons gyrate within the field.

Behind the Earth (as seen from the Sun), magnetic-field lines stretched by the solar wind form into a magnetic tail that reaches beyond the orbit

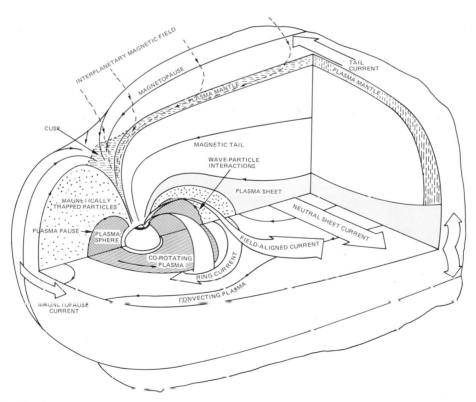

The Earth's magnetosphere seems to become more complex every year, as more spacecraft explore it. When we look out from the Earth at the starry sky, it is difficult to realize that we are not gazing through empty space but through an amazingly complex pattern of electric and magnetic fields and swarms of charged particles, some of which have enormous energies. (NASA)

of the Moon. Along the middle of the tail is another plasma sheet, which also reaches from the Earth's nightside polar ionosphere (electrically charged upper atmosphere) to beyond the Moon's orbit. This plasma is not as dense as the plasma of the solar wind but is nevertheless denser than that of the magnetosphere between the current and the solar wind.

North and south of the Earth, the electric field of the solar wind appears to enter the Earth's magnetosphere in two regions, called the cusps. Through these cusps particles stream in to produce the aurorae.

Between the magnetically trapped particles of the inner magnetosphere and the Earth itself there is a region of plasma that rotates with the Earth once each twenty-four hours. Between this co-rotating region and the magnetically trapped particles is a boundary called the plasmapause.

The whole region of the terrestrial magnetosphere and the interaction of it with the solar wind are extremely complex. Not only are there the radiation belts but there are also many different current sheets and streams of particles. Gradually, spacecraft have been unraveling this complexity to find that the activities in the magnetosphere produce magnetic effects at the Earth's surface. They have discovered how the auroral particles act, and how other particles heat the thermosphere and add to the heat generated by ultraviolet radiation from the Sun. The thermosphere, in turn, controls conditions closer to the Earth's surface.

The contact zone between the Earth's magnetosphere and the solar wind is referred to by space scientists as the bow shock. This is a very apt description, for it is the region where the high-velocity solar wind first interacts with the magnetic field of the Earth and is disturbed from its normal flow through interplanetary space. It is somewhat like the bow of a boat pushing through water and causing the water to stream around on either side. But while the bow of a boat is solid, the bow shock of the Earth is not. As the solar wind varies in intensity, it buffets this shock region back and forth, sometimes compressing it toward the Earth and at other times letting it expand farther into space.

There is also a flow of ionospheric plasmas of the upper atmosphere moving away from the Earth at high latitudes. This flow is called the polar wind. Scientists have also discovered strong interactions between the solar wind, the magnetosphere, and the ionosphere that give rise to magnetospheric substorms, a phenomenon that is still not well understood. Clouds of hot plasma are generated deep within the magnetosphere of the Earth as a result of very strong blasts of the solar wind. Somehow these hot plasmas rain particles of high energy into the lower atmosphere, where they agitate atoms of atmospheric gases to glow and form aurorae. During this process, low-frequency radio waves are generated and the upper atmosphere in the regions of the Earth's poles is heated. Energy pours downward into the lower atmosphere, adding heat that for a short time is as much as that normally injected into the lower atmosphere by the ultraviolet radiation from the Sun. The atmosphere upwells around the polar regions and spills over into lower latitudes, thereby generating gravity and acoustic waves that travel toward the equator. High winds are generated in the upper atmosphere and strong electric currents flow at many levels. For several hours the upper atmosphere is in a turmoil, radio communications are interrupted, and magnetically induced currents flow in cables and pipelines on the Earth beneath. All this results from the vagaries of the solar wind, generated far away on the Sun several days previously.

The third type of interaction between the solar wind and a planet is where the planet possesses an atmosphere but no appreciable magnetic field, such as Venus. The effects of the solar wind are even more bizarre there, since the blizzard of electrons and protons plunges directly into the atmosphere of the planet. The result is a tremendous mixup between the plasma of the solar wind and the electrified particles of the Venusian ionosphere. The complex conditions that result have not yet been explored in detail by spacecraft.

Astronomers try to understand the details of the solar wind by using several different types of spacecraft. Some orbit planets to explore how they interact with the wind; others plough through interplanetary space to check on how the wind behaves in free space between the planets.

The space inside the orbit of the Earth has been explored by several Mariner spacecraft that flew by Venus and Mercury, and a pair of German-built, American-launched Helios spacecraft that penetrated inside the orbit of Mercury for an unprecedented close approach to the Sun. Helios 1 was launched from the Kennedy Space Center on December 10, 1974, and Helios 2 followed, on January 15, 1976. Both spacecraft performed well beyond their initial design lifetimes despite their proximity to the Sun, where temperatures were hot enough to melt lead.

Helios 1 made its first close approach on March 15, 1975, when its perihelion was 28 million miles (45 million kilometers). Helios 2 flew even closer to the Sun when it made its perihelion passage, on April 17, 1976. It was then only 27 million miles (43.4 million kilometers) away from the Sun. Solar radiation there was twelve times as intense as at the orbit of Earth. Helios 2 passed through its perihelion at a velocity of 43.5 miles per second (70 kilometers per second), which is 125 times faster than the Concorde supersonic airliner.

Many findings of scientific importance came from this program. For example, scientists discovered that protons in low-speed plasma streams travel at between 185 and 250 miles per second (300 to 400 kilometers per second), while those in high-speed streams move as fast as 370 to 500 miles per second (600 to 800 kilometers per second). At times, the flow of the plasma appeared to fluctuate much more than at the orbit of the Earth. Also, for the first time two distributions of electron energy in the solar wind were measured. One consisted of moderate-energy electrons, called halo electrons because they form a halo around the Sun and do not have enough energy to escape into the outer Solar System; the other was made up of high-energy, "Strahl" electrons, concentrated into narrow beams.

One of the most important measurements of the solar wind across the Solar System was made in 1972 by five Pioneer spacecraft. Pioneer 7 was on the far side of the Sun from the Earth, Pioneer 6 and Pioneer 8 were on opposite sides of the Earth, one a quarter way around its orbit ahead of the Earth and the other trailing one quarter orbit behind. Pioneers 9 and 10 were in a straight line from the Sun between Pioneer 8 and the Earth. Pioneer 9 was just inside the orbit of the Earth (at 0.78 Earth distance, or astronomical unit, from the Sun) and Pioneer 10, on its way to Jupiter, was 2.2 astronomical units out from the Sun.

On August 2, a huge solar storm suddenly erupted and lasted for almost a week. It was the most intense storm ever observed from space and it produced a number of unexpected events. The solar wind swept by Pioneer 9 at an unprecedented speed of 625 miles per second (1,000 kilometers per second). A short while afterward, Pioneer 10 measured the flow at a distance of 130 million miles (210 million kilometers) farther

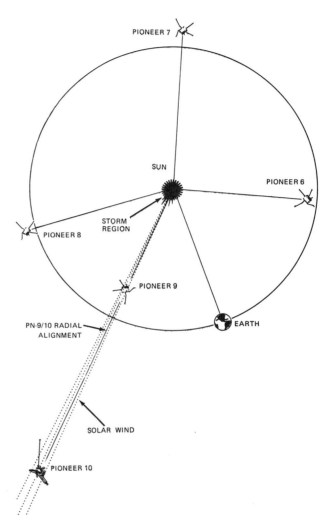

Great solar flares erupted on the Sun in early August 1972. A group of Pioneer spacecraft provided monitoring platforms in interplanetary space to measure the effects of these great flares on the interplanetary environment. A fortuitous alignment of Pioneers 9 and 10 provided an unprecedented set of measurements to show how the solar wind behaves as it climbs out from the Sun. Pioneer 10, at the greatest distance of all these five spacecraft, provided measurements to show that the solar wind smoothed out somewhat and reached a more average speed while its temperature soared to very high values. (NASA)

out in the Solar System than Pioneer 9. The speed of the solar wind had dropped to 435 miles per second (700 kilometers per second). Yet scientists were surprised to find that the temperature of the wind had increased. At Pioneer 9 it was 0.9 million degrees F (0.5 million degrees K), while at Pioneer 10 it was 3.6 million degrees F (2.0 million degrees K), almost the same as in the solar corona itself.

The instruments carried by the spacecraft recorded wild fluctuations of the magnetic fields. At times the intensity of the field reached one hundred times normal. This showed that at times the solar wind carried along with it intense magnetic fields that it had pulled from the Sun. Perhaps it was also generating its own magnetic fields as fast and slow streams collided. As mentioned earlier, such regions of intense magnetic fields scatter cosmic rays.

Equally as mysteriously, radio astronomers found that the radio waves emitted from Jupiter (which are a measure of the state of that planet's magnetosphere and its radiation belts) showed no

signs of significant change as the solar wind passed the orbit of the giant planet. Nor did the comet Schwassmann-Wachmann 1, traveling at about 5.6 astronomical units from the Sun, show any effects of the passage of this mighty solar wind. Scientists had to conclude that the wind somehow dissipated before it reached the distance of Jupiter and the comet. Only much later, when Pioneer 10 had encountered Jupiter, was it discovered that the particles in Jupiter's radiation belt do not originate from the solar wind but from Jupiter itself.

From a location point of view, the most unusual solar-environment spacecraft developed by NASA was the third International Sun Earth Explorer, or ISEE 3. Sent into space on August 12, 1978, it was the first spacecraft designed for positioning at the Sun-Earth vibration point, a location about a million miles (1.6 million kilometers) from the Earth where the solar and terrestrial gravitational pulls balance out. From this unique position, such phenomena as the solar wind, sunspots, flares, and galactic cosmic rays

1961   Oct. 12   Oct. 18   Nov. 3

A big surprise, however, was that the tremendous flare activity of August 1972 did not seem to affect comet Schwassmann-Wachmann 1, which was farther out in the Solar System than Pioneer 10 at the time of the flares. Scientists had expected to see a possible flare-up of the comet due to the increased solar wind. This comet, shown in these photographs in 1961, follows an unusual orbit. It is nearly circular and keeps the comet between the orbits of Jupiter and Saturn. With an orbital period of sixteen years, the comet is sometimes seen to increase dramatically in brightness. For example, on October 12, 1961, it was magnitude 13. Six days later it was very much brighter (magnitude 7.2), when its diameter appeared as only 0.6 minute of arc. But by November 3, its diameter had increased to 2.0 minutes of arc, while its brightness went down drastically, to magnitude 18.4. (U. S. Naval Observatory, Flagstaff) can be continuously measured. Meanwhile ISEEs 1 and 2, sent into highly elliptical, looping trajectories around the Earth in October 1977 simultaneously and in co-ordination, measure the effect of the same occurrences within the terrestrial environment.

WHERE THE WIND ENDS

As Pioneer 10 flew farther and farther from the Sun on its mission to Jupiter and beyond, it confirmed that the temperature and the density of particles in the solar wind drops off with distance from the Sun, but surprisingly the average velocity remains high. The velocity peaks and valleys smooth out, so that at great distances the solar wind does blow more steadily than it does at and within Earth's orbit. Scientists suspect that the wind blows to at least the orbit of Pluto, though they are not quite sure where it may finally end. A heliosphere has been postulated similar in shape to the terrestrial magnetosphere, enclosing not merely a single planet but the entire Solar System. Within the heliosphere the solar wind and its magnetic field predominate over the magnetic field and particles of interstellar space.

The Solar System travels through interstellar space around the center of the galaxy at a velocity of about 12.5 miles per second (20 kilometers per second). At the front of the Solar System there may be a bow shock separating the heliosphere from an interstellar wind, and behind the Solar System there may also be a wakelike magnetic tail. How far such a tail extends and how close the bow shock may be is purely speculative at this time. Fortunately Pioneer 10 is heading toward the tail and Pioneer 11, after it passes Saturn, in 1979, should head toward the bow shock. If these spacecraft continue to function into the outer reaches of the Solar System they may detect an end to the heliosphere. Scientists doubt this, however, because they do not think that these small spacecraft will be able to communicate over the immense distances expected before the solar wind ends. The task of looking for the boundaries may have to fall to two Voyager spacecraft which have more powerful long-range communication capabilities and are following the Pioneers to Jupiter, Saturn, and beyond.

WIND FROM THE STARS AND COSMIC BULLETS

Beyond the heliosphere is true interstellar space—the space between the stars that will be a challenge to mankind in the twenty-first century. This space is believed to contain about an atom per cubic centimeter, and is thus ten times less dense than the solar wind at Earth's orbit. The material from the interstellar wind has actually been detected drifting into the Solar System, because some of its particles are uncharged and can penetrate the magnetic fields of our system. These pieces of matter from between the stars are mainly hydrogen and helium, the principal interstellar gases. While the amount of material in each cubic centimeter (the approximate size of a lump of sugar) between the stars may seem infinitesimal by terrestrial standards (since there are 200 million, million, million atoms in every cubic centimeter of the air we breathe at sea level), there is so much space between the stars that interstellar gas and dust amounting to about fifty specks per cubic kilometer actually account for a substantial portion of the total amount of matter contained in our Milky Way Galaxy.

In fact, the detection of substances in the interstellar medium has been one of the most important discoveries of recent years. Astronomers have found that in addition to hydrogen and helium the interstellar medium contains the heavy isotope of hydrogen (deuterium) and other elements. Even more surprising, because ultraviolet radiation from stars would be expected to break down complex molecules into their elements, astronomers have also found that interstellar space contains molecules of chemical compounds such as hydrogen cyanide, methane, and ammonia. So many compounds have been discovered in space in recent years that some scientists have speculated that the building blocks of life originated in this cosmic dust and fell to planets to start life processes. Support for the idea that primitive biological systems may occur deep in space came in early 1978, when astronomers at the Herzberg Institute of Astrophysics, in Ottawa, detected radio emission from cyanooctaterayne of molecular weight 123 (it consists of a nine-carbon-atom chain with a nitrogen atom at one end and a hydrogen atom at the other). Meanwhile, Sir Fred Hoyle and Dr. N. Chandra Wickramasinghe, of the University College, in Cardiff, Wales, suspect that polysaccharides exist in the dust clouds of the remote Trapezium nebula. They cautiously wrote in the journal *Nature:* "Without supporting evidence, this would seem a bold conclusion, to say the least. Yet with the evidence presented here an affirmative answer does indeed seem warranted."

In places, the interstellar material condenses into dark clouds that obscure the light from more distant stars. Elsewhere, clouds shine brightly by reflected light from nearby stars. There are also clouds that are excited by ultraviolet radiation of stars to emit visible light. In fact, the spaces between the stars are now proving to be as intriguing to astronomers as the stars themselves.

In addition to the sluggishly drifting clouds of gas and dust, interstellar space is crisscrossed by high-velocity bullets, the nuclei of atoms known as galactic cosmic rays, some of which penetrate into the Solar System. Although mostly protons (nuclei of hydrogen atoms), these particles also include the nuclei of heavier elements up the atomic scale at least as far as iron. Cosmic rays are also emitted by the Sun and by Jupiter, but they do not have the enormously high energies of some of the galactic cosmic rays.

Cosmic rays were discovered by their effect upon an electroscope, a simple instrument used in many high-school physics laboratories to demonstrate that like charges of static electricity repel one another. Inside a glass chamber, two short lengths of gold leaf are suspended side by side and connected at the top. When a static charge of electricity is applied to an electrode connected to the top ends of the leaves, where they are joined together, their free ends move apart. But scientists found that no matter how carefully they designed such an instrument to maintain its charge, the leaves gradually lost their charges and fell back together. Scientists had to conclude that some kind of ionizing radiation caused the charge to leak from the leaves, for they had observed that if a piece of radium is placed near the instrument its emanations discharged the instrument.

A natural conclusion was that uranium and radium in the Earth's crust must be emitting radiation that discharges electroscopes. So an Austro-American physicist, Victor Franz Hess, placed an electroscope in a balloon to find out how the discharge might be affected by increasing height above the Earth's crust. He was surprised to discover that the electroscope discharged more rapidly at high altitudes. The ionizing radiation seemed to be coming from space, not from the Earth. At first it was assumed to be radiation, hence the name cosmic rays. Now we know that we are being bombarded by high-velocity bullets, the nuclei of atoms, rather than by rays.

Some of these bullets hit the Earth's atmosphere with energies of 100 billion electron volts. The electron volt, a term used in particle physics to describe the energy of subatomic missiles, is the amount of energy an electron acquires when it is accelerated through a potential difference of one volt. It is thus analogous to describing the energy of a falling weight in foot-pounds, the energy acquired by a weight of one pound falling through a distance of one foot in the Earth's gravitational field.

When energetic particles of cosmic rays enter the Earth's atmosphere they hit nuclei of atmospheric gases and break them apart, causing showers of subatomic particles that penetrate through the atmosphere and reach the ground. It was the effects of secondary and tertiary showers that alerted scientists to the incoming bullets from space. Later, through balloons, high-altitude rockets, and spacecraft, they ascertained the true nature of the cosmic rays in space.

The energy of a primary cosmic-ray particle is some hundred million times that of a typical proton of the solar wind. This is because the cosmic-ray bullets are moving almost at the speed of light. We still don't know how these particles acquire such tremendous energy though we think the particles may originate from the explosion of stars and are subsequently accelerated by galactic magnetic fields.

MAGNETIC FIELDS AND CURRENT SHEETS

After passing Jupiter, the spacecraft Pioneer 11 journeyed high above the plane of the ecliptic—the plane of Earth's orbit—on its way to Saturn. It was able to investigate regions of the Sun's magnetic field not previously explored. The field is believed to extend several billion miles above and below the Sun's north and south poles, respectively, and outward along the ecliptic to beyond the orbit of Pluto. Pioneer 10 has traced it out beyond the orbit of Saturn already. On its way across the Solar System from Jupiter to Saturn, Pioneer 11 reached some 100 million miles (160 million kilometers) above the ecliptic plane, which allowed it to look at the solar magnetic field at a point 16° above the solar equator.

This was important because there had been so much scientific controversy about the form of the solar field. They established that the field reverses in direction every eleven years, about the time of sunspot minimum. At this time spacecraft found that the field appeared weak and distorted. They also saw another strange effect. Because the equatorial regions of the Sun rotate faster than the poles, the field lines near the solar surface at the equator appear to wind around the Sun so that they run east-west, instead of north-south across the equator as might be expected.

In 1978, the field in general emerged from the northern hemisphere of the Sun while field lines ran back toward the southern hemisphere. Near the magnetic equator, the lines of the north and south fields were separated by a warped sheet of electric current around the Sun in the inner Solar System. In the inner system this current circles the Sun, while in the outer system it flows away from the Sun.

As the Sun rotates, the warped current sheet as seen from the plane of the Earth's orbit appears to move up and down. As a result, all spacecraft moving in this plane record a magnetic field from the Sun that is first directed toward the Sun and then, as the Sun rotates, directed away from it. Such spacecraft saw a reversal of the solar field each time the current sheet passed through their path. The effects were confusing and variously interpreted.

Pioneer 11 clarified the situation by showing that the warped current sheet was a true explanation of the current and field reversals observed by other spacecraft. When Pioneer 11 was 16° above the ecliptic, it was too far above the warps of the current sheet for them to pass through the spacecraft. As a consequence, the spacecraft's instruments observed a steady field that did not reverse in polarity as the Sun rotated.

One important question that has not been and cannot be answered until spacecraft reach the boundaries of the Solar System is whether the

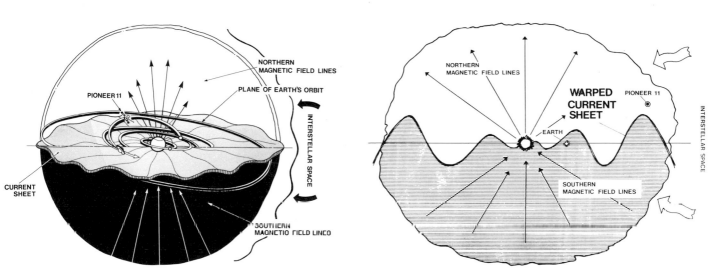

Spacecraft have generally flown in the plane of the Earth's orbit on interplanetary missions. They encountered peculiar changes in the magnetic field on the Sun. Pioneer 11, after its encounter with Jupiter, flew high above the ecliptic on a path across the Solar System to a rendezvous with Saturn. It provided a unique platform to look at the solar magnetic field some 16° above the solar equator. The first drawing shows the general field of the Sun and the path of the spacecraft from Jupiter to Saturn. The second shows how the reversals of the solar field are probably caused by a warped sheet of electric current in the equatorial plane of the Sun separating north and south field lines. (NASA)

outgoing north polar field lines link up with the incoming south polar field lines far out toward Pluto. Or do they merely join with the interstellar magnetic field? In general, though, whatever the case out at the outer limits, the solar field resembles that of the Earth, with the modification that the outgoing solar wind stretches the field away from the Sun.

The discovery that the solar field has an equatorial warped current is important because this sheet must control the motion of galactic cosmic rays as they try to penetrate the Solar System. Such high-energy charged particles must be urged into paths that run parallel to the current sheet, thus channeling the cosmic rays from the galaxy into the region of the planetary orbits. This may, indeed, have been an important factor that helped to speed up biological evolution on the Earth, because changes between progeny and parents may be the result of chance encounters between cosmic rays and genes. Terrestrial life conceivably could have been seeded from the stars by infalling cosmic dust and then speeded toward consciousness by cosmic rays channeled by the solar magnetic field to hit the planets.

INTERPLANETARY ENVIRONMENT LINKS SUN TO EARTH

Evidence continues to mount that through the interplanetary environment the Sun affects the Earth in many different ways. Also, many ancient records, both Western and Eastern, are being examined in search of links between the Sun and terrestrial events such as weather and earthquakes. While many of the alleged links have not yet been substantiated and may well never be, we are accumulating more evidence that the Sun we observe today is not a normal star; instead, it has undergone periods considerably different from what it is now, and these have significantly affected conditions on Earth.

The period from 1645 to 1715, mentioned earlier as a time when there were no sunspots, is called the Maunder period, after the astronomer who first drew attention to it. Ancient Chinese records have also been examined that tell of other periods of very low sunspot activity. There are also indications of a sixty-year cycle of sunspot activity, which corresponds to three times the period of Jupiter-Saturn conjunction, which, occurs

every 19.86 years. Scientists have also discovered that the period of rotation of the Earth on its axis may be changing in step with the solar cycles.

There have been periods of intense solar activity as well as periods of quiet. If these changes affect world climate they could be of paramount importance to mankind and we need to understand them. A period of low solar activity between 1100 and 1300 corresponded to an extremely cold period in Europe. The period from 1940 onward saw a change in global atmospheric circulation patterns that resulted in redistribution of rainfall into the south-central parts of the continents. Dust-bowl conditions of previous decades were eased and agricultural land produced bumper harvests. Predicted famine for millions in India was averted by unusually high crop yields in 1968, and United States wheat surpluses were so great that the price of the grain went to an all-time low. The importance of continued investigation of the interplanetary environment and the Sun-Earth links cannot be overemphasized, with the world population exceeding 4 billion. An aggressive program of interplanetary monitoring spacecraft may be essential for our future survival on the planet. By such spacecraft we may be warned of impending climatic changes and thus avert disaster as meteorological satellites have warned us of danger from approaching hurricanes. We haven't been surprised by a hurricane since the advent of weather satellites. We may also avoid surprises in the future from climatic changes on Earth caused by extraterrestrial events if we do the research today into the environment of space.

BUILDING BLOCKS OR DUSTY DEBRIS?

The interplanetary environment contains more than particles and fields and elusive plasmas. The more substantial bodies between the planets range in size from grains of dust to miniworlds, and in composition from solid rocks to gaseous halos of comets that are a million or more miles in diameter. These objects are not so readily classified as part of the interplanetary environment, yet they are more akin to the environment than to the larger and more substantial planets and their satellites. Depending upon whether we regard them as the raw material from which the larger bodies accumulate mass, or the debris from

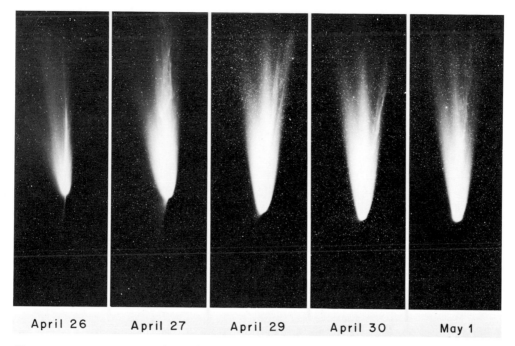

| April 26 | April 27 | April 29 | April 30 | May 1 |

Comets are very spectacular celestial objects and have mystified and awed people throughout the ages. Only in recent years have scientists been able to gain a better understanding of their true nature. When a comet approaches the Sun, some of the ices embedded in its nucleus evaporate to form a halo, part of which is pushed away from the comet by solar radiation pressure. At the same time, meteoroids are released and scatter along the orbit as a stream of small rocky bodies. This process is taking place in this view of Arend-Roland (1956h) taken in the spring of 1957. On the picture taken on April 27, meteoroidal matter seems to be fanning out away from the comet. (Hale Observatories)

the formation of the planets, we can accept them as building blocks of the Solar System or as the dust of its dissolution.

HAIRY STARS

The most spectacular of the less massive bodies of the Solar System are comets, moving starlike objects that are surrounded by nebulous halos and sprout enormous tails as they approach the Sun. The fuzzy appearance of comets has earned them the name of "hairy stars." With the possible exception of a total solar eclipse, no astronomical event causes more popular excitement than the appearance in the sky of a large, bright comet. Ancient records are filled with references to these spectacular occurrences. Sages, astrologers, and witch doctors interpreted them in two extreme ways depending on the mood at the local scene or time in history. They were said to be either special signs of divine pleasure with mankind or a special group of mankind, or omens of impending disaster, war, or other calamity: . . . "The heavens themselves proclaim the death of princes," it was once written.

Very bright comets that outshine anything other than the Sun and the Moon have always been rare, appearing no more than once or twice each century. But the advent of telescopic astronomy showed that modest comets are not uncommon in the Solar System, and even faint ones are nearly always visible; many of them arrive in the inner Solar System each year.

While most people know what a comet looks like, it took a long time for astronomers to de-

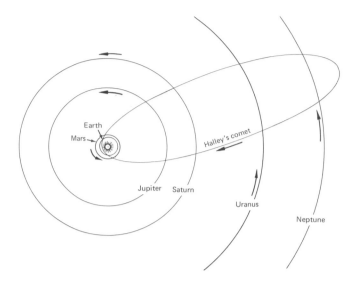

Comets follow very elongated orbits, as a rule. Some of these orbits extend far beyond Pluto, and a comet may take over a million years to traverse it. Halley's comet has a period of less than one hundred years, but even so it travels beyond the orbit of Neptune, as shown in this diagram. The comet reached its greatest distance from the Sun in 1948 and is now on its way back toward Earth for a close approach in 1986.

This is an artist's conception of how a comet's tail behaves as it approaches and swings around the Sun and then recedes from the Sun on its journey back toward the outer reaches of the Solar System. The tail of the comet always extends away from the Sun, even when it is on a receding path, a phenomenon caused by the influence of solar radiation (the solar wind).

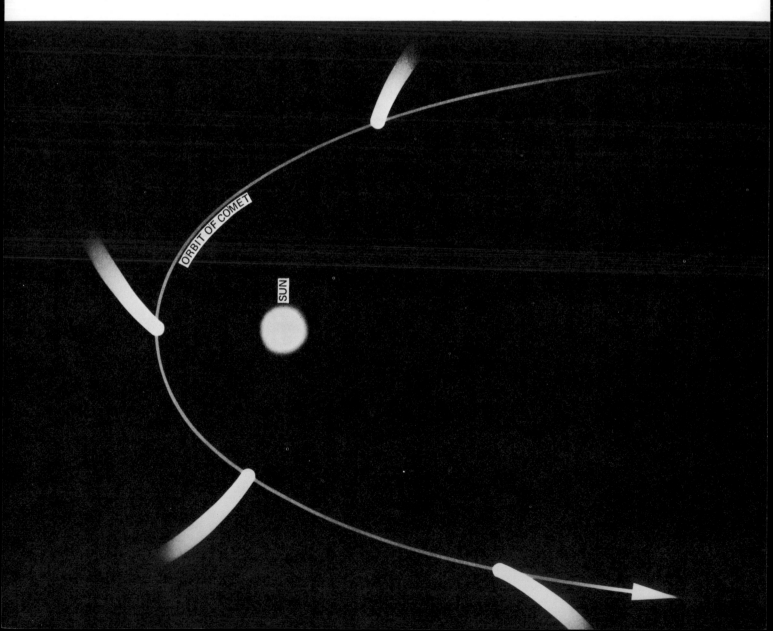

velop any reasonable understanding of their physical nature. Even today, there are many unanswered questions that most probably will not be resolved until a spacecraft is dispatched to a comet.

A comet consists of three clearly visible and quite different parts. The head, called the coma, forms a bright diffuse cloud that can sometimes be as large as the planet Jupiter. Within this coma is the second characteristic part, the nucleus, which appears as a star and is probably only 0.6–50 miles (1–80 kilometers) in diameter. It shines by reflected sunlight. The most conspicuous part of a comet is its tail, a long, tenuous appendage that the comet develops as it approaches the Sun. The tail always points away from the Sun, streaming behind the comet as it plunges into the Solar System and reaching ahead of it as it climbs out again.

Most comets travel along extremely elongated elliptical orbits. At the farthest point from the Sun a comet is in deep freeze and effectively in hibernation. Even through very powerful telescopes it appears merely as an inconspicuous star (if it is visible at all). It has neither coma nor tail. But as it swings in toward the Sun, its nucleus awakens into action. Astronomers typically discover the comet when it is between the orbits of Jupiter and Mars.

At first a faint starlike object, the comet starts to evaporate gases as solar radiation warms its nucleus. The starlike object begins to be surrounded by a fuzzy halo. Later, its tail starts to develop as solar radiation pushes material from the coma away from the Sun. The comet reaches its full brightness at its closest approach to the Sun. If the lineup of comet, Earth, and Sun is right, the comet appears as a brilliant celestial object in the evening or dawn sky. Often, though, the comet's brilliance is obscured by that of the Sun and the comet is visible only when it is before or after its perihelion passage.

One of the first to explain comets, Fred L. Whipple, suggested that the nucleus consists of a mixture of ice and rocks. This led to the popular description of comets as "dirty snowballs." Whipple suggested that the nucleus consists of a mixture of ices of ammonia, water, methane, and carbon dioxide, impregnated with particles of stony and metallic material. When the comet is far from the Sun, this conglomeration is cold and dormant. But as it nears the Sun the frozen gases vaporize to produce the nebulous coma. Some of these gases are pulled away from the comet, by the pressure of radiation from the Sun, to form the tail. Yet this theory does not account for some peculiarities in the tails of comets such as long

Comet Bennett was a spectacular naked-eye comet in the predawn skies of early 1970. The first photograph was taken in April with an f/2 Schmidt telescope at NASA Goddard Space Flight Center. It shows two distinct tail streamers trailing thousands of miles into space. The second photograph, which captures observations of the comet from Orbiting Astronomical Observatory 2 and Orbiting Geophysical Observatory 5, revealed a hydrogen cloud enveloping the comet with a size many times larger than the Sun. (NASA)

straight streamers, split tails, and spikes pointing toward the Sun.

We can answer some questions by looking more closely at the makeup of comets as we understand them today. A comet's tail must be a strongly ionized and very tenuous gas. The electrical charge of the molecules and dust particles making up the tail is affected by the charged particles of the solar wind: some are attracted and others repelled. Charged particles in the coma of a comet are dragged by the solar wind into a straight tail, while uncharged (neutral) particles form the curved tail.

Formation of a coma and a tail means that the nucleus of the comet loses matter each time it travels through perihelion of its orbit. Close passages boil off more matter from a comet than distant passages, so that comets with close perihelion lose matter quickly. Also, those comets that do not travel far out into the Solar System and make repeated passages close to the Sun soon deplete their resources.

Although there have so far been no space missions to comets, spacecraft have already added considerably to our understanding of these mysterious and spectacular objects. One of the Orbiting Astronomical Observatory spacecraft discovered that comet Bennett (1970) was surrounded by a cloud of extremely tenuous hydrogen gas fifty times the diameter of the Sun. While its hydrogen cloud was the largest known object in the Solar System, the comet was only medium-sized and contained a relatively small amount of matter, about as much as could be carried in an average ocean-going freighter.

The spacecraft also threw light on the rate at which comets lose their matter. Comet Bennett ejected its hydrogen at a rate corresponding to forty tons of water per second as the comet swung through its perihelion. The loss of material into space has another important effect upon comets. As the material boils off from the nucleus it does not emerge evenly all around the nucleus.

Instead, it spurts off in jets that act like rocket exhausts, pushing the nucleus from its path under solar gravity. So the orbit of the comet is changed in unpredictable ways. The jets may also cause the nucleus to spin wildly.

The effect of this jet action has been recognized on Halley's comet, the most spectacular comet in the Solar System of our age and one that has been returning regularly for centuries. The jet action has been changing the time of return by about four days on each of its 76.1-year orbits of the Sun.

Even more serious irregularities result when a comet passes close to a planet, particularly huge Jupiter. Several comets have indeed been captured by the giant planet and orbit the Sun with aphelion (their greatest distance) at about the radius of the orbit of Jupiter. In addition to capturing comets, Jupiter can also fling them out of the Solar System entirely. Its gravitational field has already given two spacecraft, Pioneer 10 and Pioneer 11, sufficient energy so that they are being ejected from the Solar System. As we said, one theory for the origin of comets holds that they come from storage in a cloud surrounding the Solar System far beyond the orbit of Pluto. Comets dislodged from this Oort cloud have an orbital period of about 2 million years. Nearly all of this time is spent far out in interstellar space, where they may indeed scoop up interstellar material to replace losses during their perihelion passages. So when we look at a comet's tail and analyze its gases by means of a spectroscope, we may be looking at interstellar material that it has brought into the inner Solar System.

And the materials of a comet? Scientists have identified both oxidized and reduced components. There is evidence of carbon dioxide, of nitrogen, of the amino group (NH_2) and a CH group, of water, carbon, sodium, and a rich mixture of hydrogen, carbon, nitrogen, and oxygen in various molecular arrangements.

During the 1800s, many brilliant and large

Named for Edmund Halley, of England, Halley's is the most conspicuous of the periodic comets. In studying the historic records of three conspicuous comets that had appeared in 1531, 1607, and 1682, Halley concluded that the three were, in fact, periodic appearances of one comet. He predicted that it would return again about 1758, because he believed it traveled in an elongated ellipse around the Sun. On the night of Christmas 1758 everyone was very excited when they saw Halley's comet, which made its perihelion, or closest approach to the Sun, early in 1759. The comet returned again in 1835 and in 1910, when this picture was obtained. (Hale Observatories)

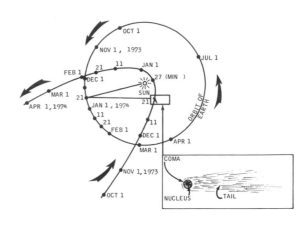

Comet Kohoutek was expected to be a bright comet. It did not perform as expected. This is the first photograph taken of the comet from the new Joint Observatory for Cometary Research, in New Mexico, on the morning of November 28, 1973. The drawing shows the orbit of the comet and of the Earth with the relative positions of these two bodies. Note how the comet became lost in the solar glare in August and September 1973 as Earth moved on the far side of the Sun. Kohoutek was again hidden in the solar glare when its speed carried it behind the Sun again, in late December 1973 and early January 1974. At no time did the comet make a very close approach to the Earth. (NASA)

comets were seen. The 1900s, by contrast, have not provided many good comets except for Halley's visit in 1910. Understandably, then, astronomers became very excited when, in March 1973, the Czechoslovakian astronomer Lubos Kohoutek found a comet when examining two photographic plates exposed some time apart in a search for an asteroid that he had discovered earlier. On these plates, of the constellation Hydra, he found a faint object that looked like a comet. It was only of fifteenth magnitude, six hundred times fainter than the faintest star visible to the unaided eye. The astronomical community became interested because the comet had been discovered when it was nearly five astronomical units from the Sun. Finding comets so far out in the Solar System is uncommon, so they anticipated that this new comet could very well become a spectacular object as it plunged toward the Sun and began to absorb energy from solar radiation. Some astronomers went so far as to predict that comet 1973f (its official name, meaning the sixth comet to be discovered in 1973), or comet Kohoutek (traditionally, comets are also named after their discoverers), would become the most spectacular astronomical event of the century.

Astronomers at observatories in many parts of the world watched the new comet eagerly as it moved toward its perihelion. It disappeared into the glare of the Sun early in June 1973 and was not seen again until September 23 when a young Japanese astronomer, Tsuomu Seki, photographed it. By the beginning of November the comet had brightened to a magnitude of 7, still just below the threshold of visibility to the unaided eye, which can see only objects of sixth magnitude or brighter. But a pair of binoculars showed it quite easily in the early-morning sky just before dawn. By late November it was a little brighter and visible to the unaided eye.

But comet Kohoutek was characteristically unpredictable. It turned out to be a disappointing object, nowhere near as bright as expected. Nevertheless, it did present a rewarding view until the motions of the Earth and the comet moved it into the dawn glare and it was again out of sight. Perihelion passage occurred on December 28, 1973, when the comet was only 13.2 million miles (21.3 million kilometers) from the Sun. It zipped past at just over 250,000 miles per hour (400,000 kilometers per hour) and then approached Earth, coming to within 75 million miles (120 million kilometers) on January 5, 1974. But it was not until a short while later that the comet again emerged from the solar glare to be visible again, this time in the sky of early evening. Astronomers and laymen alike were disappointed. Comet Kohoutek did not live up to expectations even at this time. It refused to become a cosmic spectacle, and if there was any glare in the sky, such as that from cities, it proved to be a very elusive object except with field glasses or larger optical instruments.

But despite its failure to emblazon the skies, the comet was far better observed than any previous comet, because teams of professionals and amateurs had organized well in advance for what they hoped would be a major cometary encounter. As well as being watched closely by ground-based astronomers, the comet was observed from space. America's manned Skylab space laboratory provided an ideal observing platform, free from the absorption of the Earth's atmosphere, while an unmanned Venus/Mercury-bound spacecraft provided another platform beyond the geocorona, the halo of hydrogen gas surrounding the Earth.

Skylab's third crew undertook many observations of Kohoutek, including detailed sketches of the structure of the coma and the tail, and visual interpretations of the colors of the various parts of the comet. In addition, other spacecraft turned their instruments on the comet. Thus, both, Orbiting Solar Observatory and Orbiting Geophysical Observatory satellites gathered data, as did a Mariner spacecraft on its way to explore the inner planets. Pioneers 6 and 8, stationed in interplanetary space, made their contributions. Instrumented sounding rockets, too, made breathless little forays above the atmosphere for a momentary scan of Kohoutek, and leisurely floating balloons carried heavier instruments above the absorbing layers of the lower atmosphere. Even a C-141 aircraft with a large telescope on board was pressed into service. As one astronomer commented at the time, "From a viewing standpoint here on Earth, it [the comet] is not living up to advanced billing, but . . . from the scientific point of view it is still the comet of the century."

When the path of comet Kohoutek had been calculated, it appeared that the visitor was from the outermost regions of the Solar System, possibly from the Oort cloud. Since the comet had started to develop a coma while far out in the

Solar System, astronomers concluded that it must consist of very loosely packed material, from which light gases could easily escape at fairly low temperatures. They also speculated that the comet had not visited the inner Solar System before and might be a virgin remnant of the original solar nebula from which the Sun and the planets had formed. Strengthening this speculation was the discovery that the comet emitted radio signals, the first definite evidence of this type of cometary activity. These signals indicated the presence of methyl cyanide (CH_3CN) and hydrogen cyanide (HCN), substances found (also by radio emissions) in the clouds of gas and dust of the Milky Way. It is in these same regions that stars appear to be forming, thus reinforcing the concept that Kohoutek was composed of presolar-system material. The nucleus of the comet most likely formed in a region of low temperature that was probably less than $-280°$ F ($100°$ K), probably by the accretion of small particles from gas and dust.

So far, comets have been assigned as members of the Solar System, because there is no indication that they were of interstellar origin; i.e., their orbits are all closed. Thus, comets seem to originate in the Oort cloud, which may extend to tens of thousands of times the distance of Earth from the Sun. Whether they actually form in the Oort cloud or are pushed there by perturbations of the planets is still disputed. Some theories suggest that both the comets and the outermost planets formed from the same part of the solar nebula; but, while Uranus and Neptune gradually moved into the Solar System, their gravities resulted in the comets being moved out to the boundaries of the system. Uranus and Neptune thus came closer to the Sun by forcing the comets into solar exile.

Periodically, however, something (possibly the passage of another star) dislodges a comet from the Oort cloud and it plunges deep into the Solar System. When a comet enters the inner Solar System, it has to run the gauntlet of the gravities of the planets; the greatest danger faced is mighty Jupiter. Comet Kohoutek received intermediate treatment from Jupiter by being thrust into an orbit with a period of about seventy-five thousand years, compared with a typical Oort-cloud comet period of 2 million years. The next approach of Kohoutek into the inner Solar System will be about the time when Pioneer 10 has reached the distance of the nearest stars.

The discovery of water vapor on comet Kohoutek also gave strong support to Whipple's dirty-snowball theory. Astronomers also detected carbon dioxide and carbon monoxide. At a conference convened to discuss the results of the observations of Kohoutek, at NASA's Marshall Space Flight Center in June 1974, Whipple said: "We can confidently visualize a comet as a complex lacy structure of 'whiskers' and 'snowflakes' that grew atom by atom and molecule by molecule while highly volatile molecules were trapped as clathrates."*

Six years before comet Kohoutek sped into the inner Solar System, a Belgian astronomer, Armand Delsemme, had predicted that comets might have an icy halo surrounding their nucleus. He theorized that while some materials of the coma would be derived from the breakup of parent molecules in the nucleus, others would probably come from icy grains streaming out from the nucleus when gases evolved from it as the result of solar heating. In this way, a comet should be surrounded by a halo of ices whose diameter would increase as the comet approached the Sun and would be greatest for comets that had much icy material in their nuclei. Such a halo was found around Kohoutek. It had a diameter of less than 1,000 miles (1,600 kilometers).

Skylab astronauts discovered that comet Kohoutek developed an "antitail," a spike pointing toward the Sun, just after perihelion. Whatever dust or other particles made up this antitail were clearly not being repelled by sunlight or the solar wind. Instead, the material was drifting back along or near the orbit of the comet, trailing behind it toward the Sun. Also, the material of the spike was found to be cooler than that of the coma and the main tail. It appeared to consist of particles of about one millimeter rather than the microscopic dust specks of a normal dust tail. These larger particles seemed to have been released by the nucleus before the comet reached perihelion but become visible afterward as the solar gravity pulled them inward.

A unique cosmic-ray event occurred in September 1973 which has been attributed to a nucleus of antimatter, matter in which all the particles carry opposite charges to those of ordinary mat-

ter. At this same time, comet Kohoutek was approaching the Sun. There has been some speculation that the comet itself may have been a visitor from beyond the Solar System and that it, too, consisted of antimatter. The idea is that a comet made of antimatter would not brighten as much as one of normal matter as it approached the Sun. Whether or not the environment of the Solar System is periodically invaded by objects from outside made of antimatter is highly speculative at present, although some scientists have associated the large Siberian explosion of 1908 with the impact of a small body of antimatter with the Earth.

The many unknowns associated with comets may not be resolved until they are actually visited by spacecraft. Although there have been many plans to undertake a mission to rendezvous with a comet, none have been approved and budgeted as a NASA program. The latest to fall by the wayside despite support from a prestigious working group of eminent scientists was a plan to send a spacecraft to rendezvous with Halley's comet on its next return to the inner Solar System in 1985/1986. Such missions might use conventional spacecraft or advanced systems of propulsion such as solar sails and ion engines. Many studies have been made to demonstrate their practicality. Spacecraft could be sent to penetrate the cometary coma and perhaps even land on a comet's nucleus. Since the material of comets might well date back to the origin of the Solar System (or even earlier), an expedition to explore these strange "hairy stars" might provide a clearer answer to that familiar and still unanswerable question: And in the beginning?

METEOROIDS, METEORS, AND METEORITES

Catch a falling star and put it in your pocket! Early in the evening of January 3, 1970, a flaring light from the sky pushed aside the darkness shrouding the silent, snow-covered fields surrounding the small rural community of Lost City, in Oklahoma. The light faded equally as abruptly, and few people except for a state policeman and some residents had noticed the arrival of this visitor from space, a one-ton rock that had hurtled into our atmosphere.

Scientists have always been interested in finding a fallen meteor, which is called a meteor-

---

* Clathrates are mixtures of gases in an unconsolidated structure of rocky materials. At low temperatures the gases are in the form of ices.

In the mid-1970s a group of scientists studied the possibilities of a mission to Halley's comet during its next return into the inner Solar System. The study showed that such a mission was quite feasible and would have tremendous scientific benefits. However, restricted space budgets and lack of top government support let the opportunity go by; it will not occur again for nearly one hundred years. This artist's concept illustrates one way that might have been used to reach the comet, by use of a solar sail, a lightweight structure of very thin material that would be carried into orbit by a space shuttle, extended in space, and sent to the comet by the pressure of solar radiation. (Jet Propulsion Laboratory)

ite, an aerolite, and occasionally an air stone. Before Apollo astronauts brought back samples of rocks from the Moon, meteorites were the only specimens of extraterrestrial material available for analysis. In the early 1960s a system of meteor-observing stations, called the Prairie Network, was set up with the aim of finding meteorites quickly after they fell to Earth. Speed was essential so that they could be analyzed chemically before they became contaminated by the terrestrial environment.

The Lost City meteorite was photographed by the cameras of the network. Its probable impact point was predicted, and searchers went to work. The first fragment was found in a snow-covered field about half a mile from the predicted impact point. It was recovered on January 9 and rushed to the Smithsonian Institution for examination. Other fragments were found later.

The analysis of such meteorites has revealed remarkable stories about their origin and history during their wanderings in space. Smaller than comets, sometimes related to them, and equally nomadic, these small chunks of matter travel between the planets, where they are called meteoroids. They range in size from grains of sand to chunks of matter the size of a city block. Extremely small grains, specks of dust, are called micrometeoroids.

Sometimes these objects encounter the Earth, when their energy of impact is converted to light and heat. Small objects are burned up high in the atmosphere. Large objects often penetrate the terrestrial air blanket and, like the Lost City meteorite, reach the ground. When a meteoroid contacts the atmosphere and becomes visible as a streak of light, it is called a meteor, popularly, a shooting star. If the object is particularly large it may be bright enough to deserve the name fireball. Such objects can even be seen in daylight. Sometimes the heating is so intense that the fireball explodes and becomes what is known as a bolide.

The speed at which a meteoroid enters the atmosphere of the Earth may vary greatly, from 7 to 45 miles per second (11 to 70 kilometers per second). The lower limit is the velocity attained by a body falling into the atmosphere from a great distance when it is accelerated by the force of Earth's gravity alone. Higher velocities represent a combination of this velocity and the orbital motion of the meteoroid around the Sun when it hits the Earth head on.

While scientists like to examine fallen meteorites to determine their composition, they know, through spectroscopic analysis of the light meteors emit during entry, that some of their materials are lost as heat during their passage through the atmosphere. Water has been recognized in these spectra, showing that the meteoroids in space probably contain ices in addition to the materials found in them when they reach the Earth.

Some meteoroids graze the atmosphere and rebound into space like a flat stone skipping

There are countless numbers of meteoroids in space, wandering throughout the Solar System. Some are specks of dust, others weigh thousands of tons. When a meteoroid enters the Earth's atmosphere, it becomes heated to incandescence and appears as a bright streak called a meteor, or shooting star. This picture shows the trail of a meteor near the Great Nebula in Orion. (Yerkes Observatory)

When a meteor survives to reach the Earth's surface, it is called a meteorite. This heavy meteorite was found at Hoba Farm, near Grootfontein, in South West Africa. Weighing 60 tons, it produced only a small crater, some 5 feet (1.5 meters) deep when it fell, in 1920. Although predominantly iron, it contained nearly 10 tons of nickel. (American Museum of Natural History)

Sometimes, however, when a large meteorite hits at high speed it produces an explosion comparable to a small nuclear bomb and digs a huge crater such as this one in Arizona. Meteor Crater, near Winslow, on Route 66, is almost one mile in diameter (actually 1,295 meters) and about 575 feet (175 meters) deep. It was probably caused by an explosive impact of a meteorite some fifty thousand years ago. Geologists have found crushed and pulverized sandstone beneath the sediments that now form the crater's floor, testimony to the enormous forces released by the impact. (American Museum of Natural History)

across a smooth surface of water. An Air Force satellite spotted one of these near misses on August 10, 1972. The meteoroid was over 12 feet (4 meters) in diameter, probably weighing one thousand tons. It flashed into the upper atmosphere to within 37 miles (60 kilometers) of the surface before it retreated into space. Had this celestial visitor approached Earth at a steeper angle it would have plunged to the surface and caused an explosion equivalent to a small nuclear bomb.

Man-made meteors occur when artificial satellites, slowed by the action of the interplanetary environment, finally descend low enough to be "grabbed" by the Earth's atmosphere and braked so rapidly that they are destroyed in a streak of incandescence. Soon after the end of World War II, the late Dr. Fritz Zwicky devised an experiment to fire high-velocity bullets at the apex of a V-2 rocket flight over White Sands, New Mexico, to create artificial meteors to be used in the early exploration of the composition of the upper atmosphere.

AGES AND AMINO ACIDS

In recent decades, extremely powerful new tools for analysis of materials have been developed and ways found to make chemical analyses of microscopic amounts of material. In searching for transuranium elements, for example, scientists developed techniques to analyze quantities of material of less than two millionths of an ounce in a matter of seconds. By comparing the proportions of various isotopes within meteorites, scientists can estimate the time that has passed since the material of the meteorite was last molten. Some samples indicate that meteorites are as old as 4.6 billion years.

This had led some astronomers to speculate that meteorites originated from a parent body that formed early in the history of the Solar System, probably in about 100 million years or so. Presumably, they were heated within this body and then cooled at a rate (determined by the crystalline forms within the meteorites) suggesting that it must have been no more than 60 miles (100 kilometers) in diameter. Others think of meteorites as grains or particles, left over from the original solar nebula, that have not collected together into larger bodies.

Unusual combinations and proportions of isotopes in some meteorites have been interpreted very recently as meaning that they were subjected to intense neutron flux from or within a supernova. And a special class of meteorites has been found to contain complex carbon-based molecules that are often regarded as the building blocks for living things.

Meteorites are classified on the basis of the materials of which they are composed. Stony meteorites are those in which silicates (rocky materials) predominate, while iron meteorites are those in which metallic constituents predominate. Subgroups exist within each of these two main categories, and there are sidereolites, which are transitional between iron and stony meteorites. A special class of stony meteorites is represented by the chondrites; and of these another subgroup, rich in carbon, bears the name carbonaceous chondrites. The latter have been in the limelight during recent years because they were found to contain amino acids and a great variety of organic materials. The existence of carbonaceous chondrites led to a revival of the late-nineteenth-century theory of panspermia, in which it was proposed that life was distributed naturally among worlds by spores traveling through space. Some scientists are now even questioning the generally accepted theory that life developed in the primitive oceans of Earth by a fortuitous process of radiation acting on simple chemical substances in a primordial soup. Instead, these scientists say, we should reconsider the possibility of life having arrived on Earth from space and continuing to arrive even today. They cite new pathogens as being evidence of such living things arriving on the planet from comets and meteors.

There is, however, little proof that such invasion from space has occurred or even now is taking place. But there is proof that meteorites carry amino acids, which are the building blocks of living things. Their discovery by Cyril Ponnamperuma, in 1970, was exciting and revitalized the enthusiasm of those scientists who wanted to search for extraterrestrial life. A short while later, the Viking program to search for life on Mars was given the go-ahead for a 1975 mission. One unanswered question was whether the amino acids in the meteorite resulted from biological processes out in space or on a body from which the meteorite originated, or if they were produced by non-biological processes from organic molecules, which radio astronomers had shown to be present in galactic dust clouds. We still do not know the answer, but light may soon be shed on the intriguing subject. In late 1977 and early 1978, Dr. William Cassidy, of the University of Pittsburgh, collected over three hundred meteorites in the Antarctic, of which two are of the rare, carbonaceous chondritic type. Because of their location, they are not believed to have been contaminated by terrestrial organisms. Appropriately, they were sent to be examined in the same laboratory in Houston where Apollo moon rocks were studied following their retrieval from the lunar surface.

While the search for extraterrestrial life has proved extremely disappointing, the presence of amino acids in meteorites raised anew the questions of how life originates on a planet, particularly the Earth. Some of the new theories of life originating in space and being deposited on the planets by comets or meteors run counter to many decades of accepted thinking along other lines. Yet we have been surprised many times in the past by new discoveries that shattered fashionable thinking, and it is unlikely that such surprises have come to an end. Many of today's traditional theories may have to be considerably modified if not completely abandoned as man-

kind pushes further into space and establishes permanent bases on natural and artificial worlds away from the Earth.

Astronomers believe that the meteoroid population is continuous down to microscopic-sized particles: the micrometeoroids. These are of engineering interest, because they can pose a hazard to structures in space and to spacecraft. In fact, when they were first discovered and their enormous numbers ascertained, before the first satellites flew into space, many scientists worried that they might be such a hazard to spacecraft that a manned voyage to the Moon would be impossible. This was in the late 1940s, when radar astronomy was in its infancy and had detected the first meteors during the daytime. These observations showed that some of the micrometeoroids weighed only one gram, and it was not possible to detect particles any smaller by ground-based techniques. But with the advent of artificial satellites instrumented with impact detectors and other electronic aids, meteoroids of one trillionth of a gram were detected. With such aid, engineers could figure the expected impacts from meteoroids of all sizes. Consequently they were able to design spacecraft to survive the peppering of these tiny missiles of space. Meteor bumpers were developed to vaporize all the smaller particles be-

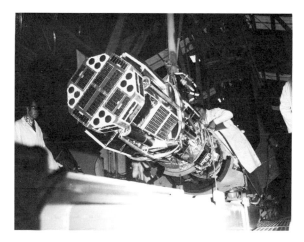 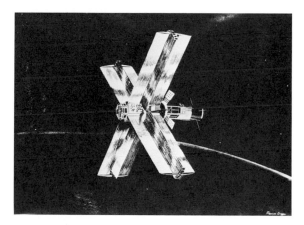

The distribution and range of sizes of meteoroids in space was not known until special spacecraft such as Explorer 46, shown here, were flown on meteoroid-detecting missions. The first picture is a general view of the spacecraft before launch; the second is a view of the spacecraft when operational in orbit after its launching, in August 1972. The meteoroid detectors extend like wings from the spacecraft; they were used to detect meteoroids. Micrometeoroids were detected by capacitor detectors. The spacecraft gathered new and valuable data on the velocities and distribution of these tiny particles moving through space in the vicinity of the Earth. (NASA)

fore they could penetrate inner, pressurized shells of spacecraft.

While some astronomers contend that meteoroids come from the breakup of larger bodies such as asteroids, others think that comets provide the material, and still others that meteoroids may ultimately accrete into larger bodies. Evidence for a cometary origin is based on the disappearance of some comets and the appearance of streams of meteors in the same orbit. Evidence for origin in a larger body, as mentioned earlier, is the composition and microscopic structure of meteorites. And evidence for meteoroids accreting into larger bodies is purely theoretical: mathematical analyses show that small bodies orbiting the Sun tend to congregate into streams and then coalesce into larger bodies.

The theory that meteoroids originated from large bodies became popular with the discovery of the asteroid belt, a region around the Sun be-tween Mars and Jupiter where there are many minor planets ranging from 635 miles (1,022 kilometers) in diameter down to the limits of telescopic vision. We would expect such bodies colliding with each other to produce a myriad smaller bodies down to grains of dust. But the theory did not hold. When Pioneers 10 and 11 passed through the asteroid belt on their way to Jupiter, they discovered that the belt did not contain a lot of dust; the only hazard within the belt was from larger bodies, and this was not great.

The streaming of meteoroids has been well known for a long time. When such streams cross the orbit of the Earth and our planet plows through them, we are treated to a brilliant display of meteors that appear to radiate from a small area of the sky. Whether or not such streams come from the breakup of comets is, as yet, undecided.

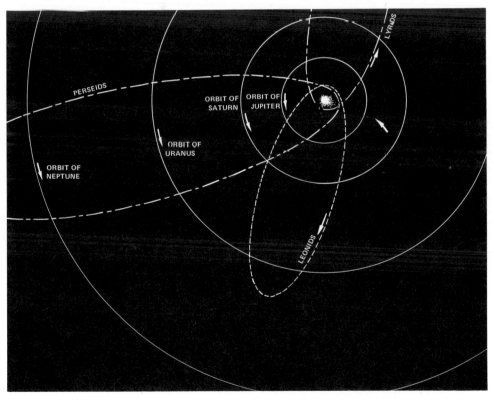

Orbits of three streams of meteoroids are shown here. These are the Leonids, the Perseids, and the Lyrids. The streams are named after the constellation from which the streaks of meteors appear to radiate when the Earth passes through the stream. These streams consist of many small bodies that orbit the Sun together. They are often associated with comets that have disappeared from the same orbit. (Arfor Picture Archives)

GLOWS AND COUNTERGLOWS

Undoubtedly there is a lot of dust in interplanetary space. You can even see it, if you know where to look. The best time is early evening in spring or the dark before dawn in the autumn. On a spring evening on a moonless night, away from the glare of artificial lighting, you can see a faint cone of light extending along the zodiac from the horizon where the Sun set. This is the zodiacal light. Pioneer 10 measured the intensity of this faint glow from space and found that it decreased in intensity as the square of distance of the spacecraft from the Sun. This light was believed to be the result of sunlight reflected from dust particles in interplanetary space. The spacecraft experiment showed that these particles are concentrated in the inner Solar System but that they also are present in the asteroid belt. Beyond the asteroid belt they appear to be virtually non-existent. The gravity of Jupiter has presumably swept the Solar System clean of them outward of about 3.3 times Earth's distance from the Sun.

Another, fainter glow seen from Earth is called the Gegenschein, or counterglow. It appears as a faint luminous patch in the dark sky opposite the Sun. For a long while, scientists were undecided whether this glow was a reflection from dust particles in space acting like full moons (i.e., showing their fully illuminated faces to Earth) or was associated with a tail of particles streaming into space from the Earth. The true nature of the Gegenschein was established by the first Pioneer to travel into the outer Solar System. As the spacecraft moved along its orbit to Jupiter, it continued to observe the Gegenschein, but its Gegenschein was in a different direction from the one seen from Earth. The antisolar glare was confirmed as being associated with particles spread around the Solar System and not confined to a tail of the Earth.

Beyond 3.3 times Earth's distance from the Sun, the Solar System is virtually free of dust particles, and spacecraft are able to measure the integrated starlight from the Milky Way Galaxy without partial obscuration by Solar System dust.

From particles and fields to meteoroids, from comets to counterglows, the environment of the Solar System has proved fascinating to a new generation of astronomers provided with the tools of the space age. Yet we are only just beginning to understand the enormous complexity of the conditions between the planets and how our lives on Earth are affected by the environment of space. Many more space missions are needed to unravel the snarled skein and weave the tapestry of space "weather." And such missions may have to have the definite objective of exploring space rather than being expeditions to other planets. Several such missions are already being planned, including swing-bys of Jupiter to send spacecraft over the poles of the Sun, high above the plane of Earth's orbit.

The exploration of the space between the planets may well prove to be as exciting and rewarding as that of the planets themselves.

5
THE INNERMOST PLANET

In 1543 Niklas Koppernigk fast approached the end of a full life. He had been a canon of the cathedral city of Frauenburg and a doctor of canon law, and had lectured in Rome, studied medicine at Padua, and used the Latin version of his name: Copernicus.

Copernicus had settled in Frauenburg in 1512, following the death of his uncle, the bishop of Ermeland. There he applied himself in his combined duties of military governor, bailiff, judge, tax collector, vicar-general, and physician. As if that were not enough, Copernicus, like Newton a century later, attempted to reform the coinage, which had become seriously debased. But he is remembered today mostly because of the great work he completed in 1530, the *De Revolutionibus Orbium Coelestium,* which swept away the belief in Earth-centered cycles and epicycles which Milton was later to describe graphically: "How gird the sphere the centric and eccentric scribbed o'er cycle and epicycle, orb in orb." Instead, he showed that the motions of the celestial bodies had fooled mankind, that appearances can be deceptive. The Earth and all the other planets revolve around the central Sun of the Solar System. Straightforward and logical today, this idea ran counter to nearly all thought five hundred years ago.

The great work was not published until Copernicus lay on his deathbed and, as some biographers claimed, was bemoaning the fact that during his eventful life he had never seen the elusive innermost planet, Mercury. Because the planet is so close to the Sun, it is nearly always obscured by the solar glare: "Can scarce be caught by philosophic eye; lost in the near effulgence of its [the Sun's] blaze," wrote an anonymous poet of the nineteenth century.

It is strange that Copernicus did not see the planet (if indeed his biographers are correct), because it is a fairly bright object at times if you know where to look: just above the western horizon after sunset or above the eastern horizon before dawn. Many astronomers in northern Europe reported having seen the planet without difficulty, despite murky skies. Denning claims he saw Mercury at least 150 times with the unaided eye from 1868 to 1905 in England. Tycho Brahe records several observations of the planet in Denmark around 1580. Gore claimed no difficulty in seeing the planet even through the rain-soaked skies of Ireland in the late 1800s.

But whether his biographers were right or wrong, it was upon the foundations laid by Copernicus that Kepler, Galileo, and Descartes built until the science of celestial mechanics was formalized by Newton and others and mankind began to take the first steps beyond the cradle of the Earth. The importance of the Copernican theory was not the strong emotional reactions that it caused among philosophers and theologians but the way in which it caused people to rethink old theories and abandon many traditional limitations. It caused astronomers to observe the planets more accurately and to question "facts" that had been almost universally accepted.

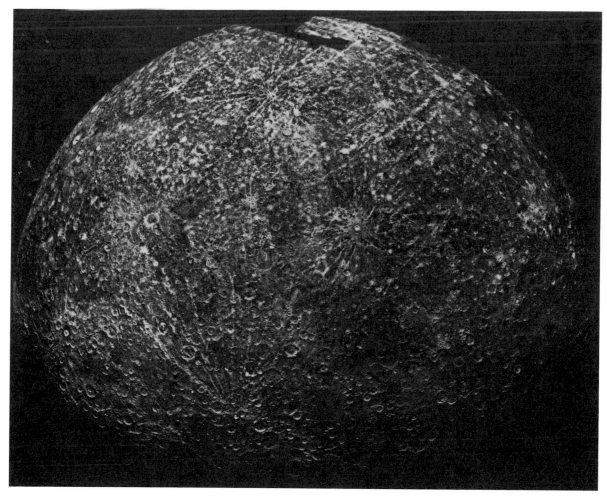

A general view of Mercury made up of photos taken by Mariner 10 on its second encounter with the planet. (Jet Propulsion Laboratory/NASA)

As a result, by 1972, the five-hundredth anniversary of the birth of Copernicus, mankind was no longer passively observing the planets from a distance but had sent seven spacecraft concurrently to explore five different planets of the Solar System: Mercury, Venus, Mars, Jupiter, and Saturn.

Darting like a moth to either side of the Sun, Mercury was known to the ancients. Isaiah was probably referring to Mercury when he wrote of "Lucifer, son of the morning," because early observers regarded the innermost planet as a false, deceitful star (*sidus dolosum*), "a squinting lacquey of the Sun, who seldom shows his head in these parts, as if he was in debt." Astrologers regarded Mercury as being malignant when in the house of Mars, while the ancient priests of Thebes associated the elusive planet with the evil god Set.

Despite observational problems, ancient astronomers and astrologers were very much aware of Mercury, though sometimes they thought it two different planets. So Mercury has two names in some of the ancient writings. The ancient Egyptians figured out that Mercury, like Venus, revolved around the Sun—a concept centuries ahead of its time. To them, Mercury was known as Sobkou, while it was known as Bi-ib-bou to the Sumerians, Ninib to the Chaldeans, Nebo to the Babylonians, Mokim to the Phoenicians, and Hermes to the Greeks. Our word Mercury is derived from the Latin Mercurius, the messenger of the gods, the god of dexterity and of eloquence, of traders and of thieves. The French word for Wednesday is *mercredi,* which is derived from the Latin *Mercurii dies,* the day of Mercury. The English word Wednesday is derived from Wodan's day, in which Wodan, or Odin, is sometimes related to Mercury in the Teutonic languages.

TWILIGHT PLANET

Because it is small and close to the Sun, Mercury still is a difficult planet to observe from Earth. With a diameter of only 3,032 miles (4,878 kilometers), Mercury is a much smaller body than the Earth. Its orbit around the Sun is more elliptical, i.e., differs more from a circle, than does the orbit of the Earth. The closest approach of Mercury to the Sun (perihelion) is 28.7 million miles (46.3 million kilometers), and the most distant part of its orbit (aphelion) is 43.3 million miles (69.7 million kilometers). The inner Solar System, which encompasses the planets Mercury, Venus, Earth and Moon, and Mars stretches from Mercury's perihelion to the aphelion of Mars (154.7 million miles, or 248.9 million kilometers) and thus covers a band around the Sun some 125 million miles (200 million kilometers) across. It takes light traveling at 186,000 miles per second (299,000 kilometers per second) only 3.22 minutes to reach Mercury from the Sun at its average distance, and 12.64 minutes to reach Mars. But to reach Pluto, the outermost of our system's planets, light has to travel 5 hours, 28 minutes.

Because Mercury's orbit is within that of the Earth, it cannot be seen far from the Sun in terrestrial skies. It appears to move from side to side of the Sun to what are termed elongations. At greatest western elongation Mercury precedes the Sun in its motion across the sky due to the rotation of the Earth on its axis. Consequently, it rises just before the Sun does and is viewed as a "morning star." At eastern elongation Mercury follows the Sun across the sky and is seen after sunset as an "evening star."

Depending on whether Mercury reaches an elongation at perihelion or at aphelion it can be separated from the Sun by a maximum of only 18° or 28° of arc, respectively.

These factors, as well as a dull surface, make Mercury difficult to study from the Earth. Even though astronomers fit their telescopes with special filters to reduce the glare from scattered sun-

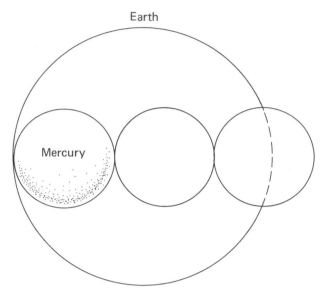

The Earth has a diameter of about 2.5 times that of the innermost planet, Mercury.

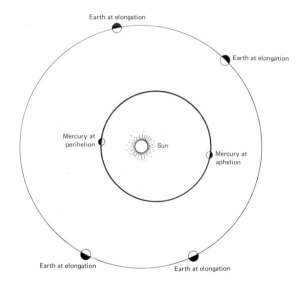

Mercury is much farther from the Sun at the aphelion of its orbit than at perihelion, as shown in this drawing. This affects our observation of the planet from the Earth. When Mercury as seen from Earth is at its greatest angular distance from the Sun, the planet is said to be at its greatest elongation. Western and eastern elongations are shown in this drawing for Mercury at perihelion and at aphelion. The aphelion elongations separate Mercury more from the Sun than perihelion elongations, and it is then most easily observed after sunset or before sunrise.

Because Mercury moves in an orbit inside that of the Earth, it exhibits phases similar to those of the Moon. Mercury is full when on the far side of the Sun. It is then at its brightest (despite a greater distance) but cannot be seen from Earth, because it is hidden in the solar glare. When between Earth and Sun, the dark side of the planet faces the Earth, but it can be seen only when it crosses the Sun, in transit. Mercury is best observed from Earth telescopically for a period of about five to six weeks around the elongations. It then appears from nearly full to a thick crescent phase.

light so they can observe the planet when it is higher in the sky, i.e., during the daytime, they still cannot see much detail on the planet.

The problem is that just as the Moon in its orbit around the Earth exhibits phases, so does Mercury in its orbit around the Sun. But there is a big difference. The size of the full Moon when the Moon is fully illuminated is about the same size as a thin, crescent Moon. But the size of a full Mercury is smaller than that of a crescent Mercury. The disc of Mercury appears fully illuminated only when the planet is on the far side of the Sun from the Earth. When Mercury is closest to Earth it presents its dark side to us.

Since Mercury's orbit lies between Earth and the Sun, the planet can occasionally appear as a small black dot crossing the bright photosphere. This is referred to by astronomers as a transit of Mercury. Such transits are relatively rare. They do not occur every time Mercury comes between Earth and Sun, because the orbits of the Earth and of Mercury are not in the same plane. Transits can occur only when Mercury passes between Earth and Sun at a point in its orbit where it is crossing the plane of the Earth's orbit. As a consequence, transits can take place only within six days either side of November 9 and three days either side of May 7. The most recent transits took place on November 7, 1960, May 9, 1970, and November 10, 1973. During the twentieth century there are only three more: on November

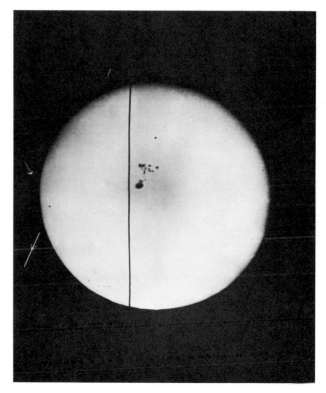

Several times each century, when Mercury comes between Earth and Sun at the right part of its orbit, namely where the orbital planes of Earth and Mercury cross each other, Mercury is seen to pass across (transit) the bright face of the Sun. The dark spot of Mercury, appearing much smaller than the sunspots, is indicated by the two arrows. (Mount Wilson and Palomar Observatories)

13, 1986, November 6, 1993, and November 15, 1999. Such transits can be seen only with a telescope, because the disc of Mercury is so small.

IMAGINATIVE MAPS OF DUSKY MARKINGS

Because of the difficulty in viewing Mercury, astronomers had trouble describing the markings on the surface of the planet. About all they could say before a spacecraft visited the planet was that Mercury was probably very like the Moon. The hazy features on Mercury when seen under exceptionally good observing conditions suggested flat, lunarlike circular plains, along with some mountains. Astronomers theorized that there would also be craters on the surface of the small world, as on the Moon, but of course, could not

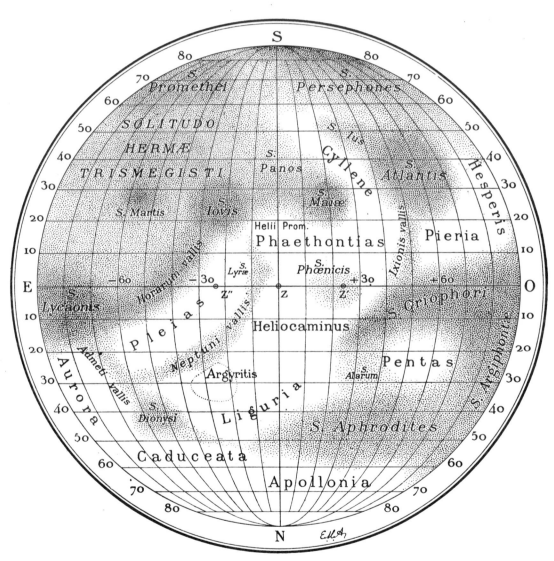

Despite the difficulties in observing the innermost planet, astronomers made valiant attempts to map the surface of Mercury. The first map was prepared by E. M. Antoniadi from observations made during 1924 through 1929 with a 32-inch (83-centimeter) telescope at the observatory of Meudon, near Paris, France. The map was published in 1934 in Antoniadi's book La Planète Mercure. *Antoniadi's studies of Mercury were entirely visual.*

see these from Earth.

Astronomers were also often carried away with their own enthusiasm for mapping other worlds, and produced maps that bear no resemblance to the surface of Mercury revealed during close inspection by a spacecraft. These represented the best that Earthbound astronomers could do, under ideal observing conditions, using the best photographs obtainable.

LOCKED TO THE SUN OR FREE AS THE EARTH?

One of the major questions that plagued astronomers for many years was the period of rotation of Mercury about its axis. Some astronomers claimed that they had observed markings that showed that the planet rotated freely on its axis in a short period compared to that of the Earth. Other, equally as astute observers said that this was not so: Mercury was locked to the Sun as our Moon is locked to Earth and turned the same face eternally to the Sun, as our Moon faces the Earth.

This uncertainty about the rotation of Mercury was not resolved until quite recently. The answer came in 1972 and it was the result of radio observations of the planet: the bouncing of radar beams and checking how they were reflected back to Earth.

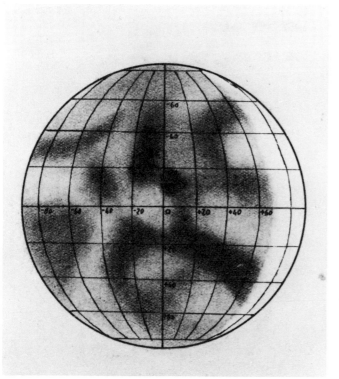

More recently, Audouin Dollfus based a map of Mercury on photographs he had taken of the planet from 1948 through 1953 at the Pic du Midi Observatory, in the Pyrenees, and on earlier observations of B. Lyot. This map of Mercury was published in As-tronomie, volume 67 (1953), by the Société Astronomique de France. It is reproduced by courtesy of Dr. Dollfus.

Generally, photographs of Mercury obtained through Earth-based telescopes are very disappointing. They show a fuzzy-looking object on which faint details are suggested. Even the best pictures of the innermost planet show little. This picture was taken by the 61-inch (1.55-meter) telescope at the Catalina Observatory, in the mountains north of Tucson, Arizona. This observatory of the Lunar and Planetary Laboratory, University of Arizona, has produced some of the finest planetary photographs. This picture was taken at elongation of Mercury in 1970. (Courtesy Lunar and Planetary Laboratory/NASA)

RADAR EXPLORATION OF THE INNERMOST WORLD

A new age in astronomy began when more powerful transmitters allowed radio astronomers to send pulses of radio energy to other worlds and later receive the echoes from the surfaces of other planets. Scientists could obtain signals from the planets that provided much new information about surface roughness, major geological features, and the planet's rotation.

Traditionally, astronomers have determined the period of spin of a planet by observing a feature on its surface and counting the time until it appears again. The elapsed time between two sightings of an identical geographic feature tells how long it takes a planet to spin on its axis so as to face the same side again to the Earth. (This length of time is known as the synodic period.) By making corrections for our change in viewpoint due to Earth and planet motions along their respective orbits during this period, the time of one revolution of the planet relative to the stars can be determined. This is called the sidereal period. The sidereal period for the Earth is 23 hours, 56 minutes, 4 seconds. The solar day is slightly longer because Earth moves a short way around its orbit as it spins on its axis. So for any point on the Earth's surface to again face the Sun requires twenty-four hours. This is the period of the solar day.

Repeated observations of Mercurian surface markings led astronomers in the past to conclude that the planet's spin period was the same as the period of revolution around the Sun. This meant that the same face of Mercury faced the Sun all the time. Such an interpretation was very popular because it seemed to confirm tidal theories that explained how the Moon continually presents one face to the Earth. The theory said that, over hundreds of millions of years, powerful solar gravitational forces acted as a brake and slowed down the rotation of Mercury until it became synchronized with its orbit around the Sun. This was generally held to be true until the mid-1960s.

In April of 1965 a major discovery took place, in Puerto Rico. In the rugged Guarionex Mountains, in the northwestern part of this Caribbean island, there is a great bowl-shaped limestone sink hollowed out by rushing streams that cut vast underground caverns and allowed the terrain to collapse. The great bowl, remote from civilization, made an ideal shape for a big radio antenna to look deep into space. In the early 1960s the Advanced Research Projects Agency of the U. S. Department of Defense provided money to build this big antenna for research into the Earth's upper atmosphere. Later it was used to probe deep into space.

The antenna consists of a featureless wire mesh, some 1,000 feet (300 meters) in diameter surfacing the natural bowl to form a concave mirror reflector for a giant radiotelescope. You approach the telescope by jeep on a winding mountain road and see a great mass of girders, dangling antennas, and cables seemingly suspended in midair over the mountains like a spaceship from another world. It is actually a support platform of 525 tons for the business end of the giant radiotelescope. And it's a frightening experience to climb to it across a catwalk stretched some 400 feet above the wire mesh. The whole assembly is supported by steel cables from three high towers at the edges of the bowl. This installation is called the National Radio Astronomy and Ionospheric Center, Arecibo Observatory.

About 20 per cent of this radiotelescope's time is spent looking at the planets. During one of these periods, on April 8, 1965, Cornell University scientists Gordon H. Pettingill and Rolf B. Dyce were in the control room, facing the antenna on a hillside overlooking the bowl. They had programmed the powerful transmitter to send a burst of radio waves toward the planet Mercury, which was then at one of its close approaches to the Earth.

Although other radio astronomers, in the Soviet Union and the United States, had previously bounced radar echoes off the innermost planet, the returned echoes had not been sufficiently strong to perform the experiments that the Cornell astronomers wanted to do. They were trying to use the Doppler effect, of a change in frequency of radio waves reflected from a moving surface, to measure the rotation period of Mercury. The radio waves returned from the edge of Mercury moving away from the Earth because of Mercury's rotation would be shifted to a longer wavelength, while the radio waves from the edge of Mercury moving toward Earth would be shifted to a shorter wavelength.

The results suggested that Mercury did not face one hemisphere to the Sun but rotated once in fifty-nine days, rather than the eighty-eight days that would be required were it locked to the Sun.

But it is actually related to the Sun in a different way.

Shortly after the Arecibo results were announced, Giuseppe Colombo, of the University of Padua, Italy, and the Smithsonian Astrophysical Observatory, in Cambridge, Massachusetts, and Irwin I. Shapiro, of Massachussetts Institute of Technology's Lincoln Laboratory, made a very important discovery. They examined more than fifty years' observations of Mercury through Earth-based telescopes and concluded that the drawings were all suspect. They also adjusted the period determined by the initial radar observations to precisely 58.65 days and concluded that Mercury is locked to the Sun in an unusual way. Three rotations of the planet on its axis are equal to two revolutions of the planet around the Sun. This strange resonance has affected observations of Mercury and trapped earlier astronomers into concluding that the rotation was synchronous with orbital revolution.

The best times to observe Mercury from Earth are every three synodic periods of Mercury, i.e., when Mercury presents the same face to the Earth. Thus, astronomers who observed the planet for many years and made sketches of the elusive markings would find that they appeared very much the same at the same phase of Mercury. The natural conclusion would be that the planet always presented the same face to the Sun. However, if the observations are continued long enough, the fact that three synodic periods of Mercury do not exactly equal one terrestrial year begins to take effect. This is why maps of Mercury made by one generation of astronomers are quite different from those made by another generation.

A solar day on Mercury is 176 Earth days. When manned bases are established on Mercury, perhaps sometime during the next century, tomorrow's explorers will witness a fantastic spectacle: First, the Sun will appear much larger than it does from the Earth and also quite a bit larger at perihelion than at aphelion. The Sun will rise in the east and set in the west once every 176 days. Moving across the sky of Mercury, it will appear to slow its westward crawl when the planet approaches perihelion. It will then move eastward for just over eight days and then resume its path, now to the west. From some stations on the surface of Mercury an observer would be able to view two sunrises or two sunsets every day! This looping of the Sun's motion at perihelion results from the elliptical orbit of the planet. Although the spin rate on its axis remains constant, the planet's velocity in orbit varies from fast at perihelion to slow at aphelion. For most of Mercury's orbit the planet spins faster than it orbits, just like the Earth, so that the Sun appears to move from east to west across the sky even though much slower than in terrestrial skies. But around perihelion the planet moves around the Sun faster than it spins, so the Sun appears to go backward for a short while.

Because of the two-thirds spin resonance of Mercury and its orbit around the Sun, two parts of the planet, diametrically opposed, must alternatively face the Sun at perihelion. Since the in-

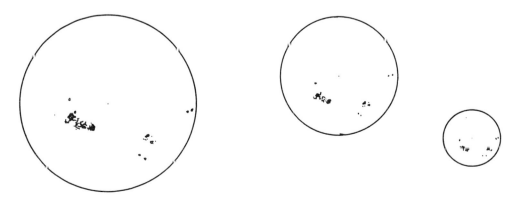

The Sun would appear awe-inspiring from Mercury, several times larger than it is seen from Earth. At perihelion (left) the Sun presents a huge, bloated globe. Even at aphelion (center) it has twice the diameter of the Sun seen from Earth (right).

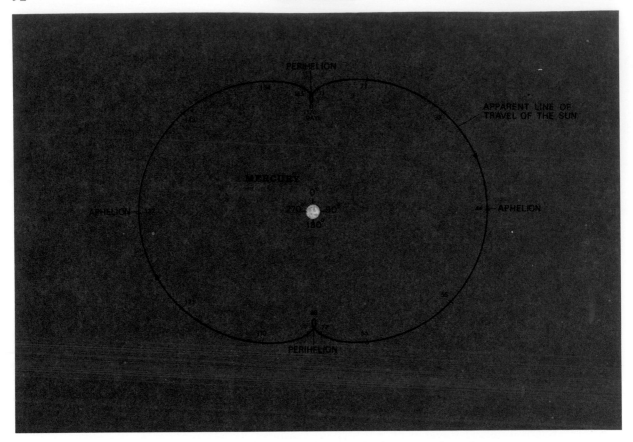

Observer on surface of Mercury sees nearby Sun rise in east and set in west over period of 176 Earth days, or more than 5.5 months. At one time or another, the entire surface of the planet Mercury is illuminated by the Sun. At perihelion positions, the Sun seems to switch direction and moves retrograde for just over eight days, then hovers, and travels prograde. Points on the Sun's apparent path are ten-day intervals. (Arfor Picture Archives)

tensity of sunlight on Mercury at perihelion is more than twice that at aphelion (because of its highly elliptical orbit), these two areas have been described as "hot poles" even though they are actually on the equator of the planet. Halfway between these hot poles around the equator are the so-called "warm poles" that face the sun alternately at time of aphelion. Their temperature is some 235° F (130° K) lower than the hot poles at midday. At perihelion the midday temperature reaches 1,500° F (700° K) while the night temperature falls to less than —343° F (100° K). Some measurements of the nighttime temperatures of Mercury made by Pettit and Nicholson at Mount Wilson Observatory in 1923 were responsible for their suggestion that the rotation of Mercury might not be synchronous with its revolution

around the Sun, even though very slow. These observations also suggested that the rotation of Mercury was in the same direction as that of the Earth. Yet astronomers were so entrenched in the idea of synchronous rotation that they invented ways in which the temperature differences might be accounted for—for instance, by the surface of Mercury being more mountainous in one part than another.

Even before radar had cleared up the problem of the period of rotation, it had shown astronomers that the surface of the innermost planet was probably much like that of Earth's satellite. A Soviet team headed by V. A. Kotel'nikov at the Institute of Radio Engineering and Electronics of the Academy of Sciences first discovered this during the summer of 1962, when they made the first radar bounce from Mercury when it was between 52 and 54 million miles (84 and 87 million kilometers) from Earth. This was confirmed in May 1963, when a team at the Jet Propulsion Laboratory under Richard M. Goldstein and Ronald L. Carpenter made contact with Mercury.

Within ten more years radar surveys of the planet were improved a great deal by new techniques of processing the data through computers, and astronomers were able to build up radar pictures of the planetary surfaces. In 1972, California radio astronomers Richard Goldstein and Shalhav Zohar discovered a number of intriguing shapes and features on the surface of Mercury. Using the big antennas and powerful transmitters of NASA's Deep Space Network station at Goldstone, in the Mojave Desert, they pinpointed at least five circular features averaging 30 miles (50 kilometers) in diameter, features that on the Moon would be prominent craters. They also located some circular features that were 185 miles (300 kilometers) across which were the size of major-impact-basin plains of the Moon. The radio astronomers also saw evidence in the returned signals of undulating hills and valleys as well as land or rock projections that were 75 miles (120 kilometers) across and 4,300 feet (1,300 meters) high.

Other than that Mercury had a high density, more like Earth than the Moon, very few other facts were known about the innermost planet before a spacecraft gave us our first view, in 1974. There had been speculation about an atmosphere, possibly one consisting of the inert gas argon produced by radioactive decay. A Soviet astronomer, V. A. Nuroz, suggested—with some reservations —that a small amount of carbon dioxide might be present on Mercury. There were even some estimates of the atmospheric pressure being about one ten-thousandth of Earth's sea-level pressure. However, astronomers didn't support such speculations, because any atmosphere of Mercury would tend to leak off quickly into space; the high surface temperatures would heat gas molecules to velocities at which they could escape the planet's gravitational attraction and rocket off into the void of space. The solar wind, too, blowing at great intensity across the inner world, would be expected to knock gas molecules off into space. Unless the planet somehow continually replenished its atmosphere from its interior, Mercury would most likely be completely airless.

FROM SPECULATIONS TO SPECIFICS . . . BY SPACECRAFT

On March 24, 1974, an ungainly-looking spacecraft with spreading panels of solar cells and a big sunshade protecting its body from the fierce solar heat was 2.7 million miles (4.3 million kilometers) from Mercury. Mariner 10 had used the gravity and orbited motion of Venus to deflect its inward-bound trajectory and plunge into the inner Solar System. It later came within less than 500 miles (800 kilometers) of the sun-drenched mysterious surface of the elusive planet. An electron beam flicked rapidly across a screen on which an image of the planet had been focused by a telescope carried by the spacecraft. Complicated electronics changed the pattern of light and shade in the image into digital numbers, which were hurtled across space at the speed of light on a radio beam to Earth. Gathered by a 210-foot (65-meter) -diameter antenna of NASA's Deep Space Network, the radio signals were decoded and the string of numbers passed to giant computers at the Jet Propulsion Laboratory in California. Within a few minutes after receipt, these numbers were converted back into areas of light and shade, which could be displayed on a TV-type screen or assembled into black-and-white photographs. The surface of Mercury was about to be revealed for the first time.

On the TV screen at the laboratory, the first picture showed a fat crescent with faint markings. Astronomers said it was about the quality that they had observed from Earth with the best conditions. But Mariner 10 was speeding toward Mercury at 23,000 miles (37,000 kilometers) per hour. By March 27, the pictures showed craters dotting the surface of Mercury, some of them only 25 miles (40 kilometers) across. And there was a bright ray crater, too, similar to the Moon. These are regarded as being fresh, because the material dug out by the impact that produced the crater appears as light streaks radiating from it over darker surrounding material. As Mariner 10 flew past the planet, it provided unique pictures of a heavily cratered surface that was extremely Moonlike. Closer inspection, however, showed that there were features unlike anything on the Moon, particularly lobate scarps that were interpreted as meaning that Mercury must have contracted after its surface had solidified.

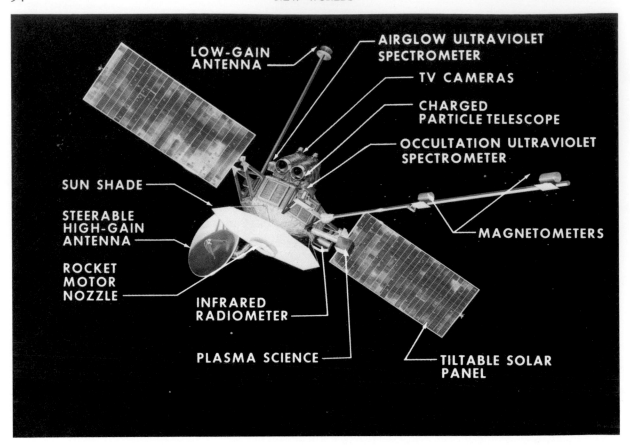

The surface of Mercury was revealed for the first time by cameras carried on a space-craft. Launched in 1973, NASA's Mariner 10 swung by Venus and made three encounters with Mercury. In addition to cameras, it carried other scientific instruments and discovered that Mercury unexpectedly has a magnetic field similar to that of the Earth, though of less intensity. (Jet Propulsion Laboratory/NASA)

In all, Mariner 10 made three swings past Mercury, mapping a considerable portion of its surface as well as exploring the planet's environment. A number of surprises were in store for the expectant astronomers back on Earth.

MERCURY IS A MAGNET

The biggest surprise of the mission was that Mercury has a substantial magnetic field. This was completely unexpected, since before its discovery geophysicists believed that a planet must not only have a liquid interior but must also rotate quickly in order to generate a magnetic field —through a process similar to that of a dynamo. The discovery that slowly rotating Mercury has a magnetic field shook up scientists, and they still lack a satisfactory explanation of how planets generate their magnetic fields.

The Mercurian field holds off the solar wind, just as does Earth's. There is also a bow shock like that of the Earth as well as a magnetosphere. There also seem to be some trapped particles within the field. But one of the big problems was why no radio emissions were observed if trapped particles were being accelerated within the Mercurian field. Scientists conceded that they were dealing with an entirely new configuration, which

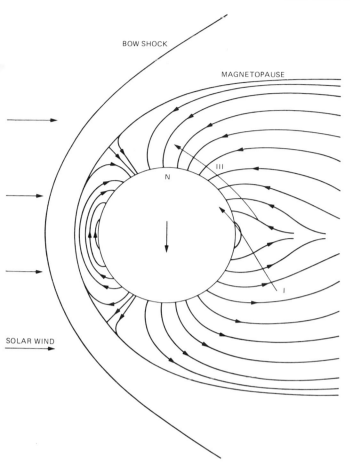

BOW SHOCK

MAGNETOPAUSE

III

N

I

SOLAR WIND

The discovery of a magnetic field of Mercury ran counter to established theory about how planets generate magnetic fields. A slowly rotating planet, such as Mercury, was not expected to generate a field. That Mercury does indeed have a field has led scientists to rethink how such fields might arise. An all-inclusive theory has not yet been devised. This line drawing shows the bow shock and magnetosphere of Mercury and the paths of Mariner through the magnetosphere during the first and third encounters with the planet. (Jet Propulsion Laboratory)

should be and what the intensity of the field should be if it is an internally generated field, like that of the Earth.

A sense of drama built up as the third encounter approached. Mariner 10 had been an ailing spacecraft nursed by its controllers through many months and millions of miles in space beyond what had originally been expected. Then it suddenly went awry, throwing one of its frequent tantrums. Like an inebriated reveler, the spacecraft had no sense of direction. And, unfortunately, there was hardly any reaction motor control gas remaining in its tanks to allow for any mistakes in bringing the spacecraft back on target in time to explore the magnetic field again. The controllers had one chance only to lock a star sensor on the bright star Canopus so that the spacecraft could be oriented correctly as it flew by Mercury. And they had to delay this maneuver as long as possible. If the spacecraft should again throw a tantrum and lose its balance, there would be no further chance to gather data on the innermost planet.

So, just a few hours before the encounter, the vital final corrections were made. Mariner 10 skimmed past Mercury, only 203 miles (327 kilometers) above the cratered surface, at 3:40 P.M. (PST) on March 16, 1975. During this fly-by the instruments carried by the spacecraft showed that the field was exactly as predicted by Dr. Ness. The magnetic field undoubtedly originates within the planet. Morever, Mercury's magnetosphere is an exact replica of that of the Earth, but only one thirtieth the diameter.

THE SHAPE OF MERCURY

The big antennas of the Deep Space Network coupled with powerful computers are able to trace the paths of spacecraft with unbelievable accuracy. Precise measurements of distance to within 5 feet at 200 million miles (1.5 meters at over 300 million kilometers) have been achieved. Such remarkable precision of measurement has permitted celestial-mechanics experiments to be conducted that determine the way in which gravity of the planets affects the paths of spacecraft flying past them. The accuracy is so great that effects on the flight path from large masses of mountains on a planet can be identified. From these experiments during the fly-bys of Mercury, scientists

appeared quite different from the interaction of the solar wind with the other planets.

At Mariner 10's first encounter with Mercury it was not clear whether the magnetic field was generated inside the planet or resulted from some interaction with the solar wind. The second spacecraft encounter took place far away from the planet, on the sunlit side to provide good photo coverage; it did not investigate the magnetic field. In preparation for the third encounter, Dr. Norman Ness, of NASA's Goddard Space Flight Center, calculated where the bow shock ascertained that Mercury is much closer to being a perfect sphere than Earth, the Moon, and Mars. This adds weight to a belief that although Mariner 10 could show us only one hemisphere of Mercury, the other hemisphere is probably very much the same. This is quite different from Earth, Moon, and Mars, on which there are great differences from hemisphere to hemisphere. One hemisphere of Earth, for example, is dominated by the Pacific Ocean basin; a hemisphere of the Moon is dominated by a volcanic plateau of gargantuan size.

The mass of Mercury was determined to one hundred times greater accuracy than previously, and it was confirmed that the planet consists of very dense materials, almost as dense as those of the Earth, and quite different from the lightweight Moon. While Moonlike on the surface, Mercury seems Earthlike in its interior. The high density of the small world suggests that it has a core of iron-rich material.

THE MOONLIKE SURFACE

Although Mariner 10's pictures covered only one hemisphere of Mercury, the surface revealed was remarkably consistent in its features. This surface has many primary-impact craters like those on the Moon, presumably caused by the impact of meteorites on a large scale. There are also many secondary, smaller craters, which are thought to result from debris, thrown up by the primary impacts, later crashing back to the surface. Many craters are also surrounded by rubble pushed out by the impact. The area covered by the rubble is called an ejecta blanket. Such blankets and the fields of secondary craters are all closer to the main craters than on the Moon. This is because the greater activity of Mercury prevented the material from being scattered as widely over the surface.

Amid the heavily cratered areas there are two types of plains. The first are comparatively smooth and lava-flooded, somewhat similar to the plains of the Moon but with less-well-defined boundaries. The second are areas of what seems to be smooth but more ancient terrain than the lava plains, ancient plains that are more heavily pock-marked with craters and may represent the original crust of Mercury, before it was bombarded by swarms of meteorites.

Many scarps appear to be wrinkles on the surface of the planet. Mercury also has one known major-impact basin comparable with those on the Moon but with an entirely different, lava-filled floor. This is the Caloris Basin, which occupies the region of one of the "hot poles"; hence the name. At the other hot pole there is a peculiar area of jumbled, mixed-up terrain that some theorists have suggested might have been caused when seismic waves from the impact that made Caloris focused around to the opposite hemisphere.

THE GEOLOGY OF MERCURY

Planetologists classify the surface features of Mercury into seven major geological types.

Intercrater plains are the most widespread on the side of Mercury photographed by Mariner 10. Gently rolling territory between the large craters and the impact basins, these plains have many small craters. The elongated shapes of these craters suggest that they are secondary rather than primary impacts. And since these plains do not anywhere damage the walls of the craters in the heavily cratered terrain, astronomers conclude that the plains predate the heavy craters.

Heavily cratered terrain looks very much like the highland regions of the Moon. Some of the big craters stand alone; others are in groups; yet others overlap each other. Their sizes are similar to those on the Moon, the biggest being several hundred miles across. Many of the craters have their floors filled with smooth plains of lava-like material.

Some craters on Mercury have bright rays. These are called *rayed craters;* they appear to be the youngest features on the planet. The rays cover other features. Instead of rays, some craters have *dark halos* that may also cover nearby features and thus indicate a younger age.

Smooth plains are the fifth type of surface feature. These level tracts appear to be younger than the heavily cratered terrain and older than the ray craters. They are crossed by many ridges and scarps. Scientists are still undecided whether these plains originated from a fluid flow of rocks as a result of heat generated by impacts that formed craters, or from an internal heating of Mercury and the upswelling of magma from the interior.

Mariner 10 produced hundreds of pictures of Mercury, which scientists later assembled into mosaics like these shown here. The first picture is a general view of Mercury taken by Mariner 10 as it flew away from the planet after its first encounter; the second is the general view as the spacecraft approached Mercury for this first encounter. Just below halfway down the first picture and near to the curved limb (edge) of Mercury, a bright-rayed crater appears. A bright ray coming from a crater near the south pole is shown near the bottom, and to the left of this ray there is a prominent light-colored crater with a dark halo and a small bright ray system. About the middle of the second picture there is a bright crater with rays superimposed on an older and larger crater. This is Kuiper. (Jet Propulsion Laboratory/NASA)

This picture shows a vast plain close to the northern limb of Mercury. Running along the edge of the plain is a major scarp, Victoria Scarp, named after Magellan's ship that sailed around the Earth. (Jet Propulsion Laboratory/NASA)

Another picture of a scarp on Mercury. This one is 185 miles (300 kilometers) long. It passes through craters, thereby indicating that the scarps were formed after the craters by a wrinkling of the crust of the planet. (Jet Propulsion Laboratory/NASA)

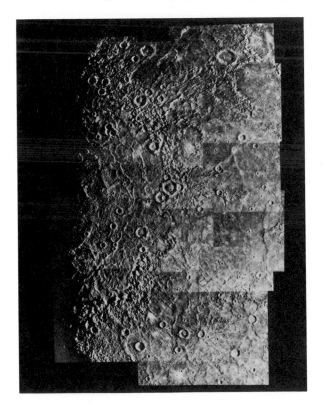

At the edge of the hemisphere of Mercury seen by Mariner during its three visits to the planet there is a huge impact basin, which has been named Caloris Basin because it is on a part of the surface that faces the Sun when Mercury is at its perihelion. Caloris is believed, however, to have been formed by the impact of an asteroid-sized body possibly 3–4 billion years ago. There are similar impact basins on the Moon and Mars. This picture, from Mariner 10, shows half the floor of the huge basin illuminated by the Sun. The rest is on the night side of the planet. The picture shows the mountain rings surrounding the basin and the ridges and fractures on its floor. At the top there are many long valleys radiating from the basin. These are thought to have been formed by projectiles gouging the surface when they were hurled out of the basin by the impact. (Jet Propulsion Laboratory/NASA)

At the other "hot pole" of Mercury (which faces the Sun at perihelion)—180° around the planet's equator from Caloris—is a region of jumbled terrain. It looks as though the whole surface has been shaken violently into jumbled blocks, breaking down the walls of craters. One theory is that this destruction of the surface came from the impact of the body that produced Caloris, that the seismic waves were focused to the antipodal point and their concentration there produced this jumbled mass.

The lower photograph shows a close-up of part of the terrain. (Jet Propulsion Laboratory/NASA)

Some areas of Mercury are heavily cratered, like areas of the Earth's Moon. The first picture shows the south-pole area of Mercury. The pole is inside the large crater on the lower limb of the planet at the lower center of the picture.

The crater stretches along the limb for a distance of 100 miles (160 kilometers). The floor of the crater, in shadow, and its far wall, illuminated by sunlight, appear disconnected from the edge of the planet in this view. Just above and to the right of the south pole is another frequently seen crater formation on Mercury, a double-ringed crater. It is about 125 miles (200 kilometers) across. A bright ray system splashes out from a 30-mile (50-kilometer) -diameter crater at the upper right.

The picture at right shows a heavily cratered region of Mercury with fresher craters on the top of older craters. This is typical of these regions. (Jet Propulsion Laboratory/NASA)

The southwestern quadrant of the planet is shown as Mariner 10 first approached Mercury, in March 1974. Several bright-rayed craters are visible. These are thought to be recent craters, similar to the bright-rayed craters on the Moon. (Jet Propulsion Laboratory/NASA)

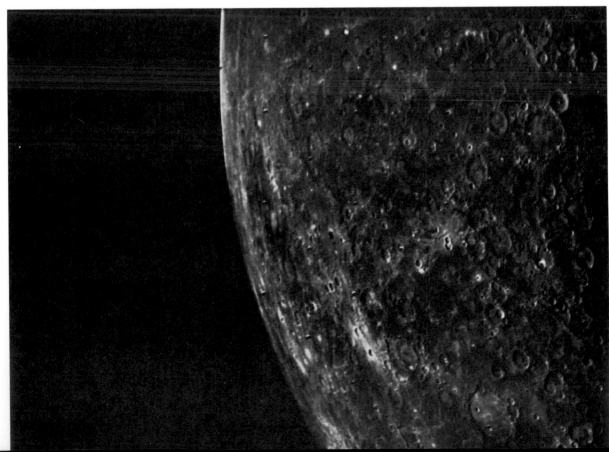

A dark, smooth, relatively uncratered area on Mercury shows surfaces similar to the mare material of the Moon. This material appears to have flowed across the surface to form bays in older terrain and in parts to have covered it. A history of heavy cratering on Mercury seems to have been followed by a period of lava flows that filled many of the old craters. The bright crater with a central peak at the center of the picture is about 19 miles (30 kilometers) across. (Jet Propulsion Laboratory/NASA)

There are also *young craters* that are somewhat older than the rayed craters. These are characterized by blankets of ejecta surrounding them and very well-preserved structure to their walls.

A *lineated terrain* consists of hills and valleys surrounding the Caloris Basin. It extends some 185 miles (300 kilometers) from the edge of the basin. The lines across this terrain must have been gouged out by massive projectiles the size of mountains that were hurled across it at the time of the impact that produced the Caloris Basin.

Scientists think the Caloris Basin was formed by collision of a minor planet (asteroid) with Mercury. The largest structure seen by Mariner 10, it is some 800 miles (1,300 kilometers) in diameter and straddled the terminator during all fly-bys of Mariner 10, so that only half of the basin could be seen. (The terminator is the line between day and night). The edge of the basin consists of a line of immense cliffs, some rising 1.25 miles (2 kilometers) high, the inner edge of great mountains ringing the whole of the visible basin. Between some of the mountains there are smooth lava plains peppered with small craters. The mountains probably represent bedrock thrust up by force of the impact of an asteroid hitting Mercury billions of years ago.

The floor of the Caloris Basin is laced with

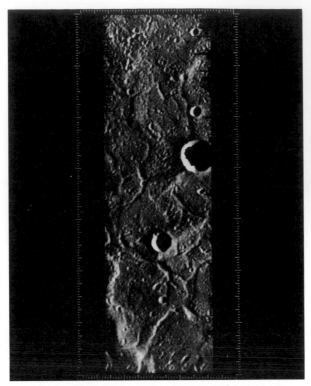

The third encounter of Mariner 10 with Mercury produced some very highly magnified views of small parts of the rugged surface. This is part of the floor of the Caloris Basin, showing small craters, ridges, and fractures. This picture is about 0.6 mile (1 kilometer) wide.

The photograph at right shows the location of the largest crater in this general view, of the basin. (Jet Propulsion Laboratory/NASA)

fractures and ridges quite unlike the basins on the Moon. These features make both radial and concentric patterns over the entire floor. Their widths vary from 1 to 7.5 miles (1.5 to 12 kilometers), and the highest ridge rises 0.5 mile (0.8 kilometer) above the floor. The deepest fractures are about 1,000 feet (300 meters) deep, and they have flat floors. These fractures could have been formed by tensional stresses similar to mud cracking when a pond dries up. Since they are not affected by the ridges, they are believed to have developed after them. First the floor of the basin generally collapsed to form the ridges, and then later it rose to form the fractures.

A TENUOUS ATMOSPHERE

From Earth-based observations, astronomers had concluded that any atmosphere Mercury possessed would be extremely tenuous. Mariner 10 flew behind the planet as seen from Earth. As the radio signals from the spacecraft were cut off by the bulk of the planet the change in the radio signals provided scientists with information about any possible atmosphere. This experiment confirmed that for all intents and purposes Mercury, like the Moon, is an airless world. The spacecraft also carried an ultraviolet spectrometer, which looked for emissions in ultraviolet light from any gases held to Mercury. This experiment detected very minute quantities of helium, amounting to a pressure only one million billionth that of Earth's sea-level pressure. This helium could have originated from decay of uranium and thorium in the crust of Mercury or might have been captured by Mercury from the solar wind. Scientists are not sure. They also found traces of hydrogen, argon, and neon.

Surface photography by Mariner showed no signs of atmospheric erosion. So if the planet ever had an atmosphere it must have disappeared into space before the time when most of the observable features formed on the surface. If the time scale of events on the Moon can be applied also to Mercury, this would mean that the appreciable atmosphere, if any, would have to have been dissipated at least 4 billion years ago.

THE COLOR OF MERCURY

A photometry team led by Bruce Hapke, of the University of Pittsburgh, confirmed that most of the surface of the innermost planet is covered by a dark, fine-grained soil similar to the unconsolidated material that overlies much of the lunar surface. But they saw more variations of color than on the Moon. Bright fresh craters and their ray systems appear somewhat bluer than surrounding areas, while the floors of some basins are redder than the darker rims. But no strong color differences could be seen on the smooth plains, nor across the large scarps. This latter observation suggests strongly that the scarps are faults, rather than lava flows, since if a scarp were the edge of a lava flow it might be expected to have a different color from the material over which the lava flowed.

WHAT DOES IT MEAN?

The Mariner fly-by of Mercury and its revelations about the nature of the surface of the planet greatly enhance our understanding of the Solar System. An earlier discovery by another spacecraft of craters on Mars similar to those on the Moon had led scientists to speculate that the inner planets had all been bombarded from space during their past history. This would be especially so if the Moon had obtained its cratered surface while in the vicinity of the Earth, and there was no reason to suppose that this was not true. The source of the bombarding bodies was thought to be the asteroid belt, between the orbits of Mars and Jupiter. If this were so it was expected that the innermost planet, Mercury, would be less heavily cratered than the Moon and Mars.

It turns out that the surface of Mercury is very heavily cratered. As a consequence, whatever bodies created the craters must have been as common in the innermost part of the Solar System as at the orbit of Mars. Bodies from the as-

teroid belt seemed unlikely: the bombarding objects had to follow extremely elliptical orbits that would dip to the orbit of Mercury and cross the orbits of all the inner planets. A more likely place was way out in the Solar System, possibly as far as the orbit of Neptune. The recent discovery of a mini-planet in these outer regions has given support to the theory that there is an outer belt of asteroids from which bodies were dislodged into the inner Solar System in the remote past.

The features on the surface of Mercury as revealed by the Mariner 10 pictures, taken in conjunction with the ages of the lunar samples returned by Apollo astronauts, allow us to estimate the probable sequence of the molding of the surface of the innermost world. This sequence most likely applies to the other inner planets, including Earth.

In general, the shaping of the surface of Mercury appears to be remarkably similar to that of the Moon, though until a spacecraft actually lands on Mercury and samples the surface materials we cannot be sure that the ages of the rocks on Mercury are the same as those on the Moon. But from what we know, the planet-forming sequence appears to have begun about 4.5 billion years ago, when material from the solar nebula started to fall together in a process of accretion that formed the main bulk of the planetary masses. Possibly this material was of differing composition at various distances from the Sun, thereby accounting for the different densities of the planets. Mercury, for example, has a density of 5.4 grams per cubic centimeter, (5.4 times that of water), which is close to the Earth's over-all density of 5.5 grams per cubic centimeter. The Moon, by contrast, has an over-all density of 3.36; Mars, 3.9; and Venus, 5.2 grams per cubic centimeter. Thus, it appears that planets closer to the Sun may have received more heavier materials than those farther out, though there are anomalies. On this basis, Earth may have formed in a region of the Solar System where water could easily condense, so it ended up as a water-rich planet.

After the accretion had taken place and the main bulk of the planet had formed, there was a period, evidenced on Mercury but not on the Moon, when the original surface became smoothed. On Mercury, this is believed to be represented by the intercrater plains, which appear to have been eradicated on the Moon because debris from subsequent impacts could be spread more widely, owing to the Moon's lesser gravity. The smoothing might have been caused by the bulk of Mercury melting under radioactive heating, during which the planet differentiated; that is, lighter materials came to the surface and later solidified into a crust, while heavier materials gravitated toward the center and became a core. Such a process of differentiation appears to have taken place on the Earth, because surface rocks have a much lower density than the planet over-all, indicating that heavier materials have accumulated at the Earth's center.

There is, however, another school of thought that prefers to think of the planets as forming from the nebula in stages: heavy material condensed first from the nebula to form the core, and then subsequent layers of lighter and lighter materials condensed later. Thus, the planets would have initially been in the form of layered shells of decreasing density surrounding a metallic core.

After millions of years the planets were bombarded again by asteroid-sized bodies known as planetesimals. These plunged through the inner Solar System and crashed onto the planets, covering much of their surfaces with craters of widely ranging sizes. Some of the impacting bodies were huge and produced vast circular impact basins, hundreds of miles across. Such basins are prominent features of Mercury, the Moon, and Mars. They probably existed on Earth, too, but have been obliterated by subsequent volcanic churning of the crust.

As the rain of planetesimals subsided, a new period of melting surface materials began, of lava flows and the formation of huge plains. The floors of craters and impact basins were filled with lava.

Up to this time the development of the planets was very similar. But afterward, except for a few straggling impacts, the planets proceeded along different evolutionary paths.

On Mercury and the Moon the internal heat engine was abruptly shut off. No more significant amounts of heat were generated. This is a mystery in itself, because Mercury and the Moon are internally so very different in size and density. Except for a few impacts from space to produce the fresh-looking ray craters, some of which may be only a few million years old, the surfaces of Mercury and the Moon have remained almost unchanged for some 3.5 billion years.

Following this concept of how the inner planets formed, we assume that the surfaces of Venus, Earth, and Mars were similar to those of the Moon and Mercury about 3.5 billion years ago. But on Mars, Earth, and Venus the internal heat engines continued to work. On Venus, a thick atmosphere was produced. It trapped solar heat and led to a planet with a tremendously dense atmosphere of carbon dioxide and a surface hot enough to melt lead. On Mars, half the planet was churned by volcanic activity to produce the biggest volcanoes known. Water flowed over the surface and eroded channels and dug immense river beds. Then Mars, too, began to slow in its planetary activity; water froze into ice and the atmosphere thinned until today it is only one hundredth the pressure of the Earth's atmosphere.

On Earth, volcanic forces have been active throughout its history. A relatively thick atmosphere formed and water condensed into oceans that covered most of the surface. The internal heat engine is still actively changing the terrestrial surface, pushing continents apart and raising volcanoes.

A peculiar feature on Mercury that is not seen on the Moon is the planet-wide system of lobate scarps. These long cliffs range from 12 to 300 miles (20 to 500 kilometers) in length and from a few hundred to 10,000 feet (3,000 meters) in height. They are relatively steep but have rounded crests and indicate that Mercury, unlike the Moon, passed through a period of global compression like the shrinking of a drying fruit producing wrinkles on its skin. This shrinkage of the crust of Mercury took place sometime after the final bombardment from space. This is apparent because the lobate scarps disturb the walls of craters. The shortening of the crust to produce these scarps could have been caused by a change in the iron-rich core from liquid to a solid state.

A RETURN TO MERCURY?

Despite Mariner 10's resounding success, the United States has no clearly defined plans for another mission to Mercury even though there are many unanswered questions that are important to understanding the genesis of our Solar System. Until we have sampled directly the surface materials of Mercury, all our estimates of ages must be purely relative.

A mission to have a spacecraft orbit Mercury would be a logical next step in the exploration of the innermost planet. One suggestion has been for twin orbiters, the first to follow an elliptical orbit that would dip to within 125 miles of the surface, the second to pursue a circular orbit about 280 miles above the surface (200 and 450 kilometers, respectively). These spacecraft would map the whole surface of Mercury much as Mariner 9 mapped Mars. Depending on the success of the orbiter missions and the availability of public support, one or more landing craft could be sent to Mercury in the 1990s. They would perform experiments on the surface and might even return some samples of material for detailed analysis here on Earth. A major aim would be to determine the composition of the crust of Mercury and the absolute ages of the materials there.

Mercury can also provide an ideal location for observing the Sun at close hand. In the next century, permanent stations may be established on the innermost planet for close-in monitoring of solar activity. And if such stations were manned,

astronomers during the two and a half months of night would witness a celestial spectacle the like of which can never be seen from our own planet. In the night sky, approximately every four and a half months they would observe a most brilliant sight: the planet Venus. Fully illuminated and casting distinct shadows, it would be many times brighter than it ever is from Earth. Observers on Mercury would also see Earth and Moon, a brilliant twin system undergoing monthly gyrations that would dominate the night sky approximately every three and a half months, Earth would appear as a blue-green jewel, the Moon as a rather dull star by contrast. The observers on Mercury would also see Mars, but it would never reach the brilliance it attains in Earth's skies.

So, having visited the innermost of the planets, we can now look outward into the Solar System through all its planets and review what we knew before the space age and the prodigious advance in our knowledge and understanding after the visits of our spacecraft. The next planet outward from the Sun is that most brilliant object in the night sky of Mercury, the shrouded beauty that is sometimes called the queen of the night: Venus.

6
THE SHROUDED BEAUTY

At the beginning of the nineteenth century, Napoleon Bonaparte was riding to the Luxembourg Gardens where the French Directory was to celebrate him. As he rode along Rue de Touracour he was surprised to find that many of the people lining the street were not looking at France's hero. Instead they were gazing into the blue sky, pointing fingers heavenward. Napoleon turned questioningly to one of the brilliantly attired staff accompanying him. He was informed that the people were looking at a bright "star" in the sky, which they associated with the conqueror of Italy. Napoleon turned to look into the bright noon sky, where he, too, saw the object. It was indeed the planet Venus.

When at its brightest, Venus is easily seen in broad daylight, if you know where to look. And at night it is so brilliant that if you hold up your hand between the planet and a light-colored wall you will see the shadow of your hand cast by Venus. When seen suddenly through a break in clouds in the evening sky it has often been mistaken for a UFO.

Being the brightest of the planets, Venus must have been the first celestial body other than the Sun and the Moon to attract the attention of ancient peoples. As early as 3000 B.C., Babylonian astronomer-priests recorded the movements of the brilliant "star" they called Nin-dar-anna, or Mistress of the Heavens. About the same time, Chinese records were noting the appearances of Tai-pe, the Beautiful White One, in the morning and evening skies. Both referred to Venus, whose name comes to us today from the goddess of love and beauty of Roman mythology.

Like Mercury, Venus is a planet that revolves around the Sun within the orbit of the Earth. As a consequence, it always appears relatively close to the Sun as seen from Earth. But as its orbit is bigger than that of Mercury, Venus appears at greater angular distance from the Sun. Even so, at its normal brightness it is still an object that is visible only for a short while in a darkened sky. This circumstance led the Egyptians to regard Venus as two separate objects, to which they gave the names Tioumoutiri and Quaiti. The ancient Greeks did the same: they called the two objects Phosphoros (light-bearer) and Hesperos (western). The former was the morning star, the latter the evening star. These names, morning and evening star, are still used today in popular writings. By the sixth century B.C., the Greek mathematician and philosopher Pythagoras had discovered that the two objects were, indeed, only one.

As with Mercury, Venus attains its greatest angular distance from the Sun at times of eastern and western elongations. When between Earth and Sun, the planet is said to be at inferior conjunction, and when opposite to Earth, on the far side of the Sun, it is said to be in superior conjunction. The orbit of Venus is close to being a circle. At superior conjunction the planet is at an average 160.2 million miles (257.8 million kilometers) from Earth. At its closest approach (inferior conjunction) Venus comes at an average within 25.7 million miles (41.4 million kilo-

This photograph is a reproduction in a blue tone to simulate the view of Venus seen by the cameras of Mariner 10. But such a view could never be obtained by the unaided human eye, which cannot see ultraviolet light. Venus to the unaided eye appears completely blank. (NASA)

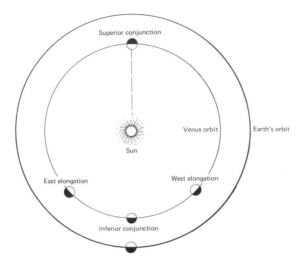

Like Mercury, Venus travels around the Sun within the orbit of the Earth. It appears at its greatest angular distance from the Sun when at east and west elongations. When between Earth and Sun, Venus is said to be at inferior conjunction. When on the far side of the Sun from Earth, it is said to be at superior conjunction. Venus orbits the Sun in 225 Earth days. Coupled with the Earth's own motion around the Sun, this leads to the apparitions (elongations and conjunctions) repeating after a period of 584 Earth days, which is called the synodic period of Venus. This period was well known to ancient astronomers of the Middle East.

meters) of the Earth. Thus, except for the Moon and a few asteroids, Venus is the nearest celestial body. Venus keeps close to its average distance of 67.24 million miles (108.2 million kilometers) from the Sun.

Venus travels around the Sun in 224.7 days. But because Earth is traveling in the same direction, the period from one inferior conjunction to the next is longer than one Venusian* year. It takes another 360 days for Venus to overtake the Earth in its orbit and gain the position between Sun and Earth. Thus, the synodic period of Venus, the time between apparitions, is 584 days.

When Venus passes through superior conjunction it cannot be seen from Earth, because it appears close to the solar glare. But as the days pass, the planet gradually pulls to the east of the

* Venusian is commonly accepted as the adjective for this planet, though sometimes astronomers use Cytherean instead. The latter is derived from Cytheria, an island in the Ionian Sea onto which Aphrodite (the Greek goddess of love) was supposed to have landed.

Sun and appears as a bright star in the evening sky. As weeks pass, Venus moves farther and farther from the Sun and sets later and later. About 220 days after superior conjunction it reaches greatest eastern elongation and sets several hours after sunset. Seen in a telescope it appears as a half-moon shape. From this point it starts to move westward toward the Sun, and because it is now closer to the Earth it moves more quickly, achieving inferior conjunction in the shorter time of 72 days. After a while, it emerges from the solar glare as a morning star and moves westward to greatest western elongation, rising several hours before the Sun. From western elongation it starts to move east again, toward the Sun, and reaches superior conjunction in another 220 days. The cycle then starts anew.

During its path around the Sun, Venus exhibits phases to an observer on Earth in the same way that Mercury does. But the change in observed size from crescent to full phase for Venus is much greater than for Mercury, because greater distances are involved. Two opposing tendencies are at work in determining how bright the planet will appear in the skies of Earth. The first is for the planet to get larger and brighter as it comes closer to the Earth. But because of the phase change from full to crescent during this same period, progressively less of the illuminated disc is presented to the terrestrial observer. As a result of these two effects, the planet is brightest when it is between elongations and inferior conjunction and appears in Earth's skies about 39 degrees from the Sun.

Since it is an inner planet, Venus, like Mercury, can sometimes pass across the face of the Sun as seen from Earth. Again, such transits are not common even though the orbit of Venus is inclined less than that of Mercury to the orbit of the Earth. As with Mercury, a transit across the face of the Sun can take place only when Venus is at inferior conjunction at a point on its orbit where the planes of the orbits of Earth and Venus cross. Since Venus takes longer to orbit the Sun than Mercury there are fewer chances of such occurrences, and transits of Venus are much less frequent than those of Mercury. The passages of Venus across the solar photosphere take place in June or December, only four times in every 243 years. In the past few centuries, since the first observation of a transit of Venus was made by Jeremiah Horrocks and William Crab-

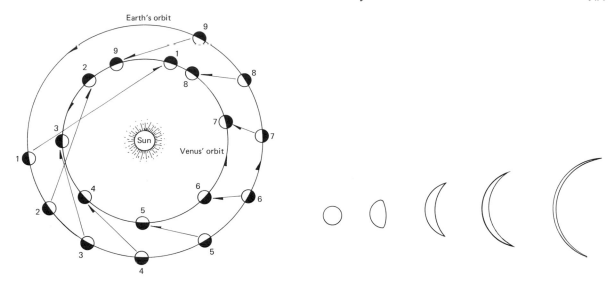

The relative motions of Earth and Venus are shown in the first drawing. The second shows the relative size of the planet as seen from Earth at various phases of Venus. Note that the size changes much more for Venus than for Mercury. While Mercury attains its greatest apparent brightness at its full phase, Venus appears brightest in Earth's skies when it is a fat crescent.

tree in 1639 at Manchester, England, they have been taking place in pairs. The next transits are due on June 8, 2004 and June 6, 2012. After that, transits will occur singly at 105.5- and 121.5-year intervals.

In size and mass, Venus is very like the Earth. Its diameter of 7,521 miles (12,101 kilometers) compares with Earth's 7,926 miles (12,753 kilometers); its mass is 0.82 that of the Earth; and the planets are almost the same density, with Venus 5.26 grams per cubic centimeter and Earth 5.5. Almost our twin physically, Venus has, however, evolved into a very different world.

THICK ATMOSPHERIC BLANKET

Transits of Venus were very important to astronomers in the eighteenth century, for by observing the transits they hoped to determine more accurately the true distance of the Earth from the Sun. So expeditions were sent all over the world to obtain long base-lines for trigonometrical exercises in calculating this distance. Unfortunately, the observations did not work out as the astronomers expected. They wanted to time very precisely the instant when the edge of Venus was just touching the edge of the Sun, but the contrast in brightness between the photosphere and the black

body of the planet produced optical distortions that prevented exact timing. In addition, astronomers found another effect, a luminous ring around the planet as it moved onto the face of the Sun.

Several astronomers observed this effect and concluded that Venus must have an atmosphere to account for the ring. The great chemist Mikhail Vasilevitch Lomonosov was probably the first to observe and interpret this effect, at the transit of June 1761. But nothing about his discovery appeared in print until the latter part of the nineteenth century, when the collected works of Lomonosov finally appeared.

The other independent discoverer also experienced a long delay in publication. David Rittenhouse of Philadelphia observed the transit of 1769 and reported to the American Philosophical Society that he had discovered an atmosphere on Venus. But it was not until the history of the society was published, in 1884, that his discovery became part of the published record.

Some years later, another astronomer, Johann Hieronymus Schröter, was observing Venus when the planet was approaching inferior conjunction, in 1790. He noticed that the pointed ends of the thin crescent—what astronomers call the cusps—extended more than would be expected on an airless world. He reasoned that the extension must be caused by a thick atmosphere. Furthermore, he reported a faint, bluish light, or aureola, occasionally outlining the circumference of Venus beyond the extended cusps when the planet was seen as a very thin crescent. He rightly interpreted this as a halo effect produced by sunlight diffusing through the atmosphere of the planet, somewhat analogous to twilight effects on Earth.

Yet all astronomers were not convinced that Venus had an atmosphere. As late as 1899, Henry Norris Russell, who had presented cogent arguments to dispute Percival Lowell's theories of

intelligent beings living on Mars, wrote in the prestigious *Astrophysical Journal* that Venus could have an atmosphere of only small extent and density. He claimed that there was no satisfactory evidence that the atmosphere of Venus was more than one third as dense as that of the Earth.

But during the first quarter of the twentieth century it became clear to most astronomers that the atmosphere of Venus was indeed extremely dense compared to that of Earth, and that its composition was also quite different from ours. Apart from its brilliance and its phases, Venus is one of

Because of the heavy clouds that completely cover the surface of Venus, astronomers relying on observations using the spectrum of light to which the human eye is sensitive have been unable to see any definite details on the planet. This picture was taken in blue light with the 200-inch (5-meter) Mount Wilson and Palomar Observatory telescope. It is typical of the best pictures obtainable by Earth-based observatories in visible light. (Mount Wilson and Palomar Observatories)

the most disappointing of telescopic objects. The bright surface of clouds seems completely unblemished by any markings. The thick atmospheric blanket guarded Venus' secrets well; astronomers had no idea what the surface was like, and speculation ran rife.

WHERE IS THE WATER?

At first some astronomers thought that the brilliant clouds were composed of water, like those of the Earth. They were disappointed when no trace of water could be found in spectra of light reflected from Venus, even with the best instruments available in the 1920s. At the end of the decade they could say only that if there was any water on Venus it was less than 5 per cent of that in the Earth's atmosphere. Even today, with space probes gathering data directly from within the atmosphere of Venus, the question of water remains of paramount importance and has by no means been answered completely. Scientists are still asking: If Venus once had water in quantities like those on its sister planet Earth, where is it now? And if Venus never had oceans, why?

If Venus is as dry as it seems, did it have oceans to begin with? One speculation is that because Venus is close to the Sun, primitive water would have steamed into the atmosphere and risen to high enough levels for it to be broken down into hydrogen and oxygen by solar ultraviolet radiation. The hydrogen could have readily escaped into space because, being a lightweight gas, individual molecules achieve high enough velocities to escape from the gravitational field of Venus, as they also do on Earth. Oxygen would, however, remain. So that its absence from the Venus atmosphere has to be explained. Some scientists doubt that Venus could have lost all its water in this way and that at least 10 per cent would have remained on the planet. They suggest that perhaps Venus formed so close to the Sun that water never condensed on the planet from the solar nebula. They suggest that Venus was always without much water.

THE CASE FOR CARBON DIOXIDE

In 1922, Charles B. St. John and Seth B. Nicholson, of Mount Wilson Observatory, wrote about the paucity of water and oxygen on Venus and speculated that the atmosphere might consist of gases such as nitrogen and carbon dioxide. Two years later, Edison Pettit and Nicholson measured the temperature of the outer atmosphere of Venus by the infrared radiation from it and found that there was hardly any difference between the night and day hemispheres of the planet. They reasoned that the atmosphere of Venus must be well mixed to equalize the temperature on the planetary scale observed.

The problem of determining the content of the atmosphere of another planet by observations from Earth is that the molecules we are looking for may be the same as those in our own atmosphere. The best way to get over this difficulty would be to place our instruments above the atmosphere of the Earth. But that was impossible until the space age, though astronomers did erect observatories on high mountain peaks and even carried instruments aloft in balloons.

As a result of flights in balloons to altitudes of 15 miles (25 kilometers) or so, some water vapor was detected in the outer atmospheric envelope of Venus, along with carbon dioxide and small amounts of carbon monoxide, hydrochloric acid, and hydrogen fluoride. But these constituents were in the upper atmosphere; astronomers could not find out from Earth the composition of the atmosphere below the clouds of Venus. In analogy with Earth, nitrogen was assumed to be a constituent of the Venusian atmosphere, but its presence could not be detected by Earth-based observations. The biggest mystery was that no traces of oxygen were found in the Venusian atmosphere.

A HOTHOUSE PLANET

While infrared radiation allowed astronomers to ascertain the temperature of the upper atmosphere of Venus, this radiation did not penetrate through the clouds. The temperature of the surface and of the lower atmosphere of Venus was unknown. Then astronomers started to use a new tool: Certain frequencies of radio waves would be expected to penetrate from the surface of Venus to space, and if these could be detected at Earth an analysis of their characteristics would allow scientists to estimate the temperature of the surface that emitted them. In June 1956 three scientists at the Naval Research Laboratory, Cornell Meyer, T. P. McCullough, and R. M. Sloanaker, obtained radio data that could be interpreted only on the basis of a very hot surface of Venus, at least 710° F (650° K), far above the boiling point of water and high enough to melt lead. Such high values were surprising to most astronomers. True, Venus is closer to the Sun than the Earth, but not that much closer. The answer seemed to lie in the atmosphere itself. While it admits solar radiation, which warms the lower atmosphere and the surface, it blocks the reradiation into space of the longer, infrared wavelengths from the hot surface and lower atmosphere. In effect, the atmosphere of Venus seems to act like the windows of a greenhouse or of an automobile left in the sun. The interior becomes much warmer than the exterior, because the heat has no way of escaping.

There were some alternatives to this greenhouse hypothesis, but they have not stood up well to further investigation. One was a suggestion that the surface of Venus was that of a dried desert, and high winds raised clouds of dust which, suspended in the atmosphere, abrasively developed large amounts of heat by friction as they blew at high speed across the surface features.

Another suggestion was that Venus might have a very active region in its upper atmosphere where molecules and atoms of gases have electrons knocked from them, by solar radiation, to become charged particles. These particles form a region of the upper atmosphere that scientists call an ionosphere, because the gases there are ionized, or electrically charged. With a violently active ionosphere, heat might be generated that would pass downward and heat the lower atmosphere. Radar beams reflected from Venus told us that the planet's ionosphere cannot be dense enough for such a theory to hold. The radar beams passed through the ionosphere and bounced back to Earth from the surface of Venus. In addition, this ability to probe to the surface of Venus with radar showed that there cannot be much water vapor in the planet's atmosphere, because it would absorb the radio waves used.

Today, we generally accept that the surface's high temperature is the result of a greenhouse effect.

PRIMEVAL SWAMPS AND HYDROCARBON OCEANS

Until modern radar and spacecraft could be used as tools to explore Venus, ideas of what it might be like on the surface of the planet were colored by astronomers' concepts of the Venusian atmosphere.

In 1918, Sweden's Svante Arrhenius, who in 1903 had received the Nobel Prize in Chemistry for his electrochemical discoveries, wrote about a wet planet in his book *The Destinies of the Stars*. He thought that the humidity on Venus would be six times that of the Earth's average and that everything would be dripping wet. The surface, he wrote, would be filled with swamps like those of Earth's carboniferous era, and there should be luxuriant vegetation, greatly accelerated in growth by high temperature and copious amounts of water. Today we can smile at these thoughts while realizing how they affected a couple of generations of science-fiction writers.

Once scientists learned that carbon dioxide was a major constituent of the Venusian atmosphere, but before they knew how high temperatures were on the planet, new speculations changed Venus to an ocean-covered planet. The reasoning was that a chemical process must have been responsible for the abundance of carbon dioxide in the atmosphere of Venus. On Earth we know that this gas reacts with silicates to form carbonates and quartz sand. If there were an ocean covering the whole of Venus it would prevent contact between silicates on the ocean floors and the atmospheric carbon dioxide. Great quantities of carbon dioxide could thereby stay in the atmosphere of Venus, whereas on Earth it has been lost to the rocks of the continental crusts.

Such eminent astronomers as Harvard's Donald H. Menzel and Fred L. Whipple once subscribed to the planet-wide-ocean theory. But Britisher Fred Hoyle disagreed. He preferred a Venus covered with oceans of oil. He based this on the possibility that all the water once on Venus had been used up in oxidizing hydrocarbons to form carbon dioxide. What remained had formed vast oceans of oil over which lay heavy clouds of hydrocarbon smog.

Yet other scientists speculated that Venus was a hot, dusty world, a vast dry desert. And they were closer to the truth.

SEEING BY RADAR WHAT WE CANNOT SEE BY LIGHT

The art of "seeing" by radio waves bounced off targets was developed during World War II, and radar was soon applied to astronomy as transmitters became more powerful and receivers more sensitive. Radar astronomy has several great advantages over astronomy at optical and near-optical wavelengths. The first is that it can be used anytime, day or night, and whether or not the sky is overcast with clouds. The second is that it can penetrate through the clouds of other planets and allow us to see their surfaces.

Radar probing of Venus put to rest all the theories about oil oceans, dripping swamps, planet-wide water oceans, and others. The first radar contact with Venus took place on February 10, 1958. It was made by scientists of the Lincoln Laboratory of the Massachusetts Institute of Technology using a large radar system at Millstone Hill. To begin with, contacts were used to establish more precisely the distance of Venus from the Earth and thus the distance of the Earth from the Sun. But subsequently, with better equipment, astronomers used radar to find out more about the surface of Venus. For example, D. B. Campbell and his associates, using another of MIT's radars, that of the Haystack Observatory, and Cornell University's National Radio Astronomy and Ionospheric Center, at Arecibo, Puerto Rico, found evidence of mountains on the planet. In particular, they discovered a peak that was nearly 2 miles (3 kilometers) high, and the radar indicated that the slopes of this mountain were gentle on the west but precipitous on the east. On Earth, such changes of slope are often associated with mountain-building processes that result from movement of plates of the Earth's crust. On the Moon and Mercury they are found in the mountain chains around large impact basins. The peak was part of a general region of high elevation that covered an area of about 3,700 miles (6,000 kilometers) east-west and 300 miles (500 kilometers) north-south. There were other peaks and valleys in this highland area.

Radar experiments also showed that the surface of Venus reflected the radar signals more effectively than the surface of the Moon and Mercury. In fact, Venus turned out to have twice the radar reflectivity of these other worlds. At the same time, the Venusian surface seemed generally smoother than the other worlds of the inner Solar System. Differences in elevation did not appear to exceed 3 miles (5 kilometers), compared with differences of about 12.4 miles (20 kilometers) on Earth and 18.6 miles (30 kilometers) on Mars.

Major discoveries were made by radar in 1972. Richard A. Goldstein and his coworkers at the Goldstone station of NASA's Deep Space Network used 85-foot (26-meter) and 210-foot (65-meter) antennas together to obtain radar pictures of the Venus surface in greater detail. The larger antenna first transmitted at high power toward Venus. A few minutes later, the radio signal that had bounced off the planet was received by both antennas. This "stereo" reception enabled Goldstein to pinpoint each area touched by the radar beam on Venus. He was thus able to detect features on the surface as small as 6 miles (10 kilometers) across and to measure elevations to within 650 feet (200 meters).

As Goldstein analyzed the data carried by the radar echoes, some surprising details of the surface of Venus were revealed. He used a computer to assemble the data into maps of Venus showing about 500,000 square miles (1,300,000 square kilometers). A big surprise was that the surface appeared to have many craters, ranging in diameter from 20 to 100 miles (35 to 160 kilometers). Though numerous and large, these craters did not seem to be very deep. The largest of the craters was 100 miles (160 kilometers) across and only about 1,300 feet (400 meters) deep. So the craters on Venus seemed more akin to those on Mars, partially filled with debris, than the deeper craters of the Moon and Mercury. While some astronomers had not expected to find craters on Venus, believing that the dense atmosphere of the planet would have prevented the impacting bodies from reaching the surface, the real surprise was more philosophical. Actually, a body able to dig a crater only 1 mile (1.6 kilometers) in diameter would have no difficulty in penetrating a dense atmosphere like that of Venus and reaching the surface. We have examples of such craters on the Earth. But, philosophically, Venus had been regarded as a twin of Earth, and Earth was characterized by the absence of large ancient craters. The discovery of such craters on Venus meant that they were either of recent, volcanic origin or that there had been little molding of the surface for billions of years. It also meant that of all the inner planets only the Earth was free of large craters now but was probably covered with them in the distant past. The presence of these large craters on Venus indicated that there could not have been widespread erosion of the surface either by

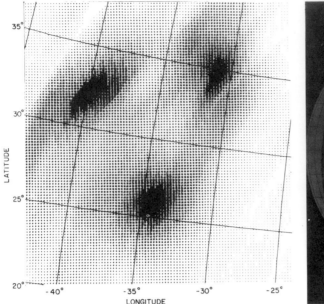
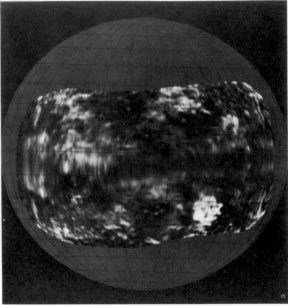

First clues about features on the surface of Venus were obtained by bouncing radio waves from the planet and observing their pattern when received back at Earth. The left is one of the early radar pictures of Venus taken with the 210-foot (65-meter) antenna at NASA's Deep Space Facility in the Mojave Desert of California in 1968. It covers an area of about 160,000 square miles (400,000 square kilometers). The dark areas are thought to be rough areas, thereby accounting for their poor radar reflectivity. In contrast the bright area at the center of this map may be a mountain. It has been named Beta and is approximately 17,000 square miles in area (60,000 square kilometers). A later radar chart of Venus, produced in 1970 (right), covers a much larger area of the planet, about one sixth of its surface, 30 million square miles (78 million square kilometers). The bright spot in the southern hemisphere, named Alpha, is another possible mountain range. It is about 30° south and on the prime meridian. Beta is about 30° north and at the opposite side of the region mapped in this picture. (Jet Propulsion Laboratory/NASA)

weathering as on Earth or alternatively by volcanic processes.

In subsequent years radar astronomers made other intriguing discoveries about the surface of the planet. In 1976, the Arecibo radar revealed a lava flow as large as the state of Oklahoma and a basin similar to a mare basin on the Moon. The bright lava area had been seen on radar maps earlier and had been named Maxwell. It seems to lie on top of an older and smoother surface. The lava sheet has long, parallel ridges extending for hundreds of miles. This might be evidence of tectonic processes similar to those on Earth that led to the building of mountains as big areas of crust move relative to one another and crinkle into mountains under the forces of compression.

On one of the most recent radar maps, there appears to be a chasmlike feature running north and south across the equator for a distance of about 600 miles (1,000 kilometers). The chasm is about 1 mile (1.6 kilometers) deep and about 100 miles (160 kilometers) wide. At its southern end, it divides into two branches, which are each about 60 miles (100 kilometers) wide. At its northern end, it breaks into many branches. Another high-resolution picture shows a raised area of the surface of Venus that may be something like the Tharsis Ridge, a vast raised area on Mars associated with large volcanoes. On Venus the raised area also shows a high, mountainous feature with a cratered top that may indeed be a volcano.

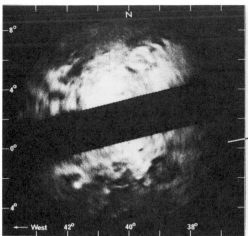 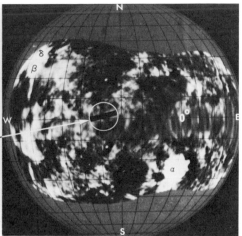

Radar pictures continually improve. This dramatic map at left was produced in 1973 to show a 910-mile (1,500-kilometer) -wide strip of the equatorial zone of Venus up to 100 miles (160 kilometers) across. The black area through the center depicts a belt that cannot be accurately mapped from the Goldstone Tracking Station of the Deep Space Network. The map on the right shows a large-scale radar chart of Venus with the view covered in the high-magnification picture shown as a circle. The exciting thing about these pictures is the discovery of huge, shallow craters on Venus, some up to 100 miles (160 kilometers) in diameter.

Below, a high-magnification radar image of Venus. This picture resolves objects as small as 6 miles (10 kilometers) across and reveals differences in elevation of 500 feet (150 meters). A crater in the lower half of the picture has a raised rim with an elevation of about 1,800 feet (540 meters). (Jet Propulsion Laboratory/NASA)

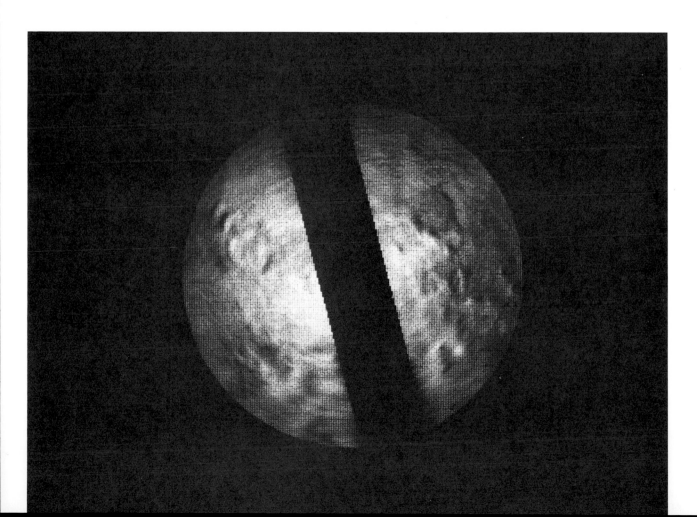

HOW MANY DAYS IN A VENUSIAN YEAR?

As with Mercury, radar solved one of the major questions about Venus that had frustrated astronomers for centuries. How many days in a Venusian year? Does Venus rotate in a period measured in hours, like the Earth, or is the planet locked to the Sun so that it turns one hemisphere sunward all the time?

In 1892, Sir Robert Ball, the famous Cambridge professor of astronomy, wrote, in *In Starry Realms,* that: "It seems hard to resist the conclusion that Venus, like Mercury, revolves so as always to show the same face to the Sun." But this was based on the then current excitement of having accounted for the Moon's facing the Earth as a result of tidal forces, and Schiaparelli's observations, later proved erroneous, that suggested Mercury turned one face to the Sun. The problem in finding out the period of a day on Venus was the ubiquitous cloud cover, which prevented any observations of surface markings. Although astronomers often claimed to see markings on Venus, these were so indistinct that it was impossible to have any faith in them.

The spectroscope had been applied to Venus to see if measurements could be made of the rate at which the atmosphere on either side of the disc as seen from Earth was moving away from our planet on one side and toward it on the other side. Results were inconclusive, but they seemed to confirm a long period of rotation, possibly the 224.7 terrestrial days that coincided with the revolution of Venus around the Sun. In 1956, Robert S. Richardson, of Mount Wilson and Palomar Observatories, suggested that Venus might slowly rotate oppositely to the Earth and the other planets, in what astronomers call a retrograde direction. Others agreed that this might be so, but the shifting of spectral lines by the rotation of the planet was so slight that astronomers could not be sure. In the meantime another problem had arisen. Photographs taken in ultraviolet light showed markings that were not visible in the region of the spectrum to which the human eye is sensitive. This series of pictures suggested that Venus was rotating in a period of only four Earth days and in a retrograde direction.

Astronomers in America, Russia, and Britain tackled the question with radar during the early 1960s. They found that the planet rotates very slowly, taking about three hundred Earth days to turn on its axis. Later, this time was reduced by more accurate observations. In the summer of 1964 the big antenna at Arecibo, Puerto Rico, narrowed the uncertainty to a Venusian day of between 242 and 252 days. By 1969, the true period of 243.09 days had been established.

This period is strangely similar to the time between successive close approaches of Venus to the Earth. Each time Venus passes through inferior conjunction between Earth and the Sun, it turns the same hemisphere toward Earth, or very nearly so. The time between inferior conjunctions is, within the limits of error, 1.7 hours longer than the spin period of Venus.

While Venus rotates backward in 243 days and goes around the Sun in 225 days, the combination of these two motions results in a solar day on Venus of just slightly less than 117 Earth days. But if a person on the surface of Venus could see the Sun, its rising and setting would be different from those on the other terrestrial-type planets, including Earth. On all the other planets the Sun rises in the east and sets in the west. On Venus it would rise in the west and set in the east. Sunrise would be painstakingly slow, and it would take a whole Earth month for the Sun to rise to its highest point, at local noon on Venus. Then another month would pass before sunset. This would be followed by two months of night. And because Venus rotates on an axis that is very close to being perpendicular to the plane of its orbit around the Sun, there are no seasons there.

The anomalous rotation of Venus is enigmatic. Mars spins in almost the same time as the Earth. Mercury became locked in its 2:3 resonance with the Sun by action of the solar gravity. The Moon is locked to the Earth by tidal action. So how did Venus get to spin the wrong way around?

S. Fred Singer, of the University of Virginia, offered a solution. He suggested that Venus once rotated like the Earth and Mars. But the planet captured one of the big bodies that entered the inner Solar System about 4 billion years ago. These bodies blasted great basins on other worlds, and some may have done the same on Venus. But suppose Venus captured one of them into an orbit the opposite way around to the normal spin of Venus, and later this body crashed to the surface. Singer suggests that the energy of impact might have been great enough to stop the normal rotation and jar the planet into a slow rotation in the opposite direction, the rotation that we measure today.

Another theory is that Venus accreted from a region of grains that were moving in circular orbits around the Sun. Such accretion would be ex-pected to produce a planet rotating in a retro-grade direction, whereas planets that accreted from particles moving along elliptical orbits should rotate in a prograde direction; i.e., like the Earth.

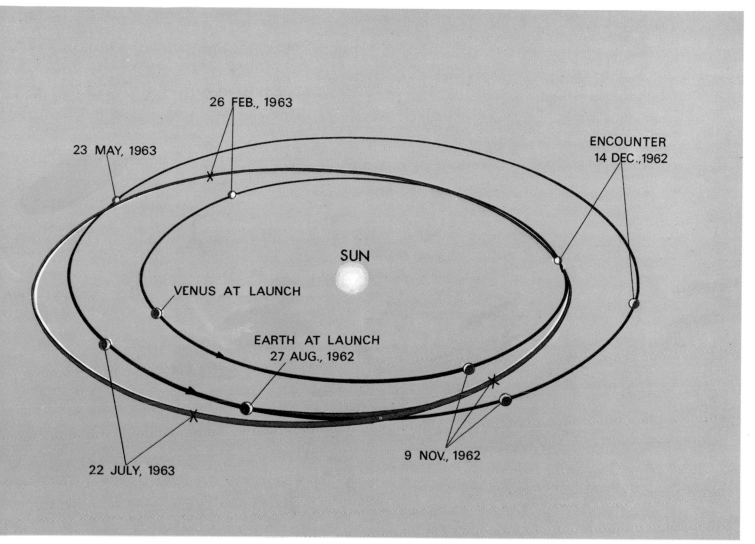

America's first successful Venus probe, Mariner 2, approached to within 21,750 miles (34,800 kilometers) of the planet on December 14, 1962. The journey lasted approximately three and one half months. Among other discoveries, the spacecraft's instruments showed that our neighboring world does not trap high-energy radiation particles, as do Earth's Van Allen belts, and hence possesses no appreciable magnetic field. Infrared radiometers detected an area in the southern part of the terminator some 18° F (10° K) lower in temperature than the surroundings, giving rise to speculations that a high mountain was situated there. (Arfor Picture Archives)

VENUS REVEALED BY SPACECRAFT

As with Mercury, the really big jumps in our understanding of Venus came when spacecraft were used to explore the planet. Although attempts to send them to Venus started in early 1961 it was not until December 14, 1962, that scientists obtained data from a spacecraft flying by Venus. On that date America's Mariner 2 flew to within 21,625 miles (34,795 kilometers) of the hidden surface of Venus. This spacecraft did not carry cameras to photograph the planet, but it made two important discoveries. The first was to confirm that because of the very slow rotation of Venus, it did not possess a magnetic field of any appreciable strength. As a consequence of there being no magnetic field, Venus does not have radiation belts like those of the Earth. The second important discovery was to confirm that the temperature differs only slightly between day and night hemispheres. Mariner 2 data indicated that the surface temperature of the planet was about 800° F (700° K).

Flying past a remote planet with a spacecraft is one thing; landing on it is quite another. Following the success of Mariner 2, the United States almost abandoned Venus. It was left to the Russians to try for a landing. They pioneered surface exploration of Venus with their Venera 4 spacecraft, which was launched on June 12, 1967, and landed on October 18 of that same year.

The capsule that was to make the historic landing, the first on another planet, separated from the main spacecraft and penetrated the atmosphere on the night side of Venus. Encountering high forces of braking as it plunged into the dense atmosphere, the tiny emissary from Earth probably glowed bright red. By the time it was 16 miles (26 kilometers) above the surface, its speed had been slowed sufficiently for parachutes to be released and further break its fall. For ninety-three minutes instruments aboard the spacecraft radioed information back to Earth on the pressure, density, and composition of the strange atmosphere of our sister planet.

Venera 4 found that the atmosphere of Venus is mainly carbon dioxide—around 90 per cent. It also discovered that there was nitrogen in the atmosphere, but only a small amount. And this applied to water vapor and oxygen. There is some doubt still about the accuracy of the measurements that indicated about 1 per cent by volume of oxygen. There is also doubt that Venera reached the surface intact. The instruments stopped sending their information at a pressure of eighteen times that of Earth's sea-level atmospheric pressure and at a temperature of 520° F (544° K), which were both much lower than had been expected from the radar results and the fly-by of Mariner 2. In addition, America's second Venus spacecraft, Mariner 5, flew by the planet a day after the Venera 4 landing. Its instruments made measurements that suggested that the surface temperature must be 980° F (800° K) and the surface pressure must be one hundred times that of Earth's atmosphere at sea level. While the American spacecraft did not measure these values directly, they were believed to be accurate. The spacecraft flew behind Venus as seen from Earth, and as it did so its radio signals were interrupted by the Venusian atmosphere. By studying the way in which the frequency, the strength of the received signal, and the way in which the radio waves vibrate (their phase change) varied as the radio signal had to pass deeper and deeper through the atmosphere to reach Earth, and using the composition of the atmosphere determined by the direct measurements of Venera 4, scientists could estimate what conditions must be on the surface.

A temperature of 980° F (800° K) would make life as we know it impossible on Venus.

In subsequent years, Russian spacecraft again landed on Venus, two of them obtaining pictures of a hostile, rock-strewn desert. And an American spacecraft on its way to Mercury, Mariner 10, obtained fantastically detailed ultraviolet photographs of the swirling clouds of Venus, showing that although the planet itself takes 243 days to rotate on its axis, the upper atmosphere appears to rotate in only four days. These spacecraft have painted an entirely new picture of Earth's sister world, a picture that has led to much speculation about why Venus should be so different, and if Earth will ever go the same way. More important, scientists are asking if there are things that we might do, or might already be doing, that could cause the Earth to move along the evolutionary path to an incredibly dense atmosphere of carbon dioxide and a surface of such intolerable heat. If so, we would like to know about it, to avoid such a calamity befalling the blue jewel of Earth.

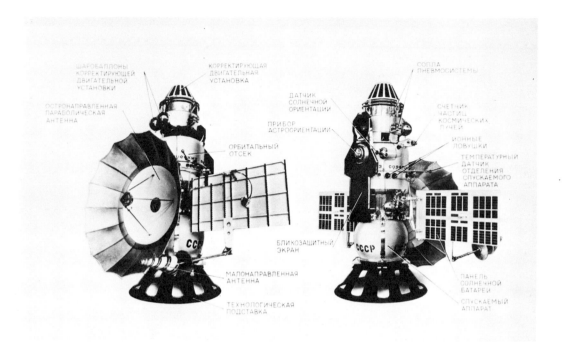

The Soviets have pursued an aggressive program to explore Venus, far surpassing the United States effort. This program has been rewarded with many space firsts and other successes, including the first landing on Venus and later the first pictures from a lander on the surface. Venera 5, typical of the Soviet spacecraft used to explore Earth's sister planet, is shown here in its flight configuration, with its solar panels extended and its directional antenna beamed toward Earth.

The lower picture shows the landing capsule that plunged through the dense Venusian atmosphere to land on the scorching surface. (USSR Academy of Sciences)

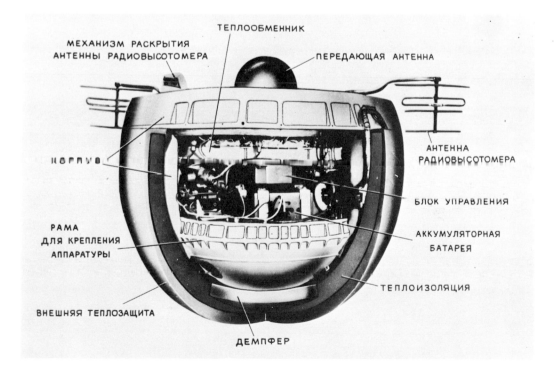

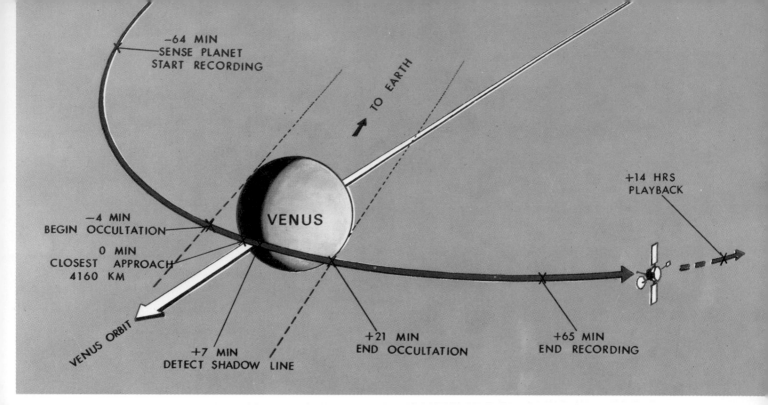

THE ROCK-STREWN SURFACE OF VENUS

Two Russian spacecraft, Venera 9 and Venera 10, have provided us with most of our information about the surface of Venus, supplemented by data from two earlier landers, Venera 7 and Venera 8.

Venera 9 was launched June 8, and Venera 10 June 10, 1975. They arrived three days apart in October of that year. Venera 9 landed on October 22 at 33° north latitude and 293° longitude in an area of Venus that appears dark on the radar maps; i.e., it has low reflectivity at radio wavelengths. The picture returned shows heaps of rocks, mostly about 12 inches (30 centimeters) or larger. These rocks are characterized by their sharp edges, which suggests that they were formed by volcanic processes associated with faulting of the crust of the planet. The picture gives the impression that the spacecraft landed on the side of a hill down which some of the rocks have been moving. Scientists suggest that they may have originated from broken lava flows.

Venera 10 landed on October 25, 1975, at 15° north latitude and 295° longitude in an area of

high reflectivity on the radar maps. As expected from the radar brightness, the surface is much smoother than the Venera 9 site. It appeared to be a plateau or a plain that is older than the Venera 9 site. On the picture can be seen rocky elevations which are covered with a dark soil that looks fine-grained. Scientists believe that this implies the rocks have been weathered, which may have been by chemical action of the atmosphere. Only gentle winds were expected on the surface, and these were indeed measured there. Such winds would not be strong enough to weather the surface rocks. There are outcrops of lighter-colored rocks projecting through the dark soil. These rocks are generally smooth, but they have some small pits in their surfaces, and depressions on them are filled with the dark soil. The edges of the rocks have been blunted and rounded, which again points to weathering.

The spacecraft measured a surface density of only between 2.7 and 2.9 grams per cubic centimeter, which is about the same as basaltic rocks here on Earth. This implies that a crust has been formed on Venus and that the planet probably has a dense core to cause the calculated over-all

This first picture of the hot surface of Venus was obtained by the Soviet spacecraft Venera 9 in 1975. The rocky surface is shown extending to a clear view of the horizon, its curvature resulting from the way in which the camera panned (scanned) around the spacecraft. The rocks appear young and seem to have been fractured in fairly recent times. (TASS)

density of 5.3 grams per cubic centimeter. This differentiation of the planet into a heavy core and a light crust is supported by spacecraft measurements of high concentrations of uranium, thorium, and potassium in the Venusian rocks, greater concentrations even than those found in terrestrial surface rocks.

To Soviet paleontologist Mikhail Marov, the rocky landscape is clear evidence that from an evolutionary standpoint Venus should be placed with the young, still active planets. Dmitryi Gregoryev, chairman of an international space mineralogy commission, commented: "It looks . . . [like] some unknown force has scattered the rocks over the planet's surface. Perhaps they fell or slipped down the surrounding rocks. It could also be that these are outbursts from giant meteorite craters."

Prior to the Venera 9 and 10 landings, astronomers often stated that only about 1 per cent of sunlight shining on Venus could penetrate through the thick cloud cover and reach the ground. The surface of Venus would be extremely gloomy, hardly the place to take photographs. For this reason, both the Venera spacecraft were equipped with powerful floodlights to aid their surface photography.

As it turned out, pictures might have been taken without them. According to optics engineer Arnold Selivanov, conditions on the ground were as bright as a cloudy June day in Moscow. Not only were the immediate surroundings of the spacecraft photographed, but also the pictures showed details of the horizon about 330 yards (300 meters) away.

These two spacecraft made excellent measurements of the surface temperature and pressure; 860° F (733° K), and ninety times Earth's sea-level pressure, respectively. The high temperatures of the rocks could lead to decomposition of many volatile-bearing minerals that are quite stable at the lower temperatures on Earth. The volatiles released would include gases that are normally trapped in the rocks of Earth.

LIKE LIVING BENEATH THE SEA

The pressure at the surface of Venus is equal to that about 33 feet (10 meters) beneath the sea. It is the weight of the dense atmosphere of carbon dioxide, extending some 130 miles (200 kilometers) to the top of the ionosphere. The Venusian atmosphere is about one hundred times as massive as that of the Earth. While Earth's atmosphere contains only 0.03 per cent carbon dioxide, that of Venus is almost all carbon dioxide.

As further information was gathered by spacecraft, the atmosphere of Venus became of increasing interest because of its great difference from that of Earth. Scientists were both mystified and intrigued by these strange characteristics of the Venusian atmosphere and how they evolved.

Venera 4 and Mariner 5 findings led some astronomers to suspect that the clouds enshrouding Venus are about 1.25 miles (2 kilometers) thick and consist predominantly of ice crystals. Others speculated that instead the clouds might be dust, possibly minute particles of quartz, and if this were so they would extend all the way to the surface as a very thick blanket. The Soviet spacecraft and the radio probing of the atmosphere by American spacecraft showed that the clouds were in layers and there was a clear space beneath the bottom layer and the planet's surface.

It became clear that the atmosphere of Venus could be divided into three distinct regions, with quite different characteristics. The relatively clear lower atmosphere, between the surface and the bottom of the cloud deck, is some 30 miles (50 kilometers) thick. The next region is that of the clouds, and these are 12 miles (20 kilometers) thick and topped by several haze layers. The third region, of the upper atmosphere, extends from the top of the cloud and haze layers out to interplanetary space.

THE LOWER ATMOSPHERE

Information about this region has been derived mainly from the Venera spacecraft. Wind velocities have been measured from the base of the clouds to the ground. At about 30 miles (50 kilometers) altitude, the winds blow at about 160 miles per hour (260 kilometers per hour). At the landing site of Venera 9, the local wind was only 1.3 miles per hour (2 kilometers per hour). Wind velocity at the site of the Venera 10 lander was only slightly higher. Thus there is essentially little wind at the surface of Venus, no dust stirred up from the ground, and a relatively clear atmosphere.

The big change in wind velocity occurs just below the clouds and probably leads to a very ragged base. There is also a layer of fine particles —aerosols or dust—between 12 and 18 miles (20 and 30 kilometers), where the wind velocity increases. Generally, however, the lower atmosphere of Venus appears to be relatively calm.

THE CLOUDS

The nature of the thick cloud layer of Venus was not understood until comparatively recently. Some astronomers had suspected a four-day rotation of the clouds as seen in ultraviolet light, meaning that winds must be blowing there at 260 miles per hour (420 kilometers per hour) relative to the surface of the planet. But photographs taken from Earth were not conclusive, because they lacked the necessary detail. While some scientists had wanted Mariner 5 to carry a camera to Venus, others had prevailed in not making that mission a photographic one. Some Russian spacecraft had attempted photography from orbit but had not been successful. So as recently as 1974 we had no idea what the clouds of Venus really looked like.

Abruptly, this ignorance was changed. Mariner 10, a NASA spacecraft equipped with twin television cameras, was to use the gravity and orbital motion of Venus to urge it closer to the Sun for a rendezvous with Mercury. This allowed its cameras an opportunity to photograph the atmosphere of the shrouded beauty. There were some problems in doing this, because Venus would be approached from its night side, so the spacecraft would have to turn its cameras and look back at the planet as it flew past. Also, and more important, the cameras were designed to be of very long focal length, to give high magnifications of the features on Mercury. While this would be fine for photographing a world that was expected to be much like the Moon, it was not suitable for photographs of vast areas of clouds on Venus.

Members of the imaging team for Mariner 10 were able, almost at the last minute, to provide an additional optical combination to a filter wheel that was part of the optical systems of the two telescopes. They added a small wide-angle lens that would permit them to take the desired pictures of Venus.

It was an exciting period in the Space Flight Operations Facility at the Jet Propulsion Laboratory when the first pictures of Venus were displayed on the TV-type viewing screens. The very first showed a narrow, pointed cusp, the northern end of the crescent. There were no details, and it was obvious that the top of the cloud deck was not broken from the towering cumulus clouds similar to those we see when flying above the cloud layers of Earth in a jet airplane. As Mariner 10 swiftly passed, only 3,600 miles (5,800 kilometers) from Venus, the pictures started to show more and more of the illuminated hemisphere, until the whole was revealed. It was a great disappointment. From the spacecraft, project scientists were able to see no more than we see from Earth. The cloud layers of Venus showed as much detail as a great, blank bank of fog. On the edge of the planet, scientists were,

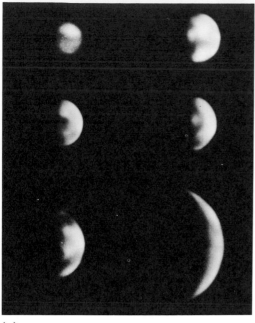

(a)

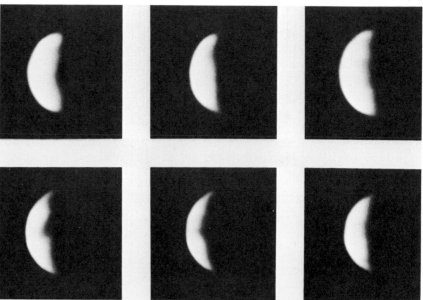

(b)

The first pictures of Venus from space showing any
details were obtained from the American spacecraft
Mariner 10. They were taken in ultraviolet light and
they confirmed the presence of marking suspected
from earlier, Earth-based photographs such as those
shown here.

(a) *Photographs of Venus taken by the 61-inch
(1.5-meter) telescope of the Catalina Observatory,
on Mount Lemon, near Tucson, Arizona, between
March and July 1967. Cloudlike markings are seen
on the images.* (Lunar and Planetary Laboratory,
University of Arizona)

(b) *The six views of Venus were obtained with
the larger telescope (100 inches, or 2.5 meters) of
the Mount Wilson Observatory, again showing some
faint markings in ultraviolet light.* (Mount Wilson
and Palomar Observatories)

(c) *A comparison of Venus photographed in red
light (bottom) and violet light (top) in April 1950.
Note the faint detail in the violet pictures and its
absence in the red pictures.* (McDonald Observatory)

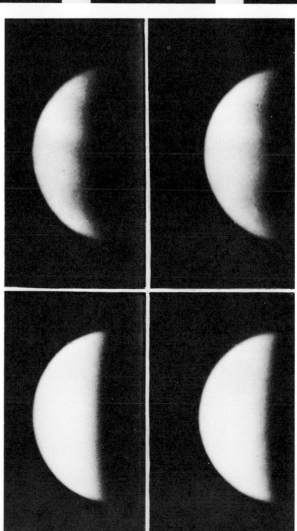

(c)

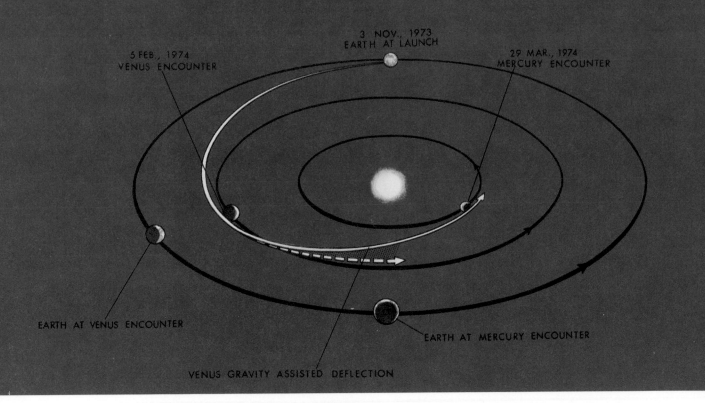

Launched on November 3, 1973, Mariner 10 first flew by Venus (on February 5, 1974), and then, using the planet's gravitational field as a trajectory deflector, continued inward toward Mercury. The historic encounter with the innermost world in the Solar System occurred on March 29, nearly two months later. (Arfor Picture Archives)

however, able to detect some haze high above the dense cloud layer.

It seemed that those in the scientific community who had opposed photo missions to Venus were justified. But a short while after the fly-by, as Mariner 10 sped away from Venus at nearly 25,000 miles (40,000 kilometers) per hour, the imaging team commanded the cameras aboard the spacecraft to start a picture sequence in ultraviolet light.

Excitement rose to fever pitch as these pictures were displayed. They showed intricate markings on the Venus clouds; swirls, streamers, polar hoods, equatorial jet streams, and rising cells. And over the course of the following days, they showed that the four-day rotation of the upper atmosphere is indeed a fact. The flow of air was found to be symmetrical between northern and southern hemispheres and to increase in velocity toward the poles, so that a spiral pattern results.

The dark markings could be related to the peculiar forms seen in the ultraviolet pictures from Earth: horizontal Y-shaped features that sometimes have a tail, Ψ-shaped features with an extension of the equatorial bar through two arms that are sometimes straight and at other times

curved, and features that look like a reversed letter C. The arms of all these features are open in the direction of the retrograde motion, which varies between 112 and 290 miles per hour (180 and 470 kilometers per hour).

The daily effects of solar heating are significant in Venus' upper atmosphere. These features are all associated with what astronomers call the subsolar point, which is the point on the surface of the planet at which the Sun appears to be shining down on it from directly overhead. But the big question is whether the apparent motions of the ultraviolet markings are real; i.e., if they are an actual movement of atmosphere around the planet or merely wave motions. The evidence favors an actual mass movement of the atmosphere, such as by winds, but there is also some evidence of wave motions that may be daily tides in the atmosphere.

Venus' snail-like rotation means that the subsolar point changes very slowly from one day to the next. In the ultraviolet markings, it appears as a very disturbed region extending some 1,200 miles (2,000 kilometers) north and south of the equator and 4,000 miles (6,500 kilometers) along the equator. This region contains cells shaped like polygons within which atmospheric gases boil to high altitudes. They are about 300 miles (500 kilometers) across and have short lives, only a few hours. These cells are surrounded by dark edges, whereas some smaller cells, of about half the size, have bright edges.

In many respects this disturbed area of the Venus clouds looks like a giant eye. From it, streaky bands of light- and dark-colored clouds reach to about 50° north and south. Beyond them is a polar ring consisting of a dark band around each pole of the planet. Scientists are still trying to unravel the circulation patterns exhibited by these ultraviolet cloud markings; their difficulty is that they possess the results of only a few days of Mariner 10 observations. In addition, scientists have not yet been able to identify the constituent of the Venusian atmosphere that causes the patterns visible in ultraviolet light. But they have a good idea of what the clouds consist of, and this was discovered before Mariner 10 flew by Venus.

By looking at the way in which the clouds reflected sunlight back to Earth, particularly in how the light waves vibrated—what scientists call the polarization—astronomers concluded in 1971 that the particles making up the clouds were little spheres, about one millionth of a meter in diameter, and they were unlikely to be grains of dust. It was also unlikely that they could be droplets of

It took a spacecraft to reveal the full details of the intricate ultraviolet clouds of Venus. In black and white, with the contrast enhanced by computer, fine structure can be seen within all the clouds, as described in the text. With these pictures scientists were able to compare the meteorology of Venus with that of Earth. (NASA)

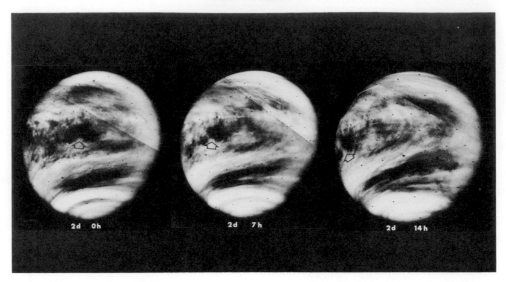

The three pictures from Mariner 10 show the rotation of the ultraviolet marking in a period of fourteen hours. The size of the feature indicated by the arrow on each picture is 620 miles (1,000 kilometers) across. (Jet Propulsion Laboratory/NASA)

water or crystals of ice; the characteristics of these just did not match observations.

A breakthrough came in 1973. NASA has several jet airplanes instrumented with telescopes to make observations as high as possible in our atmosphere. Above most of the water-vapor layers in Earth's atmosphere, astronomers are able to use the airborne instruments to obtain infrared spectra of celestial objects. In 1973, J. B. Pollack, of NASA's Ames Research Center, obtained some good infrared spectra of Venus with instruments carried in a Learjet. He compared them with the spectra of substances that might be candidates for the particles of the Venusian clouds and discovered that sulfuric-acid droplets matched most closely with actual cloud spectra. This composition had actually been suggested earlier by A. T. Young and G. T. Sill, who had pointed out that sulfuric acid is very effective in drying an atmosphere. Its presence in the clouds of Venus could account for there being so little water vapor above the clouds.

Everything seemed to be fitting together. The size of the droplets was about right, and they would remain as droplets over a much wider temperature range than water droplets. It was calculated that each droplet would consist of about 75 per cent sulfuric acid and the rest water. Thus, they are more concentrated than the acid in an automobile's battery.

There is evidence that the reactive gas fluorine is present in the Venusian atmosphere, and it, too, probably combines with water to form droplets of a very stable hydrofluoric acid. This acid can, however, combine with sulfuric acid to produce an extremely stable but very highly corrosive fluosulfonic acid. As these substances were becoming accepted as most likely constituents of the clouds of Venus, scientists began to appreciate the tremendously corrosive nature of the atmosphere, and the task of designing spacecraft to operate on Venus was seen to be not just that of overcoming high temperatures and pressures. Venus landers had to fly through a rain of corrosive acids on their way down to the surface of the planet.

While the clouds of Venus appear very opaque, they are actually quite tenuous compared with the clouds of Earth. Their opacity derives from tremendous depth. Veneras 9 and 10 determined that visibility within the clouds is between 0.6 to 1.8 miles (1 to 3 kilometers), somewhat like that in a thin haze on Earth. But they are more than twice the thickness of the cloud layers of Earth.

The process of formation of these bizarre acid clouds is quite complicated and not yet thoroughly understood. We do not know at what level the sulfuric acid is produced, though it is probably in the topmost layer or above it. In the laboratory, for example, chemists produce sulfuric acid by adding water to sulfur trioxide. Very diffuse layers of sulfuric acid in the high atmosphere of Earth are believed to form in this way. One way in which sulfur might reach the high atmosphere of Venus is by a molecule consisting of carbon, oxygen, and sulfur, formed by the action of carbon dioxide with sulfur at the high surface temperature on Venus. High in the atmosphere, this COS molecule would be broken down by solar ultraviolet into carbon monoxide and sulfur. At the same time, water and carbon dioxide would be broken down by solar ultraviolet light to provide oxygen atoms that could combine with the sulfur and with water to form the sulfuric-acid droplets.

The droplets probably fall slowly through the cloud layers, but when they reach a low altitude, close to the bottom of the clouds, their temperature has become high enough for the water to be evaporated from them and then for the sulfuric acid to break down into sulfur trioxide again. Some of this may then react with gases of the lower atmosphere to form hydrofluoric acid, which is stable enough to withstand the high temperature and to erode surface rocks. Scientists believe that the chemistry of the atmosphere of Venus is extremely complex. Above the clouds the reactions are photochemical, deriving their energy from solar radiation. Below the clouds there are thermochemical reactions deriving their energy from the hot surface and hot lower atmosphere.

ABOVE THE CLOUDS: THE UPPER ATMOSPHERE

On the pictures returned by Mariner 10 there are at least two layers of tenuous haze extending from near the equator to high latitudes. They may be associated with temperature inversions in the high atmosphere trapping layers of aerosols similar to those in the stratosphere of the Earth. At the height of the haze layers of Venus the atmospheric pressure is about one thousandth that on Earth at sea level. The temperature there is about $-103°F$ ($198°K$). The presence of such haze layers had actually been suggested in the nineteenth century by H. R. Russell, who proposed in 1899 that a haze some 0.6 miles (1 kilometer) above the main cloud deck could account for the extension of the pointed ends of the crescent of Venus.

We know most about the region above the cloud tops, for it has been examined both from Earth and from spacecraft. Above 90 miles (150 kilometers) the atmosphere of Venus is actually more rarefied than that of the Earth at the same altitude. This is because the composition of the two atmospheres is so different. But, like the atmosphere of Earth, the high atmosphere of Venus is affected by solar radiation, which knocks electrons from atoms to produce layers of electrically charged atoms and molecules referred to as an ionosphere. The ionosphere of Venus is thinner and closer to the planet than Earth's ionosphere. While the major charged particle in the atmosphere of the Earth is the ion of atomic oxygen, that in the atmosphere of Venus is the molecule of carbon dioxide which has lost one electron; i.e., it is singly charged.

Mariner 10 found that there are two clearly defined layers in the nighttime ionosphere of Venus: a main layer at 87 miles (142 kilometers) above the surface, and a lesser layer at 76 miles (124 kilometers). On the day side of the planet there was a single layer at 87 miles (142 kilometers) and several minor layers at lower and higher altitudes.

From the practical standpoint Venus has no magnetic field, so that the solar wind blows directly onto the upper atmosphere, causing a sharp boundary to the planet's ionosphere at about 250 miles (400 kilometers). But on the night side the ionosphere extends far into space. In fact, as Mariner 10 approached Venus from the dark side, it found that the planet creates a tremendous solar-wind wake, almost like the tail of a comet. This wake seemed to be even longer than that produced by the Earth.

Another important experiment performed by Mariner 10 was to measure temperature in the upper atmosphere of Venus. This measurement was made by the effects of the upper atmosphere on radio waves from the spacecraft as it passed behind the planet as seen from the Earth. The experimenters were surprised: The temperature at the topmost regions of the Venusian atmosphere was $260°F$ ($400°K$) implying that the atmosphere could not all be carbon dioxide, but must contain a much lighter gas. This gas is assumed to be helium, since it could not escape from the planet at the temperatures measured in the high atmosphere. Possibly, Venusian helium has originated from radioactive processes in the planet's crust. It is also possible that helium atoms have been captured from the solar wind. Considerable quantities of carbon monoxide were also found, suggesting that the upper atmosphere is very stable and is not mixed with the lower atmosphere as it is on Earth and Mars.

Way above the upper atmosphere is a hydrogen corona, first detected by Mariner 5 and Venera 4 and later confirmed by Mariner 10. It starts at an altitude of about 480 miles (800 kilometers). A possible origin for this hydrogen cloud around the planet is the breakdown of water vapor. Yet this should have occurred long ago. Another possible source is the solar wind. More speculative is the thought that a comet might have collided with Venus and provided the hydrogen in comparatively recent times.

WHAT DOES IT MEAN?

The greenhouse effect explains the high temperatures present on Venus, but not immediately apparent is why the Venusian atmosphere should be so different from that of the Earth. After all, the two planets are similar in size. And both were presumably condensed from materials making up the same presolar nebula.

Assuming that Earth and Venus originally possessed approximately the same amount of carbon dioxide and that this gas continued to be released into the atmospheres at approximately the same rates through volcanic and hot-spring action during the course of geologic history, why is it now tens of thousands of times more prevalent in the atmosphere of Venus than in the atmosphere of the Earth?

Taking a broad estimate, about as much carbon occurs in the carbon dioxide of Venus' atmosphere as there is carbon tied up in the crustal rocks of the Earth. Geologists believe that the carbon now in terrestrial rocks was originally present as carbon dioxide in our atmosphere. But in the presence of water this carbon dioxide was able to react with the rocks and be incorporated into carbonates. Marine animals also contributed to the extraction of carbon from the atmosphere and their dead bodies were compacted into marine deposits, which became part of sedimentary rocks.

On Venus, it is unlikely that large quantities of surface water and marine creatures ever were present. So this removal of carbon from the atmosphere of Venus could not have taken place.

But the question, then, is why the two planets of almost identical size should have followed such different evolutionary paths. Being closer to the Sun, the temperature of Venus would of course be higher than that of the Earth. The difference is, however, only a matter of about 55° F (130° C). It was enough, however, to prevent the conversion of carbon dioxide into rock carbonates and to set the planet on an accelerating course of trapping solar radiation through the greenhouse effect.

What about water? Again, both planets should have started off with roughly the same quantities. Whereas on Earth, it gradually separated from the rocks of the crust and settled into vast oceans in the basins between the continental masses, on Venus temperatures were too hot for water to accumulate as a liquid, so that it would have all been vaporized into the atmosphere. Earth holds its waters in its oceans because it is not only cooler than Venus but also has a cold trap in its atmosphere. If we trace the temperature variation with altitude, we find that at about 12 miles (20 kilometers) the Earth's atmosphere has a temperature of well below the freezing point of water. Water vapor cannot rise through this region, because it is frozen from the atmosphere. The atmosphere of Venus does not get so cold until an altitude of 50 miles (80 kilometers) is reached. Now solar ultraviolet radiation has a disastrous effect on water vapor by breaking it into its constituent elements: hydrogen and oxygen. On Earth, solar ultraviolet reacts with upper-atmospheric gases and produces an ozone layer which prevents the ultraviolet from penetrating to levels where it can break up the water-vapor molecules below Earth's cold trap. This is not so on Venus, where water can rise much higher in the atmosphere.

During the course of billions of years, the water on Venus has probably been broken down into hydrogen and oxygen and the hydrogen has leaked off the top of the Venus atmosphere into space. The present-day hydrogen corona may be evidence of continuing leakage. Meanwhile the oxygen has probably reacted chemically with the rocks of the Venusian surface. Scientists have calculated that most of the water on Venus could have been lost in this way in a time span as short as 30 million years. But there would have been about 10 per cent of the water left on the planet, which may have been trapped in the sulfuric-acid clouds.

There is, however, another problem associated with this explanation, which assumes that the oxygen reacted with the rocks and was thereby lost to the atmosphere; yet, without the presence of running water on Venus to expose fresh rocks to oxidation, it is difficult to account for the removal of so much oxygen.

OUTSTANDING QUESTIONS

Venus is still a planet of mystery, which scientists hope may be cleared up by forthcoming missions. The Soviet Union will probably continue with its exploration of the surface, perhaps ultimately using roving vehicles to gather information over wide areas. So far the United States has not developed plans for any landing on Venus comparable with the Viking Mars program. NASA's thrust has been to develop orbiters and atmospheric probes in an attempt to learn more about the atmosphere and the meteorology of Venus before working out ways to land on the surface. The big questions to be resolved include finding out how the weather systems on Venus work and precisely how the global circulation patterns seen by Mariner 10 are generated and sustained.

Also, scientists would like to be sure that the greenhouse effect is indeed the cause of the hot surface and lower atmosphere. So far, it is only a theory, not yet proved by experiment. They would like to know if the surface of Venus was ever cooler than it is now.

As for the atmosphere itself, scientists want an exact determination of its composition, particularly the relative amounts of isotopes of the various gases, because such information is vital to understanding the evolution of the planet and its atmosphere. They would also like to know the substance that is responsible for the dark markings in the clouds seen in the ultraviolet pictures.

While we have learned much about Venus from the Veneras and Mariners as well as from optical and radar observations, many unknowns remain. Some of these are certain to be cleared up by two American spacecraft launched on May 20 and August 8, 1978, from Cape Canaveral, in Florida. Despite the eleven-week spread, the two craft reached Venus only five days apart, in early December of the same year.

The first, Pioneer Venus 1, traveled along a 300-million-mile (480-million-kilometer) trajectory that first took it outside and then inside the Earth's orbit. This carefully preselected, wide-ranging path resulted in a relatively slow approach to Venus, greatly easing the entry of the spacecraft into orbit. The orbiter's mission: to take pictures in ultraviolet light of Venusian cloud cover and to make a variety of measurements of the atmosphere and planetary environment.

The companion, Pioneer Venus 2, followed a much faster trajectory to our sister inner world, permitting it to arrive a mere five days behind Pioneer Venus 1. The second craft consisted of a base platform referred to by NASA as the "bus," a large probe, and three smaller ones. All were designed to enter the atmosphere at widely scattered points to undertake measurements all the way down to the surface.

These craft may help shed light on how the bizarre conditions on our neighbor world relate to Earth, to its past and to its future. And this naturally leads us to our own Earth/Moon system, which is the next planetary object as we move outward from the Sun through the Solar System.

7
THE BLUE PLANET AND ITS MOON

A CREATION AND A CATACLYSM

The time was about 3.5 billion years ago. The world was Earth. Disintegration of heavy elements of uranium and thorium had heated its materials so that lighter elements had boiled to the surface and metals had sunk to form a heavy, nickel-iron core. Water had been released in large quantities from the heated rocks and collected on the planet's surface, forming shallow seas. Rocks eroded by the water had produced deep muds, clays, and silts. Searing radiation poured in from the Sun through an atmosphere of ammonia, methane, and carbon dioxide.

Within the metallic clays of the ocean floors strange things were happening. Shores of clays emerged from the ancient seas, but not so high that the waters did not wash over them with the low solar tides. Lightning flashed down from the turbulent clouds, and ultraviolet radiation poured down from the Sun. Ammonia, water, and methane reacted chemically under the influence of these energy surges and built bigger molecules, some of which we now call amino acids.

As the water containing the amino acids washed gently over the nickel clays, the acids were attracted by the clays and concentrated within them. The alternate wetting and drying of the clays had another strange effect. The amino acids formed long chains. Concurrently, other clays, containing zinc, worked on other molecules of the primitive oceans and became concentrated into nucleotides, which later combined into long, stringlike molecules today called deoxyribonucleic acid, or DNA. Somewhat later, these molecules developed the ability to pick up an oxygen atom to form a molecule that assembled amino acids into complex structures of proteins.

From a simple mixture of molecules formed in the ancient oceans and concentrated in the clays of wave-washed shores, a new form of matter had appeared. The long molecules were not only able to build structures around themselves but also to divide into two and build again, and again, and again. Life had struggled into existence. This event, perhaps unique in the Solar System, placed Earth apart from all the other planets, because several billion years later this life would develop into complex systems of billions of molecules with an ability to become aware of itself and the most distant stars and to question its past and its future.

Another event occurred about this same time: While the transition from chemical evolution in the ancient seas to a biological evolution that spread everywhere on the planet might be regarded as one of creation, the other event was cataclysmic. Had intelligent life been around to observe, it might have been recorded as follows.

Night and day alternated quickly on this ancient Earth because it spun on its axis in about five hours. When the night sky cleared to show the stars, five very bright objects, brighter than any other stars, could be seen wandering from night to night among the fixed stars. They were the natural moons of Earth, born soon after the

This picture of a crescent-shaped Earth and Moon—the first of its kind ever taken by a spacecraft—was recorded on September 18, 1977, by NASA's Voyager 1 when it was 7.25 million miles (11.66 million kilometers) from Earth. (NASA)

planet formed from the solar nebula. The closest orbited about 40,000 miles (60,000 kilometers) above the Earth, and the most distant was at 160,000 miles (260,000 kilometers). They were small bodies, the largest being only about 400 miles (640 kilometers), the smallest barely 50 miles (80 kilometers) in diameter. Earth seemed to be settling down after its formation. Its oceans and land masses were forming, and chemical evolution was moving on toward the emergence of living things.

From farther out in the Solar System, another planet approached. Several times it waltzed around Earth at a distance and then plunged off again around the Sun on its own orbit. But each encounter with Earth brought conditions closer to the inevitable. The other planet was finally cap-

tured; it was pulled by Earth into a highly inclined retrograde orbit. This other planet was some 2,160 miles (3,480 kilometers) in diameter, and about eighty-one times less massive than Earth. Its body was scarred with many impact craters whose pristine sharpness had not been blunted by lava flows or water erosion. Its heat engine had shut off a billion years earlier and to all intents it was dead.

But in the grip of Earth's gravity, its interior creaked and shifted; mighty tidal forces braked its spin as its own gravity raised ocean tides on Earth. The stresses and strains developed internal heat again and magma welled up to the surface, enveloping some of the ancient craters. As thousands of years passed, the alien world's orbit around Earth changed. It moved closer to the plane of Earth's equator as it approached possibly to within 24,000 miles (38,000 kilometers). And the increasing tides raised in Earth's oceans pulled energy from the planet and made it spin more slowly on its axis.

At this point the newcomer suddenly changed its path. Its orbit flipped over the poles of the Earth until the smaller world orbited in the same direction as the spin of the bigger Earth. And then it began to enlarge its orbit, gradually moving farther and farther away. Like a huge broom, it swept through the Earth's satellite system, flinging some of the smaller and larger satellites completely away from the Earth. But it did not get away scot free. It had been bombarded earlier before it reached Earth, but now it met with one of the medium-sized satellites of Earth.

From Earth the scene would have been terrifying. On the face of this intruder there erupted without warning a blinding surge of light. Around the light, the material of the Moon writhed and wrinkled until waves moved out across the solid rock like ripples around a stone thrown into a pool. A huge cloud of glowing fragments rose high into the blackness of space. Waves of debris followed the shocks across the intruder world, cutting mile-wide valleys through the wrinkled terrain. And when the debris finally settled, great masses fell back, hours later, to explode into myriad smaller craters, and the face of the intruding world had in it a vast pit, over 300 miles (500 kilometers) across, in the bottom of which a great lake of hot rock glowed redly.

A few days later some big masses of rock plunged into Earth's atmosphere to create brilliant fireballs, some of which cratered the newly formed rocks on the bigger planet.

A short while later another immense pit was opened up on the intruder world as it encountered a second of Earth's small satellites. Again Earth was bombarded.

And as the intruder moved ever more slowly away from Earth, the great pits on its surface steamed and smoked as vast lakes of lava filled them, obliterating the deep scars. Millions of years later the only evidence remaining were rings of lava fountains spouting upward around the peripheries of vast lava plains, which an emerging life on Earth would later associate with the face of a man in the Moon. For the intruder became Earth's satellite, the body we call today *the* Moon.

While highly speculative, some astronomers believe this to be a reasonable explanation of why, among all the planets, our Earth has a satellite so disproportionately large compared with the size of its primary. All other planets accompanied by moons, except Neptune, are much more massive than the combined masses of all their satellites. Neptune does possess a large satellite, but it moves in a type of orbit that suggests, it too, may have been captured.

The two most distinguishing features of the blue planet and its satellite are that Earth has evolved life and may be the only planet in the Solar System to have done so, and that the Moon is the largest satellite in the Solar System when judged in relation to its primary.

THE DOUBLE-PLANET SYSTEM

Earth and Moon are often referred to as a double-planet system. In their orbit around the Sun the Moon is at all times more strongly under the influence of the Sun than of Earth. The Moon actually weaves in and out of Earth's orbit, but the lunar orbit is at no place convex, i.e., turning away from the Sun, rather it is always concave, i.e., turning toward the Sun.

The Earth/Moon system orbits the Sun in a period of 365.25 days, at an average distance of 92.96 million miles (149.57 million kilometers). The orbit of Earth is a little more elliptical than that of Venus, so that at perihelion Earth is 91.34 million miles (146.97 million kilometers) from the Sun. This occurs during summer in the northern hemisphere. At aphelion Earth is 94.45 million miles (151.97 million kilometers) distant.

The Earth spins on an axis inclined at 23.4° and this inclination gives rise to seasons. The rotation period of the Earth relative to the stars is the sidereal day, of 23 hours 55 minutes, 41 seconds. The solar day is, of course, 24 hours. Our planet is a slightly flattened sphere whose polar diameter is 26 miles (42 kilometers) less than its equatorial diameter, of 7,927 miles (12,755 kilometers). The mass of the Earth is three millionths that of the Sun; it is 6.6×10^{22} tons (5.976×10^{24} kilograms), or 81.3 times the mass of the Moon.

The Moon revolves around the Earth at a mean distance of 239,240 miles (384,940 kilometers), in a fairly eccentric orbit that brings it at apogee (greatest distance from Earth) to

252,000 miles (405,720 kilometers) and at perigee (the closest to Earth) to 226,000 miles (363,630 kilometers). The Moon completes an orbit relative to the stars in 27 days, 7 hours, 43 minutes, and 11.5 seconds. But since the Moon is moving with the Earth around the Sun, the period from one new moon to the next (i.e., the phase when the Moon is between Sun and Earth) is longer than the sidereal period; it is 29 days, 12 hours, 44 minutes, and 3 seconds. This is the synodic period, or the lunar month. The phases of the Moon, dependent upon the relative positions of Earth, Moon, and Sun, repeat each lunar month. The Moon spins on its axis in a period equal to its orbital period, so it continuously presents one hemisphere to the Earth . . . almost.

Actually, we can see about four sevenths of the lunar surface from the Earth because of apparent rocking movements, which astronomers call librations. Because the Moon moves along an elliptical orbit, the rotation on its axis is not always in step with its revolution around the Earth, so we can see around each edge, to east and west in turn. The Moon's axis of rotation is tilted around 17°, so the poles of the Moon tilt alternately toward Earth and we see a little farther north and south. There are also some smaller and less important librations.

The Moon's orbit is inclined to the orbit of the Earth around the Sun. But if Earth, Moon, and Sun are lined up where the orbits cross, eclipses are produced. When the Moon is between the Earth and the Sun it blocks the Sun and sweeps a narrow band of darkness across the Earth. An observer who happens to be within that narrow band sees an eclipse of the Sun. If the Moon is near its apogee it does not, however, appear large enough to cover the disc of the Sun and the eclipse is not total but annular. When the lineup occurs with the Moon on the far side of Earth from the Sun, the Earth's shadow blankets the Moon and observers all over the hemisphere facing the Moon at that time witness a total eclipse of the Moon.

Alongside the gas giants—Jupiter, Saturn, and Neptune—our planet Earth is not a very impressive body. Measured against its neighboring inner worlds, however, it is quite respectable-sized. It is,

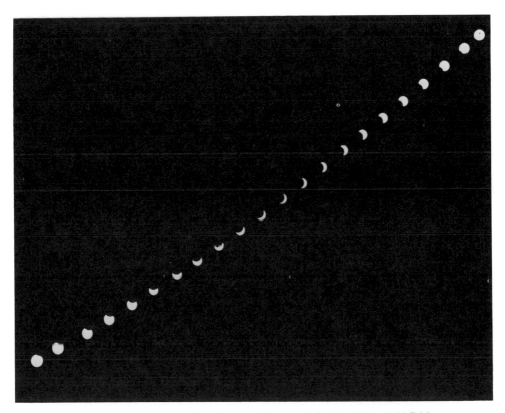

Portrait of a partial solar eclipse, taken on July 10, 1972. (NASA)

in fact, the largest planet in the inner Solar System. Since it is the home of mankind we have understandably accumulated much information about the Earth. We have poked into its land surface, exploring its inhospitable mountain peaks, polar wastes, dense tropical rain forests, and parched deserts. We have drilled deep into its crust and looked into the cold dark depths of its oceans. We have pushed into the upper atmosphere beyond the regions where we can breathe, and we have tested, probed, and monitored our planet with every imaginable device and detector. But as we explore the other planets of our Solar System, we appreciate how much we have to learn about our own.

From the epoch-making day in October 1957 when Sputnik I blazed a trail through the belt of Orion and beeped its radio message around the world we have used artificial satellites to study the Earth from space. And we have done so with brilliant success. New light has been shed on the nature and extent of the outer atmosphere, the shape and mass distribution of the materials making up the surface, the character of the gravitational field, and the planet-wide distribution of geographical, geological, agricultural, hydrological, oceanic, and cultural resources of our world.

As a result of centuries of ground-based observations and a couple of decades of space exploration, we have also found out much about the Moon. Dozens of unmanned spacecraft have gathered data about our neighboring world, some in orbit and others on landing. Between 1969 and 1972, twelve Americans visited the Moon and walked on its surface. Some of them rode a vehicle across its alien landscape. They made geological and other scientific investigations and returned with a wealth of pictures. They brought back many samples of rocks and soils from various parts of the Moon, and they left behind them automatic scientific instruments that continued to operate on the Moon until there was no more money available in NASA's budget to keep them going and to process their data. We thus gained more information about the Moon than we had about any other world except the Earth.

The study of our double-planet system permits us to compare Earth and its Moon with the other worlds of the Solar System. At the same time, we can use what we find out about these worlds to better understand the evolution, development, and likely future of our own.

APPRECIATING A PLANET ANEW

"Once a photograph is obtained of the Earth from outside . . . a new idea as powerful as any in history will be let loose," said astronomer Sir Fred Hoyle at the first lunar science conference, in Houston, in 1970. He was reminding his audience that he had made that prediction in 1948. And he asserted that his prediction had come true. "Quite suddenly everyone world wide has become seriously interested in protecting the environment. Something new has happened to create this feeling of awareness about our planet." The something new was a new perspective on our birthplace; from space a blue-green jewel floating in the dark void. Spaceship Earth!

A WATER-RICH PLANET

The Earth's solid outer crust is 71 per cent covered by oceans and seas, the hydrosphere. If it were somehow possible to smooth out the surface of the Earth, shaving off the continents to fill in the ocean depths, our world would be completely submerged beneath a 1.5-mile (2.5-kilometer) -deep planet-wide ocean. A visitor from space would see a featureless sphere of waves with frozen caps at both poles.

Fortunately for life, our world is still in a state of turmoil, its internal heat engine continuously driving great plates of crustal rocks against each other and crumpling billions of tons of rocks into great mountain chains. So the ocean is today broken by land masses of islands and continents rising above the waves.

The hydrosphere itself is characterized by motion: majestic ocean swells, surf pounding against seashores and eroding rocky cliffs, leisurely flowing currents, and deep-water circulation. All bear testimony to the dynamic nature of the liquid surface of our planet. This dynamism comes from the interaction of the ocean surface with the atmosphere, which is set in motion by uneven solar heating. There is a constant exchange of energy between the atmosphere and the ocean, energy that has its origin in the nuclear fires of the Sun.

Stand on the bridge of a great ocean liner as it plows through mountainous seas. It is hard to realize that all this energy, the heaving and swelling of the thousands of tons of steel, is derived from

When humanity first saw the Earth from space, a new awareness developed of Spaceship Earth on which no national boundaries could be seen. This view, from Apollo 17, extends from the Mediterranean Sea to the Antarctic. The cloud-free Sahara and the Arabian Peninsula are both clearly visible, as is the outline of much of central and southern Africa. Madagascar stands out at the western edge of the Indian Ocean and cloud cover builds up toward the South Pole. (NASA)

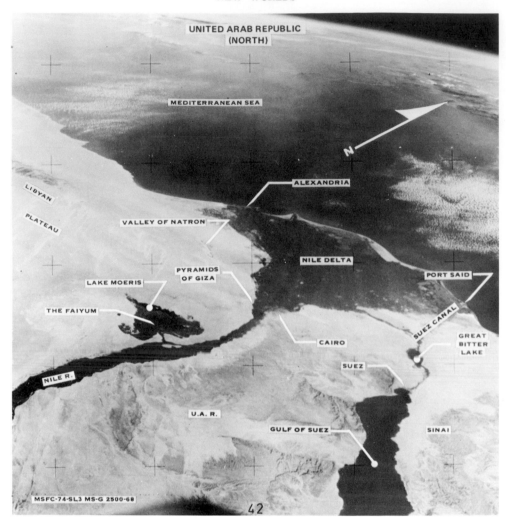

The Earth is an active planet and land masses are raised above the oceans' surface. Or-biting astronauts have repeatedly photographed the land and water surfaces of the Earth from viewpoints in space. Such pictures allow us to assess agricultural, ocean, forest, mineral, environmental, and other resources as well as the man-made geogra-phy. This view from Skylab shows the Nile Delta, the River Nile, the Suez Canal, and other well-known areas of one of the cradles of civilization. (NASA)

photons that dashed from the solar photosphere in a race to infinity and were trapped by the hydrosphere of our planet.

The vast ocean world is made up of complex solutions of salts whose composition varies with geography and season as well as with depth below the surface. The bottoms of the oceans are, for the most part, covered by sediments washed from the land surfaces and mixed with the dead bodies of sea creatures. Some regions are, how-ever, relatively free from these sediments, partic-ularly undersea mountains whose upper slopes and peaks are exposed to strong currents. Like

the land masses, the floors of the oceans have ridges, valleys, mountains, and other topographi-cal features of various elevations.

Salt water contains sixty of the ninety-two natu-rally occurring chemical elements. Every bucket of sea water consists of 96.5 per cent pure water (which is 10.8 per cent hydrogen and 85.7 per cent oxygen by weight) and 3.5 per cent dissolved salts. The salts are present as positive ions of sodium, potassium, magnesium, calcium, and strontium, and negative ions of sulfate, chloride, bromide, fluoride, and carbonic and boric acids. In addition there are traces of iron and copper and other metals, and some dissolved gases, the most common of which is nitrogen. Life can flourish within the oceans because there is some oxygen dissolved in the water too.

Water resources of our planet are concentrated in the oceans. They contain approximately 97.2 per cent of the water that has been released from the rocky crust of our planet. There is about 2.15 per cent locked up as ice in the polar regions and as glaciers. There is a mere 0.63 per cent fresh water on the land masses, and there is the very small quantity, 0.001 per cent, as vapor or cloud droplets in the atmosphere. Much of the fresh water is not readily available, for it is buried deep within the ground. About 50 per cent of all the fresh water is, indeed, below a depth of 0.6 mile (1 kilometer). Of the other half only a very small fraction is on the surface—in streams, rivers, and lakes; most is in underground reservoirs and rivers. The fresh water is derived from rainfall and melting snows. But of this, about five sevenths of the rainfall evaporates back into the atmosphere before it sinks underground or flows into the oceans.

The effect of the oceans upon the climate of the Earth is significant. Great ocean currents can carry warm and cold water and change the climate of vast regions of the land masses. Weather patterns spawned in the northern Pacific Ocean above Hawaii have been known to cause long-

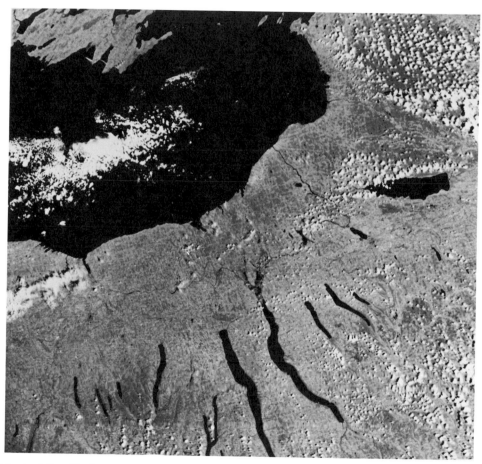

Most of Earth's free water is in its oceans. Freshwater lakes and rivers represent only a very small fraction of the available water on the planet. During periods of drought, surveys from space are invaluable in assessing water resources in a minute fraction of the time required to do so on the ground. This land and inland-water photograph (taken by Landsat 1 above the Finger Lakes region of New York on August 19, 1972) shows Lake Ontario at upper left, the city of Rochester at far left center, Syracuse at far right center, and Finger Lakes at the bottom. (NASA)

lasting cold spells in the United States and Western Europe. For example, during the winter of 1963 the North Pacific was abnormally warm. The result was a very cold winter in the East of the United States and in Western Europe, while the western United States had a relatively warm winter. The influence of the Gulf Stream in providing Europe with a relatively mild climate despite its high northern latitude is well known, as is the effect of the Labrador Current making the eastern United States much colder than European countries at the same latitude.

Abnormal warming of vast areas of ocean occurs when clouds clear and let the Sun heat the ocean for an unusual length of time. And once the area has heated, it generates storms which do not dissipate the heat of the ocean but reinforce it, so that the warm area may persist for several years and influence world climate for the same period.

A WORLD IN UPHEAVAL

Everyone is familiar with the variations in the crust of the Earth: mountains, valleys, plateaux, plains, lowlands, and so on. The highest elevations of mountain peaks reach about 29,000 feet (9,000 meters) above sea level; the deepest ocean trenches have their floors about 32,800

Satellites have provided great aid to meteorologists in assessing weather systems and their development. A great area of the Pacific north of Hawaii may be responsible for consistent weather patterns across the United States and Europe. This picture, from an Apollo spacecraft, was taken in March 1969. It illustrates how the orbital overview provides a splendid opportunity to observe major weather systems in their entirety. Here a cyclonic storm system, situated some 1,250 miles (2,000 kilometers) due north of Hawaii, is shown in the process of development. (NASA)

Unmanned spacecraft also monitor Earth's weather. In fact, since such spacecraft were placed in orbit, no one in the United States has ever been surprised by a hurricane; warnings have always been given well in advance of hurricanes reaching land. Here is a picture of Hurricane Ginger on September 13, 1971. The satellite Nimbus 4 took the picture from an altitude of about 680 miles (1,100 kilometers) when over the Atlantic Ocean east of Bermuda. At that time the center of the hurricane was about 600 miles (1,000 kilometers) east of Bermuda. (NASA)

feet (10,000 meters) below the surface of the oceans. The land masses are continuously eroded by water, wind, chemical action, and plant life. And in recent years the machines of men have added a new erosional process as civilization changes vast areas of the Earth's surface. The mechanical breakup of rocks and the transport of the resulting debris by rivers into the oceans would inevitably level the land surfaces were it not for a compensatory process of rebuilding. This results from volcanoes and from the folding into mountains of parts of the crust as great plates of rocks move relative to one another on the top of underlying hot magma.

The Earth's crust consists of rocks called silicates. Such rock includes mainly 47 per cent by weight of oxygen, 28 per cent silicon, and 8 per cent aluminum. The crust varies in thickness from about 24 miles (38 kilometers) beneath the continents to about 6 miles (10 kilometers) beneath the ocean floors. Below the crust is a region called the mantle, which is about 1,800 miles (2,900 kilometers) thick and consists of heavier silicates that contain iron and sulfur. Below the mantle is a liquid shell, part of the core of the Earth, which is estimated at 760 miles (1,220 kilometers) thick; it contains heavier elements which sank toward the center of the Earth as lighter elements floated toward its surface. An inner core, some 2,000 miles (3,400 kilometers) in diameter, is believed to be partly molten and partly solid and composed of iron-nickel alloy with perhaps some carbon, silicon, and cobalt. The high density and possibly solid state of the innermost parts of this core are responsible for nearly one third of Earth's entire mass and for making its over-all density much higher than that of the crustal rocks.

Although some 4.6 billion years have passed since the creation of the Solar System, the Earth is still thermally active. As its great internal heat engine pumps energy toward the crust, mountain-building (orogenic) forces operate to counter the mountain-depleting forces of erosion and weathering. The topmost layers, consisting of the crust and part of the mantle, are solid rocks. They are referred to as the lithosphere. This region is about 44–93 miles (70–150 kilometers) thick. The lithosphere floats upon a hot plastic asthenosphere but not as one complete shell. Rather, it consists of rafts, like ice floes grinding against each other. These rafts are called plates. The North American continent is on one of these plates, except for a part of California, which is on another. The plates are moved by upwelling of magma on the ocean floors in some areas of the world. And as the plates are forced into each other the stresses and strains are expressed in earthquakes, volcanic eruptions, faulting, and mountain building through the crumbling of the layers of rock.

While earthquakes and volcanic disasters take enormous tolls of human lives, without them life as we know it on Earth would be impossible. Earth would have become eroded to a uniform smoothness over which great ocean waves would sweep each day in lunar and solar tides.

BETWEEN EARTH AND SPACE

There is another environmental shell to our Earth which is just as important to our being here

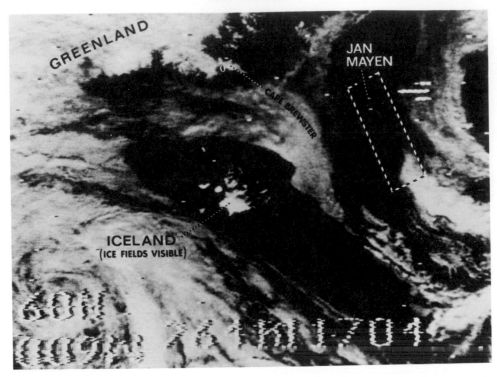

Satellites are also splendid monitors of events, natural and man-made, that occur in remote places. Here, an ash plume from Beerenberg Volcano on Jan Mayen Island was photographed by the Nimbus 4 satellite at an altitude of 680 miles (1,100 kilometers) in 1970. (NASA)

as the internal heat, the oceans, and the land. It is the gaseous envelope that extends from the surface of Earth to interplanetary space: the atmosphere. This gaseous envelope, like the oceans and the continents, is also in constant motion, sometimes pleasantly with cooling breezes on hot humid days, at other times disastrously with hurricane winds. Its restless motion is also evidenced in rising and falling masses of moist air, giving rise to an astonishing array of cloud forms. It provides air for us to breathe and moderates the temperature at the Earth's surface. And invisibly, far above our heads, it protects us from harmful radiations from the Sun.

The atmosphere today is made up of approximately 78 per cent nitrogen (an inert gas) and 21 per cent oxygen (an active gas that supports combustion). Additionally, there are small amounts of carbon dioxide, argon, neon, krypton, xenon, hydrogen, methane, and nitrous oxide. The atmosphere has varying amounts of water vapor and ozone, and it has suspended particles of dust and man-generated pollutants. Water vapor plays an important part in moving heat among regions of the atmosphere and generating weather systems. It can amount to as much as 2 per cent of the atmosphere locally but never over all. Ozone is produced mainly by incoming solar ultraviolet radiation and forms a protective layer at high altitude, where it has a concentration of about one to three parts per million.

Above the atmospheric regions is an electrified shell known as the ionosphere, where atoms and molecules are stripped of electrons by solar radiation. And far above that are the regions of plasma within the Earth's magnetosphere described earlier. These regions far above Earth's highest mountains were only explored during modern times, first with high-altitude balloons and rockets and later with satellites. One particularly fascinating type of satellite carried a propulsion system so it could dip into the atmosphere and escape again before drag could cause it to plunge to destruction. Using three of these so-called Atmosphere Explorer satellites, scientists were able to gather valuable information about a region of the high atmosphere called the thermosphere, where incoming solar radiation is changed

Satellite pictures are invaluable for studying changes to the Earth that cannot economically or physically be monitored from or at the surface. In the middle bottom of this picture taken from space is McMurdo Sound, which empties into the Ross Sea. The bottom left is Victoria Sound, and at the bottom right is the Ross Ice Shelf. This region of Earth's south polar continent is difficult to monitor from the surface because of a limited number of meteorological stations and its remoteness and harsh climate. In the picture the black areas are open water, ice-free. Land-fast ice and some ice floes are evident along the coast of Victoria Land. On the middle right of the picture the loose ice-floe pattern becomes compacted in a two-day period. Later, an irregular ice boundary can be seen in the center of the Ross Sea. (NOAA).

into heat. They discovered that this is a very complex region in which many chemical reactions take place and from which energy is transferred both upward and downward to other regions of the atmosphere.

LIFE CHANGED THE EARTH

Simple, one-celled organisms, typified by algae and bacteria, have been around on Earth for at least three quarters of the planet's 4.6 billion years. The oldest fossil algal remains appear to be 3.4 billion years old. But there may be still older life forms waiting for our discovery. The oldest rocks geologists have been able to identify on our Earth form part of a large mass in western Greenland. It dates from 3.7 to 3.8 billion years ago. Since there is evidence that the planets of the inner Solar System were heavily bombarded so as to churn their surfaces completely some 4 billion years ago, the Greenland rocks may be close to the oldest rocks still existing on Earth.

At the time when life started on Earth conditions were much different from today. The atmosphere probably consisted of ammonia, methane, water vapor, and carbon dioxide, with little if any free oxygen. This mixture, oddly enough, seems to have been essential for life to begin. Because there was no free oxygen in the atmosphere, the incoming solar ultraviolet could not be absorbed in producing an ozone layer from oxygen as it does today. It penetrated to the surface of the primitive oceans and there helped form complex molecules that became the building blocks for life as we know it.

The first, extremely positive life forms built their bodily structures from organic compounds of the oceans as long ago as 3.4 billion years. but, as the compounds became depleted, the life forms developed a way to make their own organic compounds from carbon dioxide and water by using the energy of sunlight. They evolved into what are called photosynthetic organisms, synthesizing essential materials by using light. Strangely, though life arose rapidly when the Earth had cooled off sufficiently to support it, for some 3 billion years it evolved at an extremely slow pace. Then, within a period of a mere 10 million years, most of the principal phyla of shelled invertebrates appeared. Paleontologists often refer to this as the Cambrian explosion, which took place approximately 600 million years ago. Within less than 400 million years, however, half the marine invertebrate families were extinct. And this extinction process for the most part occurred within only a few million years.

Be this as it may, the process of photosynthesis practiced by the new terrestrial life forms had far-reaching consequences. One of the products of photosynthesis is oxygen, which was of no use to the organisms and was hence discarded into the atmosphere. These early life forms thus polluted Earth's early atmosphere with the, to them, poisonous gas oxygen. Indeed oxygen probably killed off much other early life. But, later, the presence of oxygen permitted a new kind of life to develop, one that *used* oxygen is an energy-releasing process to power its biological machine. And as the amount of oxygen in the atmosphere increased over several billion years, enough of it accumulated to trigger a burst of biological evolution that led to extremely complex multicellular organisms. Some 4–5 million years ago, one of these complex creatures (that reproduced sexually to provide a greater probability of deriving offspring with better chances of survival in a changing world) produced the first of that Earth life form that we call man. Generation by generation, an awareness grew of the world that he was gradually mastering and of the bright orb that often lit the night skies above his head. But tens of thousands of generations were to elapse before humans would comprehend what the Moon is and why it behaves as it does. Against the time scale

of the universe, measured in billions of years, it was only a few seconds of the cosmic clock before this new life form on Earth had not only comprehended the nature of the Moon but had devised machines to fly to and walk upon its age-old surface.

A GREAT LEAP FOR SOME, A FALTERING STEP FOR OTHERS

While the first step of a man on the soil of the Moon was regarded as a great leap for mankind, comparable to that other great leap when early life forms emerged onto the land from the oceans, many people with less vision of our future faltered as life on Earth emerged from its womb. Some even tried to get back in the womb —intellectually if not physically. They tried to slam shut the lid of the treasure chest of a new dimension of the secrets of nature. But just as throwing Columbus into jail could not affect the ultimate development of the Americas, the post-

The Apollo program allowed men to explore parts of the Moon on the same level of detail as earlier explorers expanded human awareness of our own planet. (NASA)

Our astronauts rode over the lunar surface in a wheeled vehicle, which enabled them to explore far beyond the base of the lunar module. (NASA)

On the virtually ageless lunar surface there are now left the marks of man, which will remain there for many millions of years, irrespective of what happens here on the Earth. (NASA)

Apollo de-emphasis on space as a rightful domain to explore will not ultimately prevent the expansion of the human species into the new dimension of life beyond the limitations of planet Earth.

And the landings on the Moon were in this respect the beginning of an entirely new age, the age of interplanetary man. Again quoting Sir Fred Hoyle, long after the problems of today are forgotten, our great-grandchildren will remember that men landed on the Moon in the 1960s.

ASTARTE, QUEEN OF THE NIGHT

Early man worshiped the Moon as a goddess. It was Isis to the ancient Egyptians, Astarte to the Phoenicians, and Selene to the Greeks. But modern man sees the Moon as a steppingstone into space. Compared with the other worlds of the Solar System, the Moon is right next door to Earth, so close that it takes only 1.28 seconds for light to span the distance between the two worlds. The Moon's separation from us is only ten times the distance around the equator of the Earth. This is approximately only 1 per cent of the distance to Mars and Venus when these planets come closest to the Earth.

A quarter the diameter of our planet, the Moon has a mass of about 73 million, million, million tons, or 1.23 per cent that of Earth. The surface area of the Moon is less than one tenth that of Earth, but it is all land, because there is no water there. The density of the Moon is much less than that of Earth: 3.34 compared with 5.5 grams per cubic centimeter. This low density suggests that our satellite is poorly endowed with metals.

Over the billions of years that the Moon has

Long after the problems of today are forgotten, our descendants will remember that men walked on the Moon in the 1960s. Astronaut Scott stands on the slope of Hadley Delta, close to the awesome Lunar Apennine range of mountains. He is some 10.5 miles (17.5 kilometers) from the base of the mountains, whose rounded tops are seen in the background. A four-wheeled vehicle on which several astronauts roved about on the lunar surface can be seen to Scott's left. Apollo 15, the fourth mission from Earth to land men on the Moon successfully, was launched on July 26, 1971. The landing was almost four days later, on July 30, and the return to Earth by splashdown in the Pacific Ocean took place on August 7. (NASA)

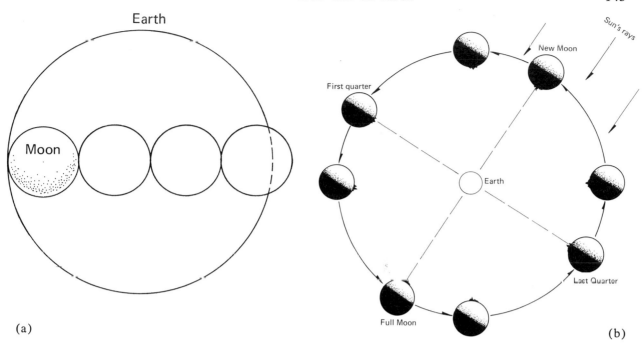

(a) *Relative sizes of Earth and Moon. The Moon is the largest satellite relative to its primary of all the satellites of the Solar System.*

(b) *Lunar day and night cycles as viewed from the Earth are seen as phases: new moon, first quarter, full moon, and last quarter. The Sun's rays come from the upper right.*

orbited Earth, the two worlds have exerted tidal forces on each other. Most people are aware that the Moon raises tides in the terrestrial oceans but not that the Moon also causes tides in the Earth's atmosphere and the crust of the Earth. Earth also has pulled the Moon out of shape. The far greater force of terrestrial gravitation has gradually slowed down the rate of spin of the Moon on its axis, until now the Moon always turns the same face toward the Earth, as already mentioned. The Moon has also been distorted in shape so that it is stretched out toward and away from the Earth.

The most conspicuous feature of the Moon visible to the naked eye is a pattern of dark areas. At one time astronomers thought that these were seas, so they gave them the Latin name *maria,* meaning seas. They are really vast lava plains filling ancient basins that were probably caused by the impact of large bodies on the Moon. They vary in diameter from about 185 miles (300 kilometers) to more than 680 miles (1,100 kilometers) and account for well over one third of the face of the Moon seen from Earth. When spacecraft orbited the Moon, scientists were surprised

to find that although there were big basins on the far side they had not been filled with dark lava like those on the Earth-facing side.

Associated with the great maria are smaller dark features, indentations along their boundaries that also received names associated with oceans and seas; early astronomers called them lakes (lacus) and marshes (paludes), and there is even one ocean (Oceanus Procellarum). These are all great lava plains.

There are countless craters on the Moon, ranging in size from the great basins of 185 miles (300 kilometers) diameter to pits in rocks that can be seen only with a microscope. These craters show that the Moon has been bombarded by bodies widely ranging in size. Most of the bombardment took place billions of years ago, but there has been a continuing bombardment over the whole of lunar history, which continues even today though probably at a much reduced rate. Young craters are characterized by radiating systems of bright rays spread over the surrounding surface. One of the spectacular examples is Copernicus, a crater about 56 miles (90 kilometers) in diameter. This magnificent young crater forms

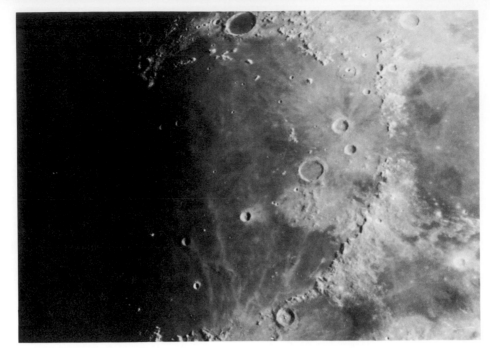

Before the space age our knowledge of the Moon depended upon our observations first with unaided eyes, then with telescopes, spectroscopes and cameras, and radio waves. This view of Mare Imbrium, a great impact basin on the Moon now filled with many sheets of lava flows, shows the crater Plato to the north, Aristillus, Autolycus, and Archimedes near the eastern edge of the basin, Timocharis lower and toward the center, and Eratosthenes at the bottom. The photograph is oriented north at top, the opposite to usual orientations in older astronomical textbooks, which showed lunar and planetary pictures upside down, as seen normally in a telescope. (Pic du Midi Observatory)

Tycho is a large, young crater in the Moon's southern hemisphere. These three photographs were taken by Lunar Orbiter 5 from an altitude of 135 miles (220 kilometers) looking almost straight down into the crater. With higher magnifications increasing detail is revealed. The third picture shows an area on the floor of the crater that is roughly 7 × 8 miles (11×13 kilometers). Fractures, flow marks in lava sheets, and protruding domelike hills with exposed rock layers are visible. (NASA)

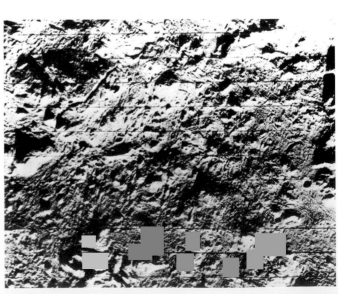

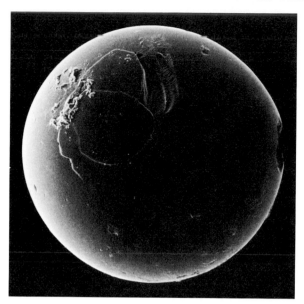

The surface of the Moon is cratered down to microscopic size. This electron micrograph shows a tiny crater in a glass bead found in a lunar sample. (NASA)

the tip of the nose of what can be imagined as the face of the man in the Moon as seen from Earth On the near side of the Moon there are more than three hundred thousand craters exceeding 0.6 mile (1 kilometer) in diameter.

A telescopic view of the Moon from Earth is dominated by the craters. And where, close to the boundary between night and day on the Moon (called the terminator) the sunlight casts long shadows, the ramparts of these craters look extremely rugged and impressive. Large craters may have walls that are 2.5 miles (4 kilometers) high. But you could stand in the center of such a crater and not be able to see the wall of mountains surrounding it. They would be beyond the horizon. And they are not as rugged as they look. The continuing rain of meteoroids has eroded the lunar surface, rounding off the rugged peaks and making the lunar mountains more like gently sloping hills, wide at their bases compared with their heights.

Other landforms seen on the Moon are ridges (strips of uplifted lava from the floors of the maria), clefts (long, groove-like depressions), rills (narrow crevices), and fields of rocks. The fields of rocks were not seen until spacecraft visited the Moon.

Lunar Orbiter also obtained oblique views of many well-known features such as the crater Copernicus, shown here. The view is looking due north from the southern rim of the 56-mile (90-kilometer) -wide crater. The mountains rising from the flat floor of this great circular depression form central peaks that reach 1,000 feet (300 meters) elevation. On the horizon, beyond the crater, a distance of about 150 miles (240 kilometers), is a 3,300-foot (1,000-meter) -high mountain named Gay-Lussac Promontory. It is part of the Carpathian Mountains, which form a section of the ring of mountains surrounding the Mare Imbrium. (NASA)

The lonely figure of an Earthman against his electric car seems insignificant against the backdrop of lunar mountains extending to a distant horizon that we have not yet been able to explore. (NASA)

AGES OF THE LUNAR FEATURES

Even before spacecraft exploration began, scientists had been able to put together a story of the Moon's history. It had been mapped assiduously since 1647, showing greater and greater detail as telescopes and methods of observation improved. Geologists studied these maps and deduced the sequence of formation of the various geological features. They divided the lunar history into several distinct phases: a pre-Imbrium period, before a giant asteroid-sized body (possibly a satellite of Earth) collided with the Moon to produce the great Mare Imbrium; the Imbrium period, when this type of basin was gouged out of the Moon and then gradually filled with lava; an Eratosthenian period, when fresh-looking craters without ray systems were formed; and a Copernican period, when the ray craters were formed.

The Copernican period extends through today.

While geologists could assign relative ages to the various features on the Moon, they had no way of finding absolute ages without obtaining samples of lunar rocks and soils. The exploration of the Moon thus aimed at four major investigations: the geochemistry of the Moon as a complete world; the composition of the soil of the Moon, the layer of rubble and crushed rocks that scientists call the lunar regolith; the composition of the lava that flowed over the mare basins; and the composition of the rocks on the highland areas projecting above the lava plains. Additionally the explorers wanted to be able to probe beneath the surface of the Moon to learn how much heat, if any, was flowing toward the surface from the interior, and to try to detect moonquakes from several stations and thereby ascertain how the material of the Moon is layered and if there is a core.

In an attempt to relate the history of the Moon to that of Earth, scientists also wanted to know if there is or ever has been water in the lunar rocks and whether there are building blocks of life such as the amino acids that have been found in some meteorites.

THE NEW MOON OF APOLLO

Both the Soviet Union and the United States followed an aggressive program of lunar exploration starting almost at the beginning of the space age, with the U. S. Air Force launching several unsuccessful lunar probes in 1958 and the suc-

Before our spacecraft went, first unmanned and then manned, to the Moon considerable effort had been spent over several centuries in mapping the Moon from Earth. This is part of one of these maps. It shows the southwestern quadrant of the Moon, in which there is the large impact basin called Mare Nectaris. (U. S. Army Corps of Engineers)

cessful orbiting of the Moon and photography of its far side by the Russian Lunik III in 1959. The U.S. program continued with major successes of Ranger, Surveyor, and Orbiter and culminated in the Apollo manned landings. The Soviet program progressed with unmanned vehicles and culminated in an automated vehicle roving over the lunar surface and an unmanned sample return.

A painstaking study of rock samples from the Moon urged some scientists to reject the capture theory for the origin of the Moon in favor of a theory that the Earth and Moon were formed from the same primordial nebula at about the same distance from the Sun. They based their conclusion on the fact that the rocks of the Moon had identical proportions of isotopes of oxygen found in the crust of the Earth. By contrast, meteorites, which astronomers believe formed elsewhere in the Solar System, have different amounts of the various isotopes.

If this theory holds, we will have to discard our scenario of the Moon's capture, a process that is dynamically very complex and relies upon many assumptions. Instead the Moon may have formed from the falling together (accretion) of mineral grains that were captured into orbit around the primordial and still forming Earth. These grains, it is reasoned, were sufficiently slowed by collisions with each other and with the remaining gases of the nebula for some to fall onto Earth and build up its crust. Others remained in orbit and added to the bulk of the Moon. In this way the material was distributed between Earth and Moon to account for the same proportions of oxygen isotopes.

The oldest rocks retrieved by the Apollo explorers are between 3.9 and 4 billion years old, a time that may have been a major turning point in lunar history, the tail end of the massive infall of bodies that built up the bulk of the Moon. These rocks are what geologists call anorthosites, granular igneous rocks that are rich in calcium and aluminum silicates. This type of rock is formed when a hot molten magma cools and as the temperature falls various materials crystallize from the molten rock in sequence. Thus such rocks may have been the first to form as the outer molten layers of the Moon cooled and solidified.

Two other important rock types were found on the Moon. One is a dark, dense, igneous rock called basalt. The other is a granular rock of lighter minerals which lunar geologists have called KREEP norite. A norite is a rock formed at high temperature. KREEP stands for the chemical symbol of the element potassium, the initial letters of rare earth elements, and the chemical symbol of phosphorous, which are present in the rock in high proportions.

KREEP norites and basalts were produced from their parent rocks, believed to be the anorthosites, by partial melting of the lunar interior. Subsequently they reached the lunar surface in lava flows that issued through fissures and cracks that may have resulted from impacts of meteoroids.

The widespread distribution of anorthosites on the Moon is deduced from three types of evidence: First, material thrown out onto the surface of the Moon as a result of cratering by planetesimals and meteoroids is mainly anorthositic. And such material could easily have been excavated from tens of miles below the surface by the bigger impacts.

Second, X-ray fluorescence detectors carried aboard orbiting command-service modules of Apollo measured X rays emitted by lunar rocks as a result of solar radiation and found that they corresponded to emissions expected of anorthositic rocks. Third, seismic experiments indicated that vibrations are transmitted through the rocks of the Moon as though these consisted of anorthosites. And experiments showed that below a crust of anorthositic rock there appear to be higher-density rocks forming a lunar mantle.

If the Moon is indeed covered with a crust of anorthositic rocks to considerable depths and if geologists are correct in asserting that this kind of rock can only form as a fully melted parent material is cooled, the lunar surface must at one time have been an ocean of liquid magma glowing red-hot and presenting an awesome spectacle in the skies of Earth if there had been anyone to observe it. What agent might have created this molten shell? The answer seems to lie in the process of accretion, the kinetic energy of the infalling bodies that built up the body of the Moon. It seems logical to suppose that this energy would have led to widespread melting of what later became the solid crust and mantle.

A cratered-highland phase of evolution appears to have followed the gradual cooling of the molten shell and formation of the crust. This took place about 4.1–4.4 billion years ago; it was the period of formation of the overlapping craters

APOLLO LANDING SITES

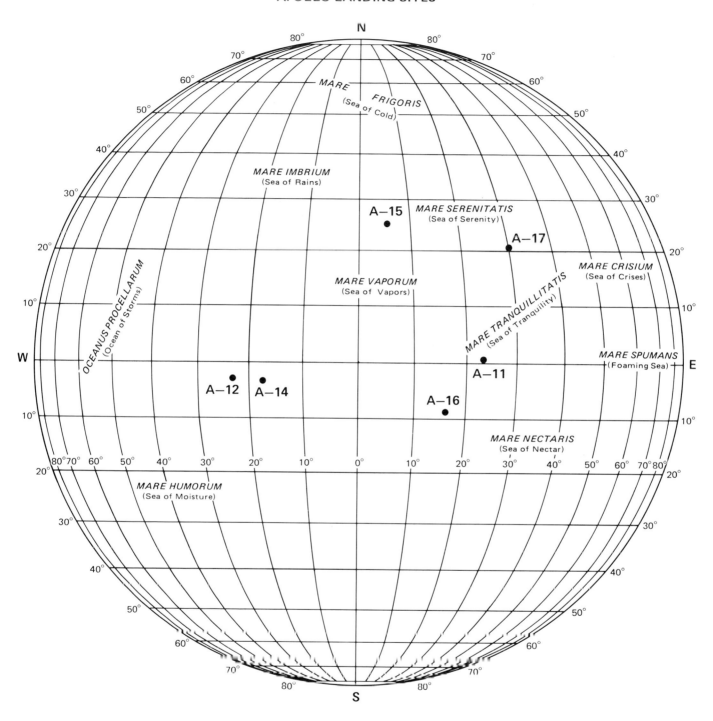

The Apollo program was one of the outstanding human accomplishments in exploration. But despite its success and the wealth of new information it gave us, it has really
only touched the fringe of lunar exploration. The locations of the Apollo landing sites
are shown on this map of the Earth-facing hemisphere of the Moon. They are concentrated in equatorial regions. And we have not yet physically explored the far side of
the Moon, even though we have photographed it from orbit. (NASA)

that distort the surface of the light areas of the Moon called the lunar highlands.

As the cooling progressed downward within the Moon and the anorthositic crust thickened, the stage was set for the appearance of KREEP norite. Sometime between 4.1 and 3.9 billion years ago the Moon, like the Earth and the other inner planets, was subjected to a surge of bombardment by planetesimals and meteoroids, some of which were large enough to excavate the enormous lunar basins. This was the Imbrium period, previously mentioned. About 3.9 billion years ago some old basins seem to have been partially filled with debris. Light-colored plains were also formed. Lunar scientists are uncertain to what ex-

tent direct bombardment was responsible for these plains, as compared to their formation by volcanic flows.

Then lava started to flow from within the Moon. Through cracks and fissures in the upper mantle and crust it welled up, filling the floors of craters and the big basins, laying down sheet over sheet for a period of 600–700 million years. The uppermost layers of lava were the last to escape from the interior of the Moon through the thickening and cooling lithosphere. But since activity continued for such a long time it is not surprising that the basalts vary in composition as well as in age. This is common on the Earth and on the volcanic piles of Mars. In general, on the Moon the

During the Apollo missions, astronauts left many automated instruments on the Moon to continue monitoring conditions there after the astronauts had returned to Earth. This picture shows the lunar roving vehicle (left) and the lunar module (right) with astronaut John W. Young preparing the rover for an excursion across the lunar surface as part of the Apollo 16 mission. (NASA)

older basalts are richer in titanium and they appear to have originated at depths of up to 90 miles (150 kilometers). Later basalts containing less titanium are thought to have originated from greater depths. The radioactive decay sources that were presumably responsible for the heat that partially melted the crustal materials to produce the lavas, appear to have moved gradually inward as the crust cooled and the lithosphere thickened.

The final phase of lunar development began about 3 billion years ago and is still in progress. Except for an occasional planetesimal impact, such as those that formed Copernicus and Tycho, the Moon's crust has been essentially dormant over this immense period. To be sure, there is still some activity, mostly the result of an almost continuous sprinkle of meteoroids and micrometeoroids, which have been detected by seismometers left on the lunar surface by Apollo astronauts.

MOONQUAKES AND MAGNETS

Seismometers erected on the lunar surface detected impacting meteoroids and the falling shells of spacecraft. Interpretations of the vibrations transmitted through the lunar rocks from impacting masses of about one ton are that the crust under the maria extends to a depth of about 37 miles (60 kilometers). It is underlain by the upper mantle, which continues down for another 125–185 miles (200–300 kilometers) and is believed to be rich in magnesium and iron silicates. Between some 185 miles (300 kilometers) depth and at least 685 miles (1,100 kilometers) there is a lower mantle, which may contain material similar to that found in a class of meteorites known as chondritic bodies, which are stony but embedded with small, rounded mineral grains. The lower part of this mantle may be partially melted. Below is the iron-rich core, believed to be some 745 miles (1,200 kilometers) in diameter. The Moon has a much thicker lithosphere than the Earth, some 625 miles (1,000 kilometers), compared with 95 miles (150 kilometers) for Earth.

The Apollo seismic network recorded several thousand moonquakes each year it was in operation. Almost all of them were extremely weak compared with earthquakes. And they appeared to originate from far deeper than earthquakes, at 435–625 miles (700–1,000 kilometers) below

the surface. This is a result of the deep lithosphere of the Moon. The rigid lunar interior fractures in response to stresses and strains, whereas the more plastic terrestrial rocks bend easily at such depths.

Other experiments by Apollo astronauts on the Moon measured the heat flow from its interior. Surprisingly, the heat rate was at first found to be double what had been expected. As data continued to be collected over many months, it was found that the heat flow was not so great after all. Still, the instruments, which may have been in abnormally hot areas (they were on the edges of the lava basins), indicated a flow of about half the average heat flow on Earth. This very high value remains unexplained.

The Apollo experiments measured two types of magnetic fields on the Moon: permanent fields due to fossil magnetism, and transient fields induced in the material of the Moon by the flow of solar wind past it. Although no general magnetic field like that of the Earth was discovered on the Moon, local magnetic fields were measured. Rock samples returned from the Moon were found to be magnetized. Scientists estimate that sometime in its past the Moon had a dipole magnetic field with a strength of 1,000 to 2,000 gammas. By contrast, the terrestrial field is 50,000 gammas, or 0.5 Gauss, and the local lunar fields about 300 gammas.

Since magnetic fields are believed to be generated by rotating planets with liquid cores, the fossil lunar magnetism implies that the Moon may have had a liquid core in its early history.

The fields induced in the Moon by the solar wind have been used as a way of exploring the interior lunar structure by estimates of the electrical conductivity. From such observations, scientists have estimated an internal temperature of about 2,732° F (1,773° K) at a depth of 625 miles (1,000 kilometers).

No atmosphere of any consequence was expected to be present on the Moon and none was found. Scientists detected only heavy gases, such as argon and krypton, that are generated by radioactive decay of still heavier elements. There were also minute quantities of hydrogen, helium, and neon, which originated from the solar wind. The temperatures on the surface of the Moon ranged from highs of 444° F (502° K) at local noon to lows of −306° F (85° K) just before sunrise.

Extremely careful and detailed analysis of the lunar samples revealed no traces of water on the Moon today nor of there having been any there in the past. No prebiological carbon compounds other than those expected on the surface from the infall of meteoroids were found.

Apollo's was a dead Moon. But it is a world blessed with immense natural resources, including metals, glasses, and other minerals, unlimited solar energy, a low gravity field, vacuum conditions, and other potentially attractive conditions.

A NEW WORLD FOR MANKIND

Although Apollo and the unmanned spacecraft preceding the manned landings yielded an incredibly bountiful harvest of scientific information about the Moon, the exploration of Earth's satellite has really only just begun. After all, it is a world whose diameter is fully a quarter that of our own planet and we have only explored a minute part of its surface.

The Moon explored by our space program seems a dead world compared with the dynamic beauty of planet Earth seen here rising over the desolate lunar landscape. This dramatic picture was taken by the last of the manned missions to the Moon, Apollo 17, in 1972. Space planners are not sure when humans will again return to visit our satellite and begin to utilize its resources. (NASA)

Almost every exploratory effort has raised more new questions than it answered. Apollo and lunar exploration was no exception. We still want to know why basalt-filled maria are present only on the near side of the Moon. We want to know how long vulcanism has persisted on the Moon. And we want to know more about the history and development of the magnetic field, to name just a few of the many remaining puzzles.

The future exploration of the Moon will most likely start with several polar-orbiting satellites to map the magnetic and gravity fields of the Moon. They will also provide information on surface composition over a wider area than that possible with the Apollo spacecraft. The gravity measurements are needed to provide more information about the internal structure of the Moon. Earlier missions discovered concentrations of heavy material (mascons) under several of the big basins. The polar orbiters would also measure the radioactivity of surface rocks all over the lunar surface and thereby determine the degree of melting during the period following the formation of the Moon as a world.

After the orbiters may come a series of lunar rovers, unmanned vehicles directed from Earth to explore large areas of the lunar surface. Some would shoot back samples to the Earth as the Russians have done from their unmanned lunar explorers. As a result of the orbiters and the rovers we may discover valuable areas of mineral resources on the Moon. And the sampling process should be extended to the far side, from which no samples have yet been returned.

Inevitably, man will return to the Moon, not only to explore but also to use its resources. Just when this return will take place is not clear today, for it depends very much upon the national spirit of adventure and exploration and of entrepreneurship on an extraterrestrial scale. Experience with the space shuttle and its payloads during the 1980s and 1990s and a possible development of solar power stations in Earth orbit may be factors that force us to go back to the Moon.

Despite many scientific, technical, political, social, and financial uncertainties, the United States space community is undoubtedly looking forward to the day when we return to the Moon. In January 1976, NASA released its long-awaited *Outlook for Space* study (SP-386), which contained sections that discussed a manned base on the Moon similar to bases that an earlier generation

There are vast regions of the Moon still to be explored and major questions yet to be answered. One of these is why the Earth-facing hemisphere (part shown at left of this picture) has lava-filled basins, while the far side (part of which is shown at right) does not. This dramatic view of Earth's satellite, a view that can never be obtained from Earth, was taken by the first manned spacecraft to orbit the Moon, on Christmas Eve, 1968. Mare Crisium, the circular, dark-colored area near the upper center, is close to the edge of the Moon's disc as seen from Earth. The large, irregular, lava-filled maria are Tranquillitatis and Foecunditatis, also on the Earth-facing hemisphere. Features on the lunar far side, which are never seen from Earth, include the large Tsiolkovsky crater, the dark splotch with central light area (a central peak) near the edge of the Moon on the right. (NASA)

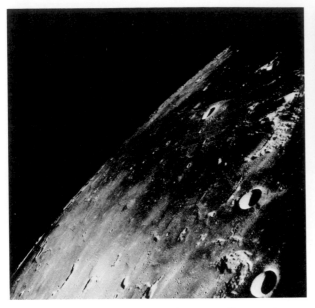 

This oblique photograph looks generally northwest from the Apollo 8 spacecraft into the Sea of Tranquillity. The lower (nearer) linear feature is the Cauchy Scarp. The upper linear feature is the Cauchy Rill. The prominent crater Cauchy lies between the rill and the scarp. (NASA)

A large lunar rock stands alone in the foreground offering its resources to the half-Earth in the skies above and waiting for the people of Earth to start mining the Moon. This prophetic picture was taken by astronauts during their final visit to the Moon. Now many are asking when we will return? What can we do with the Moon? (NASA)

established in Antarctica.

The report emphasized that space development could more easily be justified from an economic point of view if the resources necessary for such development could be gathered from space itself instead of having to be carried into space from the Earth. Among such resources are oxygen, aluminum, magnesium, titanium, silicon, and thorium, plus energy. All these materials are in plentiful supply on the Moon. And energy is almost limitless if tapped from solar radiation. Materials of the Moon could be mined and used for construction, for propellants, for farming, and even to fuel a nuclear power station by converting lunar thorium to uranium 233 in a breeder reactor on the Moon.

Many different activities are envisioned for such a base on the Moon. First, it would be used for continued scientific exploration of the Moon. Radio and optical astronomy would be boosted by observatories there. On the far side the Moon itself would shield radio telescopes from terrestrial radio interference. The Moon in general is better for optical astronomy because it is free of an atmosphere. Solar-energy converters on the Moon might be used to beam microwave energy to the Earth, and many material-processing and manufacturing operations that are inconvenient or impossible on Earth might take place on the Moon.

It is difficult to foresee clearly the scientific or broader uses to which the resources of the Moon will ultimately be put. Almost as difficult as the Europeans trying to visualize the wheat fields of the western plains, the Hollywood film capital, and Detroit's production lines at the time when the Pilgrim Fathers landed on Plymouth Rock.

Today we lay the foundations for tomorrow. People with vision see that some of those foundation stones will be laid first on the Moon and then on the other planets.

We are looking to the Moon in terms of today and even now see tremendous potential from use of its resources. One suggestion that has captured popular imagination is the building of colonies in space from the material of the Moon, which would include new settlements, towns, and industries in artificial worlds orbiting the Earth. (NASA)

8

EXPEDITION TO "BARSOOM"

Readers of *All-Story* magazine in 1912 were introduced to an unusual Earthman, Captain John Carter. He reached out his arms toward Mars, glowing redly in the night sky, and transported himself by will power to a dying planet. The keen imagination of Edgar Rice Burroughs titillated readers with Carter's rescue of a princess of Mars from primitive warring peoples. And for two generations Barsoom became synonymous with Mars in countless adolescent and many adult minds.

Burroughs' story and its sequels had followed closely on the works of the astronomer Percival Lowell, who was convinced that Mars was inhabited by intelligent beings who had constructed a planet-wide system of waterways to distribute dwindling supplies of the life-supporting liquid. Despite learned scientific statements to the contrary, many people accepted Martians as being the most likely form of extraterrestrial life.

When the first spacecraft to fly by Mars showed a world depressingly like the Moon, disappointment was keen. Even scientists seemed to lose interest for a time. And then the exploration of Barsoom began in earnest. And instead of looking for Dejah Thoris, as John Carter had so often done, the new visitors to the red planet searched for Martian microbes. Instead of mystically transporting our projected bodies to Barsoom like the sword-wielding captain, we used huge rockets to fling an electronic marvel through the void and project our senses to the red-dust plains of Chryse and Utopia.

As a consequence, on a morning in July 1976, millions of people were able to sit in front of their television sets drinking their coffee as a view of Mars unfolded before their eyes as clearly as if they had been John Carter arriving there by astral projection.

Earthmen had reached the red planet. They had demonstrated through Viking, an automated, remote-controlled machine of enormous complexity and great versatility, that they were capable of true interplanetary flight, with a landing on another planet. The leap beyond the landing on the Moon had been made. Now it would be only a matter of time before we would walk on Mars as our astronauts had walked on the Moon. Such a landing has not yet been made, but it is inevitable; we have the technology, but we still lack the will to define a manned expedition to Barsoom as a worthwhile goal.

THE FASCINATION OF MARS

Mars has always seemed to hold a tremendous fascination for mankind. The color of the planet, which, after Venus, appears as the brightest star in the skies of Earth, has for centuries been associated with conflict and bloodshed. In Babylonia, Nergal was a solar deity whose cult centered at Kutha, now merely a mound at Tell-Ibrahim. Nergal graduated to become the god of war and in doing so absorbed a number of minor solar deities. Astrological doctrine depicted him as the planet Mars, in which capacity the identification

This color picture of Mars was made by the Viking 1 Orbiter on June 18, 1976.

with war continued.

The term Mars is derived from a contraction of Mavers or Mavors, which is of Roman origin, Mavors being a poetic term for Mars. In Roman religion Mars was associated with Ares, the Greek god of war and son of Zeus and Hera, who delighted in the slaughter and destruction wrought during battle. Probably introduced into Thrace, Ares was far from a popular god and was not generally worshiped in Greece. In Rome, however, he held high honor, partly because he was believed to be father to Romulus, the founder of the nation.

The relative proximity of Mars—it comes closer to the Earth than any other planet except Venus—its warlike associations, and its conspicuous color are not the only reasons for man's fascination with this planet. In a series of memoirs published between 1878 and 1886 the Italian astronomer Giovanni Virginio Schiaparelli reported to the Reale Academi dei Lincei, in Rome, his

investigations of the physics and topography of the red planet. The detailed reports contained many drawings and maps of the planet made during the good oppositions of around 1880, when the planet was close to Earth and the clear skies of Italy revealed much detail that had not previously been recorded.

Among the features described by Schiaparelli were straight lines connecting darker features of the planet across intervening light areas. Schiaparelli referred to these markings as *canali*, or channels. Translated into English, *canali* became "canals" (many industrialized nations were only just emerging from the canal age into the railroad age and imaginative engineers were still dreaming of major canal projects).

Speculation naturally arose that the channels on Mars were canals constructed by intelligent beings . . . Martians. And these beings were obviously more advanced than mankind. This might be expected, because many people still held on to

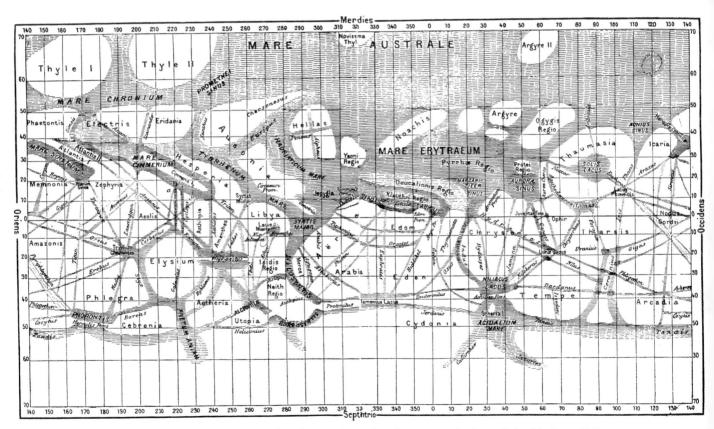

Schiaparelli's 1887 map of the planet Mars, showing many single and double "canali."
This map resulted from the Italian astronomer's observations of the red planet between
1866 and 1877. The observations began with a 9-inch-aperture telescope and continued
with the 18.5-inch telescope at Milan Observatory.

the nebular hypothesis, in which Mars was an older, and therefore probably more advanced world than Earth. This concept was strongly expressed in imaginative novels of that period.

A spate of books and articles followed Schiaparelli's publications. Camille Flammarion wrote a two-volume work entitled *La Planète Mars et ses Conditions d'Habitabilité: Encyclopédié Générale des Observations Martiennes*. The first volume appeared in 1892, covering observations up to that year. A second volume, published in 1909, covered observations after 1892. The great American astronomer Percival Lowell brought out his book *Mars* in 1892, and he, too, supported the idea of there being canals on the red planet. He wrote:

It is by the very presence of uniformity and precision that we suspect things of artificiality . . . the better we see these lines [the canals] the more regular they look; and, second . . . the eye is quicker to perceive irregularity than we commonly note. Yet more conclusive [is the fact that] the lines form a system; that instead of running anywhither they join certain points to certain others, making thus, not a simple network, but one whose meshes connect centers directly with one another . . . . . the intrinsic improbability of such a state of things arising from purely natural causes becomes evident on a moment's consideration.

Lowell seemed entranced by the idea of canals on Mars, and followed his first book with a longer one, in 1908, which he entitled *Mars as the Abode of Life*. He received great criticism. Alfred Russel Wallace, who was important in the development of Darwinism, wrote a critical review of Lowell's work which he published in 1907 under the title *Is Mars Habitable?* Wallace was sure it was not. "Mars," he concluded, "is not only uninhabited by intelligent beings such as Mr. Lowell postulates but is absolutely UNINHABITABLE." And Wallace was a Fellow of the Royal Society, etc.

But the great debate continued. Science-fiction writers stirred the pot, and newspapers carried odd reports about strange apparitions that today might be categorized as UFOs, and the Martians seemed as real as they had ever been, and they continued to be accepted on faith because it was difficult to prove or disprove their existence.

In 1930 the eminent French astronomer E. M. Antoniadi wrote a monumental book in which he reviewed observations of Mars over the period from 1659 to 1929. In this book, *La Planète Mars,* he examined afresh the question of Martian life. He compromised, saying that he felt that the absence of ice in the temperate and equatorial regions proved that the temperature of the planet was not too low to maintain mobile creatures. And he added that the presence of clouds on Mars indicated that the atmosphere was not too rarefied for human beings. But because astronomers could not find traces of water on the planet, he suggested that any life on Mars would be more restricted now than it may have been in the past. As for the canals, he did not consider them to be the work of intelligent beings. "No one has ever seen a true canal on Mars," he cautioned, "and thus the more or less straight or 'double' canals of Schiaparelli exist neither as canals nor as geometric marks."

As tools of astronomy became sharper and observations became more critical, the concept of Mars as an abode of intelligent life was gradually replaced by that of a lifeless world—a frigid, dry, desert body with the merest wisp of atmosphere that was almost without oxygen and water vapor.

And then, in recent years, the quesion of life on Mars was revived. It resulted from the discovery of prebiological molecules in meteorites and in the dust clouds between the stars, and from an emerging scientific doctrine of chemical evolution leading to biological evolution. But no longer was there any thought about canals, no longer Martian princesses to be rescued from savage warlords. Instead, a new breed of exobiologists, scientists who wanted to find a trace, any trace, of extraterrestrial life to support the hypothesis of chemical through biological evolution, emerged and began to plan experiments to search for microbes on the red planet.

MARS AS A PLANET

Mars is smaller than the Earth but slightly larger than Mercury. Its diameter is 4,220 miles (6,790 kilometers) and its surface area is a little more than one quarter that of the Earth. Because so much of the terrestrial surface is covered by oceans, there is actually more land area on oceanless Mars than here.

Mars is similar to the Earth in that it rotates on

its axis only about forty minutes longer than our planet (its day being 24 hours, 37 minutes, and 23 seconds), and its axis is inclined to the plane of its orbit around the Sun by close to 24° just slightly more than Earth's axis is inclined to Earth's orbit. As a result, Mars experiences seasons much like the Earth. There is, however, one great difference. While Earth travels around the Sun in an orbit that is close to being a circle, the orbit of Mars is quite elliptical. While the varying distance of Earth from the Sun is so small that it hardly affects Earth's seasons, the distance of Mars from the Sun varies to a much greater extent and has a pronounced effect on the Martian seasons.

On the average, Mars is 141.62 million miles (227.9 million kilometers) from the Sun, but at perihelion, its closest approach, it is only 128.44 million miles (206.66 million kilometers). And at aphelion Mars is at 154.8 million miles (249.1 million kilometers). Thus the intensity of solar radiation received by Mars at its closest to the Sun is 45 per cent greater than when Mars is at its farthest from the Sun. This causes winters in the southern hemisphere to be much more severe than those in the northern hemisphere, and summers in the southern hemisphere to be warmer than those in the northern hemisphere.

Mars takes almost 687 days to follow its orbit completely around the Sun. Because Earth and Mars are traveling around in the same direction, the two planets come close together on the same side of the Sun only once every 780 days on the average, which is slightly over two years. So Mars has moved a little farther along its orbit each time Earth comes level with it. As a result, if Earth is close to Mars when Mars is at its closest to the Sun, subsequent approaches will be farther apart until a maximum distance is reached about seven or eight years later. Then the approaches will become closer until after a total of about fifteen years a very close approach again occurs.

The close approaches occur when Earth and Mars are approximately lined up on one side of the Sun. This is termed an opposition, because as seen from Earth the red planet is very close to being opposite to the Sun in the night sky. It appears due south at midnight, when the Sun is due north but below the horizon.

Because of the ellipticity of the orbit of Mars, some oppositions bring Mars within about 34.67 million miles (55.78 million kilometers), while most distant oppositions place Mars at 62 million miles (100 million kilometers), which is nearly twice as far away.

Because Mars orbits the Sun outside the orbit of the Earth, it cannot appear as a crescent or even a half-Mars phase but is seen from Earth as gibbous or fully illuminated. Full illumination is seen close to opposition and also when on the far side of the Sun from Earth.

CLASSICAL MARS OF THE TELESCOPISTS

Mars is a disappointing object the first time you see it through an astronomical telescope, even at a favorable opposition. You see it as a reddish globe with faint gray markings, possibly with a whitish bright spot close to one of the poles. The image shimmers and dances in a tantalizing fashion as Earth's atmosphere distorts the light rays coming from the distant planet. If you keep looking, the atmospheric tremors may stop for a few seconds and you gain an impression of a tremendous amount of detail in the grayish areas of the planet, detail that eludes clear definition. If you are lucky and have a big enough telescope at a good opposition, you may see two extremely faint starlike objects close to the planet and seemingly winking in and out in the glare of Mars. These are the planet's two satellites, Deimos and Phobos, named after the mythological attendants of the god of war.

Most of the planet's disc consists of light regions that have a reddish-orange hue. The darker, gray areas were named maria because early astronomers thought they might be Martian seas. The polar caps are conspicuously white though sometimes hidden beneath hoods of clouds.

Forced as they were to examine the planet from a distance of tens of millions of miles and from the bottom of a deep atmosphere, Earthbound astronomers recognized that the true nature of the Martian surface would depend upon many factors, including the types of rocks originally present in the crust, the planet's subsequent evolution, weathering processes, changing atmospheric composition, infall of meteorites, and the pressure and temperature of the atmosphere. Astronomers speculated that the reddish areas were great regions of Martian desert. Except for periodic cloud cover, these regions seemed to change little with the Martian seasons. By contrast, the polar caps and the dark areas did show seasonal and other changes.

Astronomers were unable to find oxygen in the Martian atmosphere and they reasoned that this oxygen might have been used up in oxidizing the surface rocks of Mars, particularly to produce iron oxides, which could be responsible for the reddish color of the deserts. Comparisons with terrestrial minerals showed that the red deserts of Mars were most likely composed of an iron oxide that is relatively free of water and is called limonite, and a mineral that contains both iron and titanium that is known as ilmenite. There was also a possibility of there being a fine-grained felsite, a rock that is rich in quartz and silica and is also called rhyolite. But there was no real proof that these substances were on Mars. In fact, the presence of large quantities of iron-bearing minerals might seem incompatible with the low density of Mars, which astronomers had found was only 3.94 grams per cubic centimers as compared with Earth's 5.53 grams per cubic centimeter. However, the iron oxides might be present in only a small amount, finely divided, to account for the red coloring of the planet.

Up to the 1960s, when space probes visited Mars, there was no evidence of high mountains on Mars. Astronomers had reasoned that any mountains of four miles (six and a half kilometers) height should be visible along the border between day and night, the terminator, because their lighted peaks should stand out from the dark area of shadows at their bases. Such mountains had not been seen. This led astronomers to think of Mars as a relatively flat world.

This seemed to be confirmed by the first radar observations of Mars, made in the early 1960s by Soviet and American radio astronomers. However, by 1967 better radar revealed differences in elevation amounting to as much as six miles (ten kilometers).

In general, astronomers were not able to decide from visual observations which areas on Mars were at high elevation and which were at low. Some guesses were made. Regions such as Hellas (a circular light area) and Ogygis (an ill-defined light area), in the southern hemisphere, from time to time took on a lighter hue than usual. Astronomers thought that this might indicate high plateaux covered by frost or clouds. They were later proved quite wrong.

But the polar caps were observed not to retreat uniformly during the spring and summer. Bright areas become detached and remain white long after most of the cap has retreated southward or northward from them. This is particularly true of the southern cap. Some astronomers suggested that these isolated areas of whiteness represented high mountains; in fact, one near the south pole was called the Mountain of Mitchell.

The polar caps grow and diminish with the Martian seasons, being greatest during the winter in each hemisphere and smallest during the summer. A small residual cap lasts all year, though this is not always seen at the south pole. Scientists could not readily determine whether the caps consisted of ice or frozen carbon dioxide, though many scientific papers argued the point. Large permanent caps of carbon dioxide might be very important to climatic changes on Mars, because if they ever melted, say from a wobble of the Martian poles, sufficient carbon dioxide could be released to create an atmosphere of much lighter pressure than astronomers believe Mars has today.

The next important visible feature of Mars in terms of total surface area are the maria. Unlike the deserts, they appear to change both in shape and color with the seasons: astronomers have claimed that they pass from a dark, grayish green in spring and summer to a brownish gray in winter. Despite their name, the maria probably never were bodies of water, and this was accepted by most astronomers before the space age. Instead they were often regarded as regions of plant growth, because some observations suggested that their color, polarization, and reflectivity changed with the seasons and their boundaries against the desert region appeared not to be fixed but retreated and advanced. Many astronomers left accounts claiming they had seen a darkening wave move across Mars's surface, which they described as follows: As spring advances in one hemisphere and the winter polar cap retreats, the maria along the border regions surrounding the cap begin to darken, occasionally exhibiting reddish-brown and brownish-blue colors. The changes in the maria take place as a darkening wave that gradually moves to and crosses the equator as the season advances.

Somehow the maria seemed to be influenced by the changes in the polar caps. An obvious explanation was that water evaporating from the caps moved toward the equator and the opposite pole, and as it did so the wave of humidity affected the maria. While some astronomers claimed this was evidence of vegetation growing as the wave of humidity passed, others said that the color changes could equally as well be explained by minerals that absorbed the water and changed their color slightly. Such minerals are called hygroscopic.

However, another school of thought suggested that the changes in the maria and in the polar caps were not directly related. It was speculated that seasonal wind changes might be responsible for moving clouds of dust and sand about the planet in regular patterns, like our trade winds, and the deposition of materials on the surface or the scouring of a surface by the winds could account for the changes.

There were even supporters for a theory that the darker marking on Mars resulted from volcanic ash deposited from perennial wind systems. This theory relied upon there being active volcanoes on Mars today. Scientists had wondered why there were more maria in the southern than in the northern hemisphere of Mars. Supporters of the volcanic ash theory suggested that the southern region's longer and warmer summers and the resulting increased tendency for atmospheric circulation would lead to larger deposits of ash in these regions. By mid-1965 it was impossible to prove or disprove the volcanic theory; and some astronomers postulated that the volcanic ash theory together with simple plant life on Mars might explain the seasonal changes to the dark markings. The plants, they explained, might be obtaining nourishment from the ash and water from melting polar caps.

DRY ICE OR WATER ICE?

Depending on the season, the polar caps advance from and retreat toward the north and south poles of Mars. By analogy with Earth's polar caps, they were long assumed to consist of snow and water ice. If this were so, there would be enough water vapor in the Martian atmosphere for it to form appreciable deposits by condensation, first around one pole and then around the other.

Observations indicated that the caps were very thin, probably well under one yard (one meter), except close to the poles, where the thickness might be several yards. As a summer pole melted, the opposite, winter pole grew quickly, indicating a brisk movement of water vapor from one hemisphere to the other. Astronomers never understood why this movement should be so rapid.

Since astronomers were unable to identify any lines that could be associated with water vapor in the spectrum of the light from Mars to support their snow and ice theory for the polar caps, they looked for other interpretations. Because of the known carbon dioxide content of the Martian atmosphere, frozen carbon dioxide snow or compacted "dry ice" could form polar caps if the temperature at the Martian poles was cold enough each winter. Some even speculated that the polar caps might consist of frozen nitrogen tetroxide.

In the northern hemisphere the spring lasts for 199 days on Mars, summer 182 days, autumn 146 days, and winter 160 days. As winter advances, the north polar cap grows its largest, usually accompanied by a covering haze of clouds known as the polar hood. At its greatest push southward, the edge of the cap reaches 65° north latitude. By contrast, the southern cap extends to 50° south latitude. Both cover many millions of square miles of land. Since in the southern hemisphere summer occurs at the perihelion of Mars, temperatures are somewhat higher than during summer in the northern hemisphere, which occurs at aphelion. However, southern winters, which coincide with aphelion, are colder than northern winters, with the result that the southern cap is larger in winter than the northern cap and much smaller in summer than its northern counterpart. Sometimes the southern cap seems to disappear completely, but the northern cap never does.

THE UNCERTAIN ATMOSPHERE

Astronomers knew that Mars has an atmosphere very soon after they started observing the planet by telescope. The planet has clouds and hazes. The precipitation and later melting and evaporation of the material of the polar caps also required the presence of an atmosphere. But since astronomers could readily see surface details on Mars, they concluded that the atmosphere must be very thin. As the years went by, they gradually modified their estimates downward. In 1940 the consensus among astronomers was that the pressure was only 11 per cent that of Earth's sea-level atmospheric pressure. By the 1950s they had further reduced this to 7 per cent, which is equivalent to the pressure in Earth's atmosphere at a height of 11.4 miles (18.3 kilometers). As a result, Mars seemed an unlikely place to support life similar to that on the Earth. But worse was yet to come when spacecraft first flew by Mars, in the 1960s. Their experiments pushed the Martian surface atmospheric pressure down to 0.8 per cent of Earth's sea-level pressure, which is equivalent to the pressure in our atmosphere at an altitude of almost 21 miles (33.8 kilometers). The prospects for life on Mars looked very bleak.

Meanwhile astronomers tried with varying degrees of success to find out the composition of the Martian atmosphere. They relied upon the spectrum of Mars but had to contend with the stronger effects of the spectrum of Earth's atmosphere, which masked the details in the Martian spectrum. Carbon dioxide was the only constituent of the Martian atmosphere that was clearly identified, and this was not until 1947. But how much of this gas was present could not be ascertained from Earth. However, some calculations suggested that the amount of carbon dioxide in the Martian atmosphere might be almost twice that in the Earth's atmosphere at sea level. Other estimates were quite different. Despite persistent efforts, astronomers were unable to detect either oxygen or ozone. Nitrogen cannot be detected by Earth-based observations. Astronomers guessed that there was probably some in the Martian atmosphere, as there is in Earth's atmosphere. How much? Some astronomers thought the bulk of the Martian atmosphere might be nitrogen. But this was questionable, because it was based solely on analogy with the atmosphere of Earth.

In 1925, astronomers identified water vapor in the Martian atmosphere. However in the 1940s, with much better equipment, they could not find any evidence for its presence. By the early 1960s there was still no proof of water vapor in the atmosphere of Mars. Based upon the detection capability of their instruments, scientists concluded that if it were present on Mars it must be in quantities less than in the driest desert regions of Earth. They concluded that neither rain nor snow could fall on Mars. As a result, the idea that the polar caps consisted of frozen carbon dioxide came back into vogue.

But there were still features of the atmosphere that were not easy to explain without the presence of water. Clouds of several different types had been observed for many years on the planet. Some were very thin and appeared as hazes that covered small areas of the surface. Others were much thicker and blanketed large areas of the surface from view. Clouds also appeared and disappeared, sometimes lasting for only a few hours, at other times for days. One huge set of clouds appeared regularly each Martian afternoon over a vast area of the deserts. Shaped like a huge W, it seemed remarkably like water-vapor clouds condensing over high mountains in late afternoon.

In general, astronomers identified three main types of clouds on Mars by their color: white, blue, or yellow.

White clouds were further identified as of three main types: dense, brilliant clouds that seem to be large systems moving in temperate or equatorial regions as well as stationary systems that cover the developing polar caps; thin, misty clouds that dissipated quickly after sunrise; and orographic clouds, similar to the W cloud, that appeared to be connected with areas of high elevation.

Blue clouds are very similar to the white clouds and were identified as a separate type because they reflected blue and ultraviolet light more strongly. Sometimes they appeared only in blue and ultraviolet light and were invisible in red light. Many astronomers thought that they were high clouds, possibly consisting of ice crystals. Sometimes blue clouds are so widespread that pictures of Mars taken in blue light show a featureless disc. At other times, the atmosphere is completely clear of blue clouds, and photographs in blue light show details almost equal to those taken in red light. These periods were known to astronomers as periods of blue clearing.

Yellow clouds were generally attributed to dust storms, because they showed strong motions and polarized light in a way characteristic of large particles. There are several types of these clouds: faint yellow veils that obscure the fine details of the surface; small, local, dense clouds that move rapidly like dust storms on Earth; and major formations that sometimes envelop the whole planet and hide all surface details from view.

These yellow clouds were observed most fre-

quently during the oppositions of Mars that occurred when it was close to its perihelion. Scientists thought that high surface temperatures
would give rise to fierce winds capable of raising
the major storms. But a big question was why, if
such dust storms enveloped the planet, they did
not cover the dark areas so that Mars would appear as an almost uniformly reddish-ocher disc.
The believers in plant life on Mars explained this
on the basis of continued regrowth rising above
the layers of dust.

A CLIMATE FAR FROM BEING MILD

Astronomers have traditionally measured the
temperature on Mars by the infrared radiation
and radio waves it emits, since before the space
age there was no way of placing a thermometer
there. The first measurements made at infrared
wavelengths, in 1924, showed that the southern
hemisphere is indeed warmer than the northern.
The warmest part of the Martian year is a month
or so after the summer solstice (as on Earth),
and the warmest part of a Martian day is in the
early afternoon (as on Earth).

The average surface temperature on Mars appeared to be about —45° F (230° K), according
to the best infrared measurements.

In the 1950s, astronomers measured the radio
waves coming from Mars at centimeter wavelengths, or "thermal radio waves." These measurements showed a lower temperature than the
infrared measurements, probably because they
came from slightly below the surface and thus
represented a more average temperature of the
planet, whereas the infrared measurements gave
the average temperature of the sunlit surface
only.

The big difference between Earth and Mars is
that we possess large bodies of water, while Mars
does not. Water is extremely effective in modify-
ing the climate of Earth, as large amounts of heat
are extracted from the atmosphere when water
evaporates and large amounts returned to the atmosphere when water precipitates. Additionally,
the oceans store vast quantities of heat to reduce
temperature extremes. But Mars is an oceanless
world where the climate is far from mild. There
are wide temperature variations during the course
of each day and thoughout a year, even in the
temperate and tropical zones. Maximum equato-

rial temperatures during summer were measured
from Earth as being 80° F (300° K) on the average, but desert regions showed 68° F (293° K)
and the dark areas 98° F (310° K). This suggested that the dark areas were at lower elevations than the deserts.

No direct measurements of nighttime temperature could be made, since the night side of Mars
faces away from Earth. In equatorial regions, the
early-morning temperature was believed to be
about —99° F (200° K).

While much about the Martian wind remained
unknown to astronomers before the space age,
the patterns appeared to be less complicated than
those of Earth. Astronomers had several explanations, some which later proved unfounded. They
said that the simplicity derived from the fact that
Mars had no oceans (true), no high mountains
(false), and a cloud cover less than the Earth's
(marginal). Attempts were made to establish seasonal wind patterns like those on the Earth, but
such attempts were very speculative.

The low gravity of Mars compared with that of
Earth (38 per cent only) means that Mars cannot pull its atmosphere down toward its surface
as strongly. As a result, the density of the Martian atmosphere does not fall off as rapidly with
increasing height above the surface; the atmosphere is more evenly spread out.

As for the upper atmosphere and a possible
magnetosphere of Mars, astronomers could ascertain very little from Earth. Attempts were
made to find out if the red planet had radiation
belts like ours by checking for the radio waves
that such belts would be expected to generate.
Since we found no such belts, we could not determine whether or not Mars had a magnetic
field that would form a magnetosphere to protect
the planet from the solar wind.

FEAR, PANIC, AND AN ALIEN SPACESHIP!

Following the close opposition of Mars in
1862, astronomical textbooks continued to refer
to Mars as a planet without a satellite. But at the
next opposition Asaph Hall, an American astronomer who is said to have become tired of reading
that Mars had no satellite, decided he would look
for one. After a period of disappointing nights
during which Mars was either obscured by clouds
or so badly distorted by Earth's atmosphere that
observations were almost impossible, Hall discov-

ered a faint starlike object close to Mars, on August 10, 1877. A few days later a second moon appeared, and he was astounded to find that it was the only satellite in the Solar System to orbit its primary in less time than its primary rotates on its axis. Hall named the satellites after the attendants of Mars, the mythological Deimos and Phobos (Panic and Fear).

Deimos is the more distant of the two moons; it follows an almost circular orbit about 14,600 miles (23,500 kilometers) from the center of Mars. Its size is estimated as only about six miles (ten kilometers) across. Its orbital period is 1 day, 6 hours, and 18 minutes, so it rises in the east and sets in the west, seen from the surface of Mars. But its period is so close to that of Mars itself that if you were standing on the planet, Deimos would almost appear to be hanging in the sky.

Phobos orbits much closer to Mars, at a distance of only 5,800 miles (9,300 kilometers) from the planet's center, a mere 3,700 miles (5,900 kilometers) above the Martian surface. It is so close that from parts of the Martian surface

beyond a latitude of 68° north and south the satellite cannot be seen. Phobos orbits Mars in 7 hours and 29 minutes, so from the surface it appears to rise in the west and, five and a half hours later, to set in the east. Its diameter was long thought to be about 12 miles (20 kilometers).

In 1959 the Soviet astronomer I. S. Shklovskiy suggested that Phobos might be an artificial world, a hollow body with low density. This might then explain the slow descent of the satellite toward Mars that astronomers thought was occurring. To cause such an orbital change the satellite would have to be moving through a resistive medium such as the far outer atmosphere of Mars. But if the satellite were a reasonably dense body, similar, say, to an asteroid, the atmospheric density would not be sufficient to slowly change its orbit to the extent observed. However, all this was very speculative, since even the observations of the orbital changes were disputed by some astronomers. The idea of Phobos being some alien spaceship in the Solar System joined the mythology of Phobos and Deimos as attendants of the god of war. Instead, most astronomers regarded

U.S. MARS EXPLORATION SPACECRAFT

Name	Launched	Arrived	Results
Mariner 4	Nov. 28, '64	July 14, '65	Successful fly-by with photographs of narrow strip of surface
Mariner 6	Feb. 25, '69	July 31, '69	Successful fly-by with photographs of much wider strip of surface and closeups of selected areas.
Mariner 7	Mar. 27, '69	Aug. 5, '69	Successful fly-by, strips of surface and polar region, covered with telephoto frames of selected areas.
Mariner 9	May 30, '71	Nov. 13, '71	Orbiter; photographed surface in great detail, also satellites. Much other information on clouds, polar caps, temperatures, atmospheric structure and composition.
Viking 1	Aug. 20, '75	July 20, '76	First landing of U.S. spacecraft on another planet. Many pictures from surface and from orbit; pictures of moons; much other information including analysis of soil and atmosphere. Virulent soil chemistry discovered.
Viking 2	Sept. 9, '75	Sept. 3, '76	Repeat of Viking experiments at a second site on Mars. No conclusive evidence from either site as to whether or not soil chemistry was due to biology, but no organics found at either site.

the two Martian satellites as captured asteroids from the belt of such bodies between the orbits of Mars and Jupiter. The really big breakthrough in understanding these tiny worlds came when they could be inspected closely by spacecraft.

In fact, despite three centuries of observations of Mars, which was the best situated of the planets for us to observe, we really knew very little about it. Speculations had filled many volumes, but that was mainly what they were. Before the first spacecraft flew by Mars, in 1965, the finest detail seen with our best telescopes under exceptional conditions exceeded 60 miles (100 kilometers) in size. But in less than fifteen years, through the eyes of spacecraft, we have seen grains of dust on the Martian surface, mapped most of the planet to show details of tens of meters, and closely inspected both the Martian satellites. And to add to our knowledge, we have sampled the atmosphere, sampled the soil, probed the ionosphere, and searched for life on the planet. The six American spacecraft sent so far to Mars are summarized in the table on p. 167.

THE MARS OF MARINER 4

Automated spacecraft to explore Mars had been suggested in the literature during the early 1950s, but the first attempts were not made until 1962, when the Soviets launched their Mars 1 probe. Communications were lost on the way to Mars and it was believed that the spacecraft missed Mars by about 125,000 miles (200,000 kilometers).

In November 1964, three more probes were prepared for missions to Mars: two American spacecraft (Mariners 3 and 4) and one Russian (Zond 2). Only one of them, Mariner 4, met its objectives. On July 14–15, 1965, after a voyage of 228 days that took it 325.6 million miles (524 million kilometers), part way around the Sun, it approached within 6,119 miles (9,845 kilometers) of the red planet. At that time, the recording of photographs within spacecraft and their subsequent transmission to Earth was only at a rudimentary stage, so that the Mariner 4 was able to obtain only twenty-two pictures of Mars. And it required eight hours to send these few pictures back to Earth.

And the equipment was so limited that it was not possible to obtain over-all views of the planet showing any transition from Earth-based pictures to the closeups obtained by the spacecraft. The twenty-two pictures each showed only a very small area of Mars. Surprised astronomers gazed upon pictures of a cratered surface that at first glance seemed very much like that of the Moon. Across roughly 1 per cent of the surface of Mars that was pictured by Mariner 4, project scientists counted seventy clearly distinguishable craters between 2.5 and 75 miles (4 and 120 kilometers) diameter and many more of smaller size. Although NASA officials, supported by reports of the imaging team, wrote of Mars as being a Moon-like body with its surface molded mainly by impacts without erosional or tectonic activities, other astronomers, not connected with the space agency, pointed out that there were lineaments visible on some of the Mariner 4 pictures. These lineaments suggested that there may have been other forces at work molding the surface of Mars, particularly volcanic activities and major faulting. Nevertheless the discoveries of Mariner 4 had a strong impact on scientific concepts of Martian history and caused many scientists to think that the surface was extremely old and had not changed greatly over billions of years. The ancient sea bottoms of Barsoom changed overnight into ancient cratered terrain like that of the Moon and interest in the further exploration of Mars somewhat waned.

Additionally, Mariner 4 confirmed that Mars has no significant magnetic field and that the solar wind must blow directly into the planet's atmosphere. But, most devastating to those who still pictured Mars as the most likely abode of extraterrestrial life in the Solar System, Mariner 4 measured the surface atmospheric pressure as only four tenths of 1 per cent of that of the Earth; i.e., the same as the pressure of the Earth's atmosphere at an altitude of about 125 miles (200 kilometers). Clearly Mars could not be a place for living things.

THE MARS OF MARINERS 6 AND 7

It was another four years before American spacecraft went again to Mars. And the Soviets were concentrating on Venus during that period. Realizing the limitations of Mariner 4, imaging-team members had designed a new television system that would be able to photograph Mars dur-

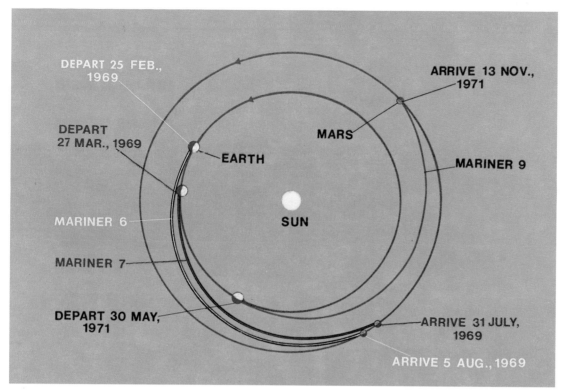

Transfer trajectories of Mariners 6, 7, and 9. The first two spacecraft took advantage of the February–March, 1969, launch window, photographing the planet Mars as they flew by, in late July and early August. Mariner 8 left Earth on May 30, 1971, and entered into orbit around its target on November 13. Surpassing all expectations, for nearly a year its cameras photographed and its instruments sensed the planet's atmosphere and surface. Among its greatest accomplishments were the discovery of the great rift canyon and the photograph of the giant Nix Olympica volcanic structure. (Arfor Picture Archives)

ing the approach, start with pictures that were about equal in detail to the telescopic views from Earth, and show increasing detail as the spacecraft zoomed toward the planet. The final pictures would be scans at high magnification covering small strips of the surface, like the pictures of Mariner 4. One of the spacecraft, Mariner 6, flew over the equator, while Mariner 7 scanned dark regions close to the equator and then over the south polar regions. Two cameras were carried by each spacecraft, one to provide wide-angle pictures, the other narrow-angle pictures of high magnification, like a telephoto lens system. The best pictures of Mariner 4 could show as a spot on the picture a feature that was 2 miles (3.2 kilometers) across on the surface of Mars (a resolution of two miles). Mariners 6 and 7 could show features that were only 0.2 mile (0.3 kilo-

meter) across (a resolution of 0.2 mile). The cameras were also equipped with filters to penetrate the Martian hazes.

The missions of the two spacecraft were highly successful. Both got off to good launches in February and March of 1969. After their long journeys part way around the Sun they flew past Mars in July and August of that same year. Mariner 6 passed within 2,120 miles (3,411 kilometers) of Mars and obtained 75 photographs. Mariner 7 passed within 2,190 miles (3,524 kilometers) and obtained 126 photographs. Both spacecraft had greater storage capacity than Mariner 4 and could return the pictures more rapidly.

The new photographs appeared to confirm the interpretations of Mars from Mariner 4: it was a highly cratered planet with few if any mountainous regions, no "canals," no big valleys, no

Interior view of the space flight network support and communications room in Pasadena, California, during the Mariner 9 orbital phase. (NASA)

Mariner 7's television cameras made this series of photographs of the south polar region of Mars on August 4, 1969, the pole being situated in the lower right portion of frame 7N17. At the top is the raw picture of the area showing the camera's compensating effect for the rapid change in contrast between bright polar cap and the darker area surrounding it. The mosaic at the bottom results from computer correction and enhancement of the individual frames. The edge of the cap is visible along a 90° span of longitude, from 290° to 20° east, and the cap itself covers a latitude range from its edge at 60° southward to and beyond the pole. The evening terminator is at the right edge of the mosaic. Craters on the edge of the cap show snow-covered southward-facing and exposed northward-facing slopes. Telescopic pictures placed within the wide-angle frames are shown enlarged slightly top and bottom of the raw-picture mosaic. South is at the bottom in these pictures. (Jet Propulsion Laboratory/NASA)

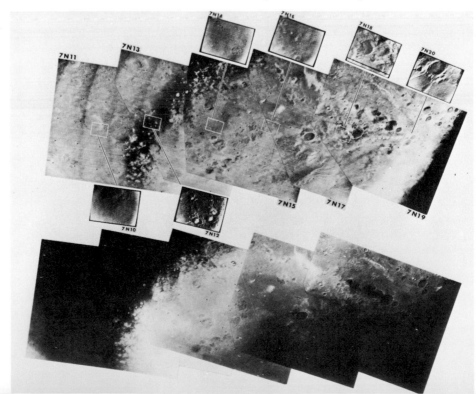

The pictures across the surface of Mars were overlapped slightly so that scientists were later able to make mosaics of strips of the surface. Four pictures taken by Mariner 6 on July 30, 1969, form this mosaic. It covers an area of about 435×2,485 miles (700×4,000 kilometers) parallel to about 15° south of the equator. The black spot on each picture originates from a blemish within the television camera. Pictures such as this appeared to confirm the Mariner 4 view of a bleak, cratered, essentially dead world. (Jet Propulsion Laboratory/NASA)

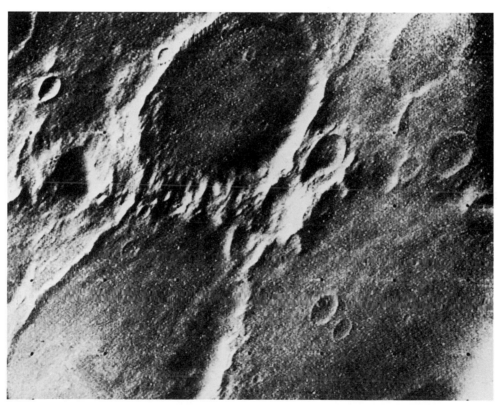

Enlargement of a telephoto frame shows two adjacent craters in the polar region, foreshortened by the oblique viewing angle from the spacecraft. Covering an area of about 200×80 miles (320×135 kilometers), this picture became known as that of the "giant's footprint." (Jet Propulsion Laboratory/NASA)

evidence of volcanic activity, and craters even into the polar regions. There were, however, two unusual types of terrain. One was a peculiar system of concentric marking at the edge of the south polar cap, and the other was regions of chaotic terrain that seemed to be depressed areas of the Martian surface on the floor of which the material was jumbled into small blocks, ridges, and depressions, and was almost free of craters.

Another discovery was that the great circular feature known as Hellas on the classical maps of Mars appeared to be surrounded by mountain chains across which the terrain changed from a heavily cratered one to a completely featureless area covering the floor of the huge depression. Scientists questioned whether Hellas was a dust-filled bowl, a flat desert, or covered with clouds.

A big disappointment was that although the wide-angle pictures gradually showed an increasingly large disc of Mars similar to that seen from Earth, the telephoto pictures did not reveal any surface features to account for the differences in light and shade. Mariners 6 and 7 did not solve the problem of what surface features of Mars produce the red-ocher "desert" regions and the gray-green maria that we see through telescopes from Earth. There were, however, some faint clues: Some of the craters near a region of Mars called Sinus Meridianus, a dark area, were seen to have dark markings over parts of their floors. And some of the dark spots on Mars that had been termed oases by earlier generations of astronomers, turned out to be large craters with dark floors.

Strangely, the two Mariners did not show Mars as a very cloudy planet. The only significant clouds photographed were a hood over the south polar cap.

Instruments carried by the spacecraft measured temperature on the planet based on the infrared radiation emitted from its surface. It was ascertained that temperatures at the south pole hovered around $-189°$ F ($150°$ K), very close to the temperature at which carbon dioxide solidifies when at a pressure about equal to that in the Martian atmosphere. Temperature along the edge of the south polar cap ranged from $-180°$ to $-45°$ F ($155°$ to $230°$ K), while daytime temperatures on other parts of the Martian surface rose from the values at the edge of the cap up to a high of $68°$ F ($293°$ K) in early afternoon on the equator. Also, measurements were made of the night-side temperature of Mars for the first time. They showed that it dipped to $-98°$ F ($175°$ K). The temperature of the south polar region appeared to confirm that the polar caps of Mars consist not of water ice but of carbon-dioxide ice.

Another instrument carried by each Mariner detected the presence of a slight fog of water vapor but showed that the atmosphere consisted of about 90 per cent carbon dioxide. There were also traces of carbon monoxide, atomic hydrogen, and molecular oxygen, but no sign of ammonia, methane, nitrogen, or nitric oxide.

Both spacecraft provided new information about the atmosphere as they passed behind Mars as seen from Earth and their radio signals were distorted by the atmosphere before being cut off by the bulk of the planet. The atmospheric pressure had to be revised upward slightly. Things looked just a little better for the possibility of life and of liquid water existing on the surface of Mars in the very deepest depressions, at least in early afternoon during the summer.

These Mariners also made new measurements of the size of Mars, showing that the planet is flattened at its poles to about 12.4 miles (20 kilometers) less than its equatorial radius of 2,109 miles (3,393 kilometers).

Mars looked interesting again. The result was that a fleet of spacecraft (six Russian and four American) was sent to Mars at the next two oppositions. Actually, opportunities for spacecraft to fly to Mars do not coincide with the oppositions, but they occur, like the oppositions, approximately every two years and two months. Of this fleet, only three Russian and one American spacecraft reached their destination; the others failed at launch or en route. The American spacecraft was an orbiter; those sent by the Russians, landers. While the American orbiter was highly successful in its orbital mission, the landers were singularly unsuccessful. They reached the surface but were not able to send back data from it.

A NEW MARS BEGINS TO EMERGE

Mars 2 and 3 were launched in May 1971 and reached Mars later that year. Each consisted of an orbiter and a lander. The landing spacecraft were separated from each orbiter before it went into orbit around Mars. While both orbiters were successful and returned much information from orbit, including some pictures of Mars, both the landers failed to operate when they reached the Martian surface. And because they had been designed to transmit the data collected during passage through the atmosphere when they had successfully landed, they did not supply their Soviet science teams with any data about the atmosphere of Mars.

The Mariners were also launched in May 1971. Mariner 8 failed in its launch, but the successful Mariner 9 reached Mars safely and was placed in orbit around the red planet on November 13, 1971. Following some maneuvers to trim this orbit, the spacecraft circled Mars in a period of 11.97 hours in an elliptical orbit, the lowest point being 862 miles (1,387 kilometers) above the surface. After some time in orbit, the periapsis (closest approach) was raised higher to allow the spacecraft's cameras to obtain broader views of the surface.

The first big surprise from the Mariner 9 mission began as the spacecraft approached the planet and before it went into orbit. During their approaches, Mariners 6 and 7 had sent back to Earth remarkably clear pictures of the light and dark areas of Mars. The pictures from Mariner 9 were indistinct, as though the camera lenses were covered with condensation. But it was not the lenses of the cameras, it was the fault of Mars itself. The planet was shrouded in a planet-wide dust storm, so intense that even the bright polar caps were only very faintly visible through the murky atmosphere of Mars. Scientists believed that it was this gigantic storm that had caused the failure of the two Soviet landers, which attempted their landing during its height. This was a lesson of a great need for adaptability in missions to Mars, a capability to change objectives, and was applied to the later American Viking missions.

Momentarily frustrated at not being able to photograph the surface of Mars when Mariner 9 was in orbit, the project scientists quickly realized that they had a new and unexpected opportunity. They could study at close hand the development of this great storm, the greatest they had so far observed on the planet. Actually it had started

When Mariner 9 approached Mars the view was very different from that obtained by the previous three U.S. spacecraft. Usually prominent surface details were obscured by a raging dust storm. This picture was taken on November 11, 1971, when the spacecraft was still 435,000 miles (700,000 kilometers) from Mars. (Jet Propulsion Laboratory/ NASA)

during September, while Mariner 9 was still on its way to Mars. The first sign of the storm was a yellowish cloud in the high desert of heavily cratered Noachis, just west of Hellas, in the southern hemisphere. Within a week it had spread to engulf the whole planet.

As Mariner 9 allowed astronomers to watch the clouds gradually clear, they were able to estimate the size of dust particles and to calculate the speed of Martian winds needed to stir up this dust. Winds as high as 100 miles per hour (160 kilometers per hour) appeared to be necessary. Once a small dust storm arises when Mars is close to perihelion and receiving maximum input from solar heating, the dust in the atmosphere traps more heat, raises the atmospheric temperature so that winds increase, and more dust is raised. The process continues almost explosively. So a major dust storm quickly covers the planet, then takes a month or so to die down as Mars moves from perihelion and the atmosphere cools and the dust settles.

As the dust of this great storm began to settle, Mariner 9's cameras sent back some strange pictures. The first that showed any detail on the generally featureless globe of the dust-enshrouded Mars were characterized by four dark spots. These were in Tharsis, a light-colored region long thought to be a high desert. It was also the region of the W cloud mentioned earlier and of several dark "oases" identified on Earth-based maps.

For these four spots to be seen above the dust clouds, they had to be extremely high features. Earth-based radar observations had actually placed the height of peaks in this area at about five miles above the average surface of Mars. The positions of these dark spots coincided with four features mapped from Earth. North Spot was at the position of Ascraeus Lacus (lake), a dark spot, Middle Spot corresponded with Pavonis Lacus, and South Spot was about the position of a feature called Arsia Silva (forest). Alongside the three spots was the fourth. Its position corresponded with that of a prominent feature known as Nix Olympica (snows of Olympus) because it was a bright area in the middle of the vast reddish deserts of Arcadia and Amazonis.

The major surprise in the whole of the exploration of Mars came when scientists directed the spacecraft to take high-resolution pictures of these dark spots. The returned photographs showed great circular depressions, a series of

(a)

overlapping craters. It had to be admitted that it was most unlikely that they could be impact craters, that meteorites had selected the tops of the highest known Martian mountains to crash onto the surface! A quick *volte-face* had to be made: from the accepted idea of a Mars molded entirely by bodies falling from space, to a Mars that possessed the biggest volcanoes in the Solar System, bigger even than anything on Earth.

As the dust cleared still further, these gigantic volcanic piles were seen to be at least 18 miles (29 kilometers) high, and Nix Olympica, which was quickly renamed Olympus Mons (mountain), spread its lava flows to form a base of 300 miles (500 kilometers) diameter. The other mountains were renamed Pavonis Mons, Ascraeus Mons, and Arsia Mons. And other volcanoes were soon identified elsewhere on Mars. Mars was looking less and less like the Moon.

The lineaments discovered at the time of the Mariner 4 fly-by were seen again; in fact, great areas of Mars were found to be broken by intricate fracture patterns. In parts of the planet these fractures had widened in huge valleys in which erosion had sculptured Martian Grand Canyons.

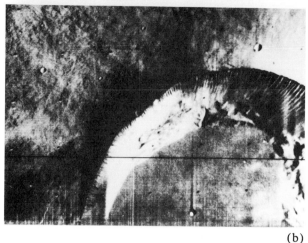

(b)

(c)

The first features seen on Mars by Mariner 9 after it entered into orbit about the red planet were dark spots rising above the dust clouds. Scientists directed the spacecraft to take telephoto pictures of each of the spots and were surprised when the pictures showed huge craters, because they would have to be on the peaks of enormous mountains. These craters were volcanic calderas, not impact craters. Completely unexpectedly, one hemisphere of Mars was found to have volcanoes greater than any on Earth.

(a) opposite page: North Spot (Ascraeus Lacus, later renamed Ascraeus Mons) looked like this. The main crater is about 13 miles (21 kilometers) across, and the complex of craters has a diameter of about 25 miles (40 kilometers).

(b) Another shield volcano, Pavonis Mons (initially called Middle Spot) has a 25-mile (40-kilometer) -diameter summit crater. The top view was taken by the wide-angle camera, the lower one is a telephoto picture. The rough volcanic flank is splattered with impact craters with raised rims. The smooth floor of the caldera suggests that of a lava lake. The caldera's walls are scarred by down-sliding materials that were probably carried away by the wind.

(c) South Spot (Arsia Mons) possesses a 70 mile (110 kilometer) -diameter caldera with multiple concentric fractures on the western rim and many rimless craterlets. (Jet Propulsion Laboratory/NASA)

But the canyons of Mars dwarfed anything on the Earth. The great system of equatorial canyons discovered by Mariner 9 stretches a distance equal to the width of the United States. The Grand Canyon is even dwarfed by some of the tributaries of the Martian canyons. And even more fascinating, Mariner 9's pictures started to show evidence of channels all over Mars, channels that resembled dried-up river beds, some of which were larger river systems than the Amazon.

Why hadn't these fascinating features been seen by the previous Mariners? Simply because they had looked at the wrong parts of Mars. It

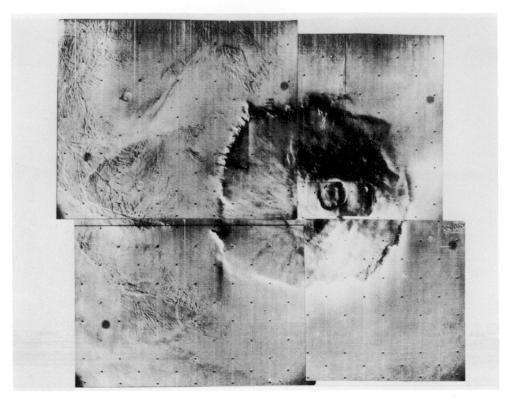

But the greatest of all the Martian volcanoes was not seen in full until all the dust had settled. The base cliffs of this mighty pile, which formed the light circular feature seen earlier, are 910 miles (1,465 kilometers) across. The peak with its great caldera rises more than 12.5 miles (20 kilometers) above these cliffs. The summit caldera is 43 miles (70 kilometers) wide. It is the largest known volcano on all the planets we have visited. Volcanic constructs such as this are made by the upwelling of lava and its pouring down the sides. In some places they overflow the base cliffs shown earlier in a picture taken some years later by a Viking spacecraft. When the magma flow subsides (as in the Hawaiian volcanoes), there occurs a general collapse around the volcanic throat, which gives rise to the caldera craters. Successive flows produce the intersecting craters. (Jet Propulsion Laboratory/NASA)

was comparable to an alien spacecraft visiting Earth and making a close inspection of the Pacific Ocean only, and the extraterrestrials then concluding that the Earth was entirely covered by water, with one or two small atolls breaking the otherwise featureless surface. The experience with Mars shows that an expedition to a planet is not enough to tell us even the gross features of that planet unless we are sure we have inspected all of its surface. As we shall see in a later chapter, even a gaseous planet such as Jupiter has quite different features, depending upon what parts of its globe we inspect.

So a new Mars emerged rapidly from the hundreds of pictures being returned by Mariner 9 to Earth. The concept of a Moon-like, largely crater-covered, dead world quickly evaporated before the eyes of astonished experimenters and scientists in general. Instead, a varied planet was revealed, a world with an extensive tectonic, volcanic, impact, depositional, and erosional history. And because of the freshness of some of the features it seemed that some of these processes might have been operating on Mars in comparatively recent times. It had not been dormant for several billion years, like Mercury and the Moon.

The greatest problem confronting scientists when they studied the new Mars was the presence of innumerable channels of all sizes. There was no easy and logical way to explain them other than that they had been produced by running water on the surface. The atmosphere today has

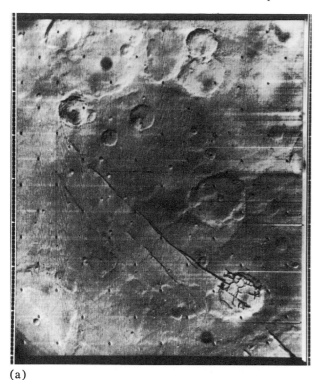

(a)

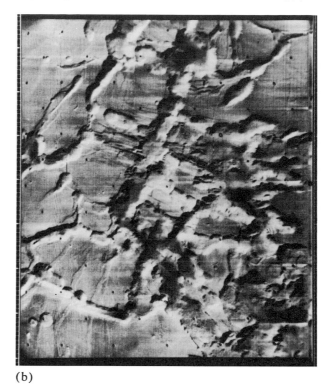

(b)

Mariner 9 in its long-term survey of Mars also rediscovered the lineaments suspected in the Mariner 4 pictures and found them on a much larger scale than imagined. (a) *(left) This fault zone extends for hundreds of miles in an echelon pattern east of the Tharsis Ridge. Each fault offsets from one above it. The crater exhibits what lunar geologists have called a "turtle back" structure. Presumably, the crust of Mars was penetrated by the body crashing into it from space, and the resulting weak floor was later molded by lava pressure from below.* (Jet Propulsion Laboratory/NASA)

(b) (right) Another type of fault structure is shown in this picture of Labyrinthus Noctis which is also on the boundaries of the Tharsis uplifted area of Mars. This web of partially closed valleys seems to have been formed by partial collapse of the Martian crust along a network of fractures with subsequent erosion of the material. The network is located at the head of the great canyon system of Mars. (Jet Propulsion Laboratory/NASA)

too low a pressure for water to exist in such large quantities. The problem was therefore twofold. How could Mars have had a dense atmosphere in the past for enormous quantities of water to flow on its surface, and what had happened to the water?

Despite further spacecraft expeditions to the red planet, these questions have not been answered satisfactorily.

A favorite speculation at one time was that much carbon dioxide is permanently locked in the polar caps. It was reasoned that when the axis of Mars wobbles there would be a period during which the major permanent cap would be trans-

ferred from one pole to the other. During this transfer, so much carbon dioxide would be released into the atmosphere that its pressure would be high enough for liquid water to be present on the surface. At the same time, the climate would be warmer, because the dense atmosphere would act like the windows of a greenhouse, and ice in the crust of Mars would melt into water.

This speculation seemed fine until a later spacecraft discovered that the permanent polar caps of Mars consist of water, not carbon dioxide.

Because of these dilemmas other scientists have sought ways to explain the channels by wind ero-

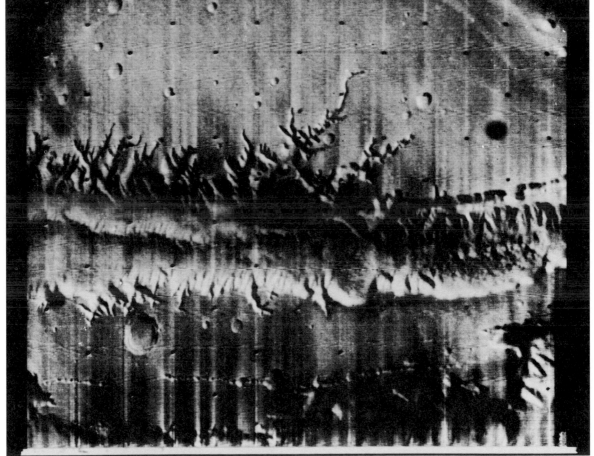

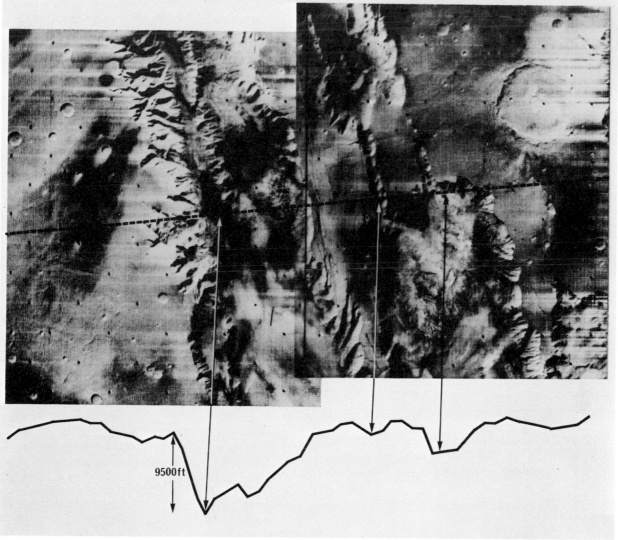

(c)

(a) *This mosaic of many photographs from Mariner 9 shows the remarkable canyon lands of Mars stretching along the Martian equator for 3,000 miles (4,800 kilometers). The great volcano Olympus Mons and the other volcanoes of the Tharsis Plateau are visible at the upper left. The mosaic, which consists of several hundred separate pictures, extends from 30° north to 30° south latitude and stretches more than one third around Mars, from 10° west longitude at the right to 140° west at the left.*

(b) *Part of the big canyon showing its many side canyons, which appear to have been eroded by water. The walls of the canyon have suffered enormous landslides and material has piled up on the floor of the canyon at the foot of the walls. The Earth's Grand Canyon would be lost in the small tributaries of this Martian canyon.*

(c) *A profile of one of the canyons gives an idea of its size. The deepest part is 9,500 feet, compared with the Grand Canyon's 5,500 feet (2,900 and 1,700 meters). Its width is 75 miles, compared with the Grand Canyon's 13 miles (121 and 21 kilometers). The line below the picture represents changing atmospheric pressure, which can be translated as shown into depth or elevation measurements. The dotted line across the photograph represents the path scanned by an instrument called an ultraviolet spectrometer, used to calculate the pressure at the surface. (Jet Propulsion Laboratory/NASA)*

(a)

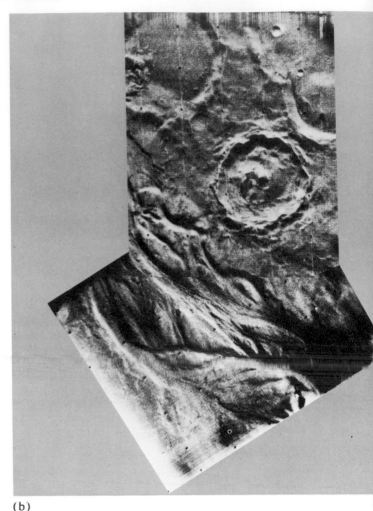

(b)

Another unexpected and important discovery from the Mariner 9 expedition was that much of the Martian surface is marked by strange, meandering channels, very different from the lava channels of the Moon, and resembling more the dry arroyos of terrestrial deserts.

(a) A sinuous channel some 450 miles (700 kilometers) long has multiple tributaries leading into the one channel. The stream deepens and widens downstream and appears to go from higher to lower elevation, like some ancient Martian river might have.

(b) It is difficult to explain the feature shown in this picture except by water erosion. The braided channel sweeps past a crater and has islands. The only terrestrial features that look like this are caused by intermittent water flows. (Jet Propulsion Laboratory NASA)

sion and lava flows, but their speculations are not very convincing.

An important result of the Mariner 9 orbital surveillance was that there is a belt around the planet's northern hemisphere which is generally beneath the mean surface level of Mars, like the floor of some ancient ocean of Barsoom. We have no proof that it ever was an ocean; in fact one of

the mysteries of Mars and its ancient water is that we see no evidence of fossil beaches. But the important thing about this low-lying band is that atmospheric pressures there are great enough, because of its depth, for liquid water to exist on the surface of Mars for part of each day.

This excited the exobiologists with the possibility that perhaps in this belt they would find microbes living today. And other scientists pointed out that from the great canyons there seemed to have been vast floods flowing northward in the past into the "ocean" basins. They suggested that the regions at the mouths of the canyons would be where fossil life might be discovered, and certainly where rocks that had been washed from the canyons might give a clue as to the ancient rocks of the equatorial highlands of the planet.

These two areas, the mouths of canyons and the low-elevation northern plains, were selected as landing sites for an American expedition of two spacecraft to land on Mars at the time of the Bicentennial. The project to do this was called Viking.

THE MARS OF VIKING

Beween Mariner 9 and the Viking project, two Russian spacecraft, out of four launched, reached Mars and sent back useful information supplementing that obtained by Mariner 9. Mars 5 entered orbit about the red planet and sent pictures and other data to Earth. Mars 6 landed on Mars but did not send any information from the surface. However, it did send information about the atmosphere during its descent to the surface.

In August and September 1975, twin Viking spacecraft left for Mars. Each consisted of two spacecraft: an orbiter and a lander. The mission plan differed in an important way from that used by the Soviets. Instead of separating the lander from the orbiter before the latter entered orbit around Mars, the two spacecraft would remain together and both enter orbit. Once in orbit, the orbiter would survey the surface of Mars and a decision could be made if the preselected landing site for each lander was indeed suitable. The orbital reconnaissance would also ensure that the lander would not plunge to destruction into a Martian dust storm.

The plan paid off.

Despite some initial difficulties in finding a good landing site for each spacecraft, successful landings were made. The first spacecraft touched down in a low basin called Chryse Planitia, while the second landed farther north, in an area called Utopia. Both survived the landings and sent valuable information from the surface for almost a whole Martian year. And the orbiters performed equally as well. Except for the failure of the seis-

mometer on Lander 1, all the equipment worked well and all the planned experiments were carried out.

Measurements were made of the Martian atmosphere—from space by the orbiters, and during entry through the atmosphere and from the surface by the landers. Winds and other meteorological conditions were measured on the surface at the two landing sites.

The composition of the atmosphere was measured. The major gas was confirmed as carbon dioxide with traces of nitrogen, argon, oxygen, nitric oxide, and carbon monoxide. Even more important, the instruments aboard the Viking landers were able to find out how much of each of several isotopes of the atmospheric gases were present. These proportions are important, because they allow scientists to estimate how the atmosphere might have evolved so that the isotopes are present in the amounts they are today. The isotopes of carbon and oxygen were found to be similar to those in the atmosphere of the Earth. But the atmosphere of Mars is enriched with more than 1.75 times as much of the heavier nitrogen 15 compared with nitrogen 14. This indicates that nitrogen outgassed from the rocks of the planet into its atmosphere very early in its history and also that a large amount of nitrogen has been released in total, probably one hundred times as much nitrogen as there is in the atmosphere today. This nitrogen probably escaped into space from the top of the Martian atmosphere.

Viking also measured relative abundances of isotopes of argon and krypton. Argon 40 is produced by the decay of radioactive potassium 40. The amounts of two other isotopes relative to argon 40 and to krypton suggest that outgassing on Mars was similar to that on Earth. If this was so and the amount of carbon dioxide outgassed on Mars was proportionate to that outgassed on the Earth, taking into account the size of the planets, then Mars must have had in its past more than ten times as much carbon dioxide as is now in its atmosphere. But this is still many times less than the amount of carbon dioxide that the Earth outgassed. The evolution of the Martian atmosphere seems to present many contradictions. The excess of krypton over xenon in Mars's atmosphere is different, too, from that found in carbonaceous chondritic meteorites, which are believed to represent early material of the Solar System. But it more closely matches the proportions in the Earth's atmosphere. On Earth, we attribute the deficiency to the absorption of xenon in sedimentary rocks. If Mars had periods of plentiful water in the past, this same absorption might have applied on that planet too.

The composition of the Martian atmosphere determined during entry and at the two landing sites is 95 per cent carbon dioxide, 2.5 per cent nitrogen, 1.5 per cent argon 40, and traces of oxygen, krypton, xenon, and water vapor. There are also trace amounts of neon and helium.

There were several surprises when the Vikings measured the temperature throughout the atmosphere from its topmost regions down to the surface. They found that the upper atmosphere was colder than had been expected from observations made by fly-bys and from orbit. There were also temperature waves moving upward and increasing in amplitude at higher altitudes. Throughout the atmosphere, however, the temperature was above that at which carbon dioxide can condense, which meant that all the clouds must be clouds of water vapor or ice.

Each day at Chryse, soon after the landing, the temperature at the surface of Mars varied from a low of $-121°$ F ($188°$ K) during the night to a daytime maximum of $-20°$ F ($244°$ K) during the afternoon. Chryse was at $22.48°$ north latitude. At Utopia, which was farther north, at $47.96°$ the daily swings were from $-115.6°$ F ($191°$ K) to $-32°$ F ($241°$ K). And at both sites the meteorological instruments detected a daily pressure change of about 1 per cent of the almost 8 millibars measured. This surface pressure of 8 millibars is about eight tenths of 1 per cent of the Earth's atmospheric pressure at sea level. As the days went by, the Martian landers measured a gradual fall in atmospheric pressure. This was caused by carbon dioxide freezing out of the atmosphere at one of the poles of Mars.

The amount of water vapor in the atmosphere of Mars is negligible compared with that in the terrestrial atmosphere. But the Martian atmosphere is very much less than that of the Earth. As a consequence, the atmosphere of Mars is really quite wet; in fact, its relative humidity is so high that the atmosphere is almost at saturation. Thus water vapor is continually moving in and out of the atmosphere and condensing into clouds.

A major Viking discovery was that the residual polar caps—the small caps that remain each summer when the caps of large extent have vanished —are of water ice. The large caps are mainly carbon dioxide and are quite thin. The permanent water-ice caps may be very thick and consist of deep layers of ice and dust. These ice caps provide considerable amounts of water vapor that also collects in lower elevations. Because there is very little oxygen in the atmosphere, incoming solar ultraviolet radiation does not produce an ozone layer and, in fact, penetrates to the surface. This solar ultraviolet radiation breaks up the molecules of water vapor and produces hydrogen and the very reactive hydroxyl radical to form hydrogen peroxide. The hydrogen escapes from the atmosphere into space and forms a hydrogen cloud around Mars. The radical helps to recombine carbon and oxygen into carbon dioxide, thus preventing the escape of those substances when solar ultraviolet radiation breaks down the carbon dioxide of the atmosphere.

While most of the clouds on Mars are water clouds, carbon dioxide ice clouds do form over the polar regions in winter. Snowfalls of dry ice from such clouds may help to build up the vast seasonal polar caps.

The landers also produced records of wind directions and speeds at the two sites. And the orbiters recorded the growth and progress of major dust storms. The W-cloud phenomenon that forms each afternoon in the Tharsis region was shown to be large cloud systems associated with the big volcanoes there. These clouds develop during the morning and grow from diffuse masses into more complex structures. Sometimes they were seen streaming off from the tops of the volcanoes into the evening darkness.

Many hazes were seen during the mission; they often filled the floors of craters and canyons. Some appeared shortly after sunrise and then dissipated as the day advanced. At times, great areas in all the low-lying regions of the planet were streaked and dotted with mist patches. There were also many cases of small fluffy clouds that are seen, under high magnification, obscuring fine detail beneath them, but are invisible in high-altitude, wide-angle photographs. Their presence could only be inferred by a general softening of the gross features of the area covered by them.

The landers sampled the material of the surface at the two sites. Although the sites looked quite different in the pictures returned from the surface, the materials at both were remarkably similar. But unfortunately the sampling device was

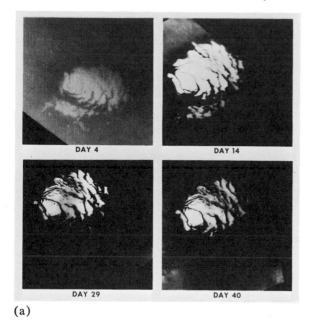

(a)

(b)

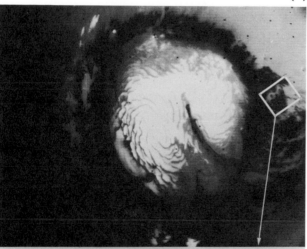

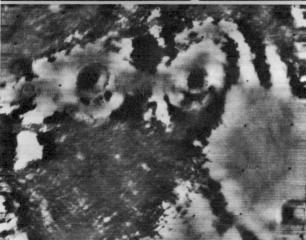

able to sample only fine material from between the rocks, and this could be deposited from a uniform mixture of fine material that is blown all over the planet by the great dust storms. All the samples had high proportions of iron plus a moderate amount of magnesium, calcium, and sulfur. There was also some aluminum and traces of alkalis and rare earth elements. The composition seems to indicate that these fine materials come from the weathering of mafic igneous rocks. Water was released from all the samples when they were heated to a high temperature. The concentration of sulfur in the fine material was a surprise. It was ten to one hundred times as great as found in Earth soils.

The surface viewed through the eyes of the cameras on Viking Lander 1 was a boulder-strewn desert on the western slopes of Chryse Planitia. The general terrain is gently rolling hills with several ridges between the spacecraft and the

Spacecraft revealed many fascinating features about the polar caps of Mars.

(a) Mariner 9 provided a history of the shrinkage of the south polar cap over the thirty-six-day period through which photographs were taken. It shows the widening of the separations and the breaking up of the large detached area which disappeared on day 29. The cap shrunk rapidly at first, suggesting the sublimation of carbon dioxide snow. Then it retreated slowly, more suggestive of the underlying residue being water ice. The picture on day 4 was obtained while the planet was shrouded with dust.

(b) The north polar cap at its maximum extent— about 620 miles (1,000 kilometers) across—was photographed on October 12, 1972. The curved patterns in the interior of the frost cover are formed on outward-facing slopes that receive more sunlight than the flat areas and thereby defrost sooner. The spiral pattern and the eroded channel running from the center of the cap toward the lower right are characteristic of the layered sedimentary complex localized in the central region of the polar region. A dark collar of rough-textured terrain surrounds the polar region. The view below is a telephoto of the area defined in a white box in the top picture. The cameras of Mariner 9 did not have the resolution to show the true nature of this textured terrain. It remained for another spacecraft, Viking Lander 1, to find that around the polar cap of Mars there is the biggest dune field in the Solar System. Mars may be a small planet but it has the most impressive geological features. (Jet Propulsion Laboratory/NASA)

Assembling the Viking Mars spacecraft in Martin Marietta's Denver facility. (Martin-Marietta)

A summer day on Mars, August 15, 1976. Toward left in this color panorama of the Martian surface is a small dune of fine-grained material scarred by trenches dug by Viking 1's surface sampler on July 28 and August 3. The sampler scoop is seen in its temporary "parked" position. The bright reddish orange surface of Mars is strewn with a variety of angular rocks of various shapes and types. (NASA)

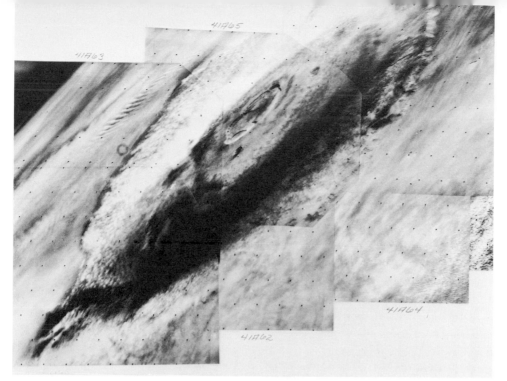

The great Martian volcano was photographed by the Viking 1 Orbiter on July 31, 1976. The mountain is seen in mid-morning wreathed in clouds that extend up the flanks to an altitude of about 12 miles (19 kilometers). The multiringed volcanic caldera pushes up through the clouds into the stratosphere. A well-defined wave cloud train extends several hundred miles beyond the mountain at the upper left of the picture. The clouds are thought to be composed of water ice condensed from the atmosphere as it cools by moving up the slopes of the volcano. In the Martian afternoon these clouds on the big volcanoes become so extensive that they can be seen from Earth during spring and summer in the northern hemisphere of Mars. (Jet Propulsion Laboratory/NASA)

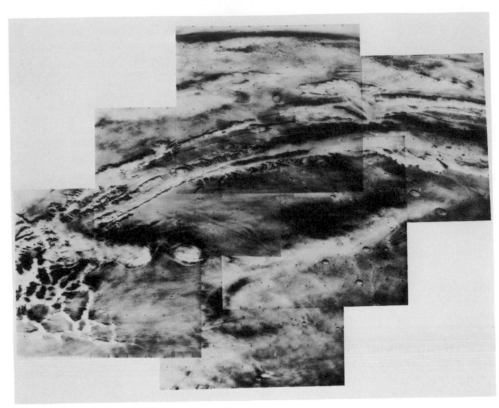

This picture, also from a Viking Orbiter spacecraft in 1976, shows the great canyons of Mars, their floors filled with early-morning mists. Labyrinthus Noctis, to the left, is also mist-filled, and on the higher ground a wavelike pattern results from a layer of tiny cloudlets like our cirro-cumulus clouds. Mist is spilling over from a large crater at the left of the picture. On the far horizon other types of clouds can be seen. (Jet Propulsion Laboratory/NASA)

This picture of part of Labyrinthus Noctis is made from several pictures assembled into a mosaic. On some of the pictures, taken at high magnification, the surface seems covered with small cottonlike balls. These are actually fluffy little cloudlets. On other pictures these cloudlets are not so clearly photographed, but their presence leads to a blurring of surface detail. Other pictures were made through clear skies and show the rugged nature of the canyons. (Jet Propulsion Laboratory/NASA)

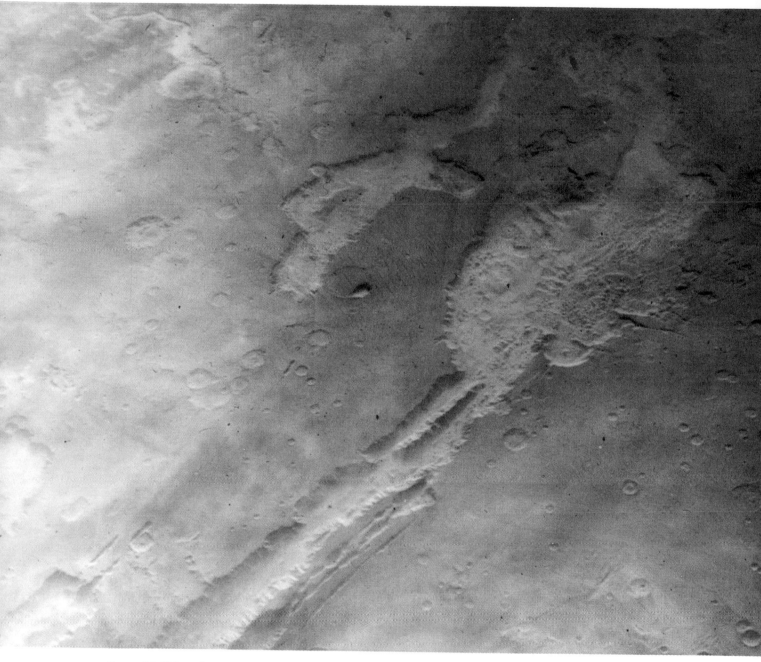

From 19,200 miles (31,000 kilometers) above Mars, Viking Orbiter 1 took fifteen black-and-white photos through three color filters—violet, green, and red—to form this color mosaic of part of the so-called "Grand Canyon of Mars." The canyon, Valles Marineris, lies just a few degrees south of the equator and parallels it for some 3,100 miles (5,000 kilometers), from east to west. The fifteen photos, taken near the highest point in the Orbiter's thirty-second revolution of Mars (on July 22), were computer-processed and combined at the United States Geological Survey facility in Flagstaff, Arizona. (NASA)

horizon. There are high features on the horizon, thought to be the rims of distant craters. The general surface is rock strewn with fine material between the rocks and drifts of the material to the northeast of the spacecraft. Some of the drifts are the size of dunes. Rocks are blocky and angular, and some are pitted with small holes called vesicles. The shade of the rocks varies from dark to light, their color is a reddish pink. The experimenters thought that this pinkish color came from a veneer of limonite. This color was on boulders, small rocks, and surface fine material. It colored the dust blown into the atmosphere and resulted in a pink sky for Mars.

Lander 2 came down on much flatter terrain, in the Utopia Planitia. Also boulder-strewn, but the boulders were of a more uniform size and color. There were no drifts of fine material, but there were troughs that were thought to be part of a polygonal network seen from orbit and similar to features in the tundra north of Earth's Arctic Circle caused by melting of the surface of the permafrost.

Small platy objects suggested that the surface was crusted, possibly as a result of water carrying salts to the surface and leaving them behind when it evaporated. The rocks were more pitted than those at the Chryse site. This suggested that they may be of volcanic origin. Flutes on some of the rocks also indicated later wind erosion.

At neither site was there any evidence of living things, no vegetation or artifacts that might be attributed to living creatures; nor was there much apparent change on the surface even after dust

(a)

(b)

storms had passed over the sites.

Exciting discoveries were made from orbit about the polar regions. Layered deposits, first seen by Mariner 9, were confirmed at both poles of Mars. They are thought to represent the deposition of dust and ice over periods of hundreds of millions of years. The layering may have resulted from changes in the rate of deposition by a cycling of the climate of Mars, perhaps caused by changes to the orbit of Mars and the wobbling of the planet on its axis. Some of the material eroded to form terraces at the poles is thought to have been deposited in the form of vast areas of sand dunes that were first discovered by Viking. The dune fields surrounding the poles, like most things on Mars, appear to be larger than any other known fields of dunes. Dune fields have also been found in many other places on Mars.

As a result of the exploration of Mars, we have some new ideas of how the planet evolved and the relationship to the evolution of other planets. There are three major types of volcanic activity displayed on Mars. Great volcanic plains are made up of extensive flows of lava somewhat similar to the lava plains of Mercury and the Moon. There are very large shield volcanoes that

(c)

Viking Lander 1 touched down in the plain of Chryse and its cameras showed a rugged, rock-strewn, and slightly undulating landscape stretching to a clear horizon.

(a) General panorama from the spacecraft.

(b) Color photograph assembled from pictures taken through various filters shows a pinkish surface and a pink sky. These are believed to be true colors because other pictures showed color charts and cables on the spacecraft against which the color processing of the pictures could be matched.

(c) Dune field taken in late afternoon. Note the variety of rock shapes and sizes seen in these pictures. (Jet Propulsion Laboratory/NASA)

The second Viking Lander touched down in Utopia, closer to the Martian north pole. The terrain here was quite different from Chryse: the landscape flatter, no distant hills, rocks of a more uniform size, and no dune fields. The site was characterized, though, by channels running across it whose courses seemed free of many rocks. The rocks, too, were more pitted here. General panorama from the spacecraft. (NASA)

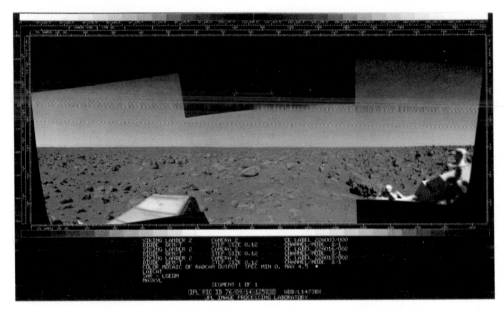

The Martian horizon as seen by Viking 2 stretches across nearly 200° in this composite of three color photos taken on as many days. Color of the surface is predominantly rusty red, caused by a thin coating of limonite (hydrated iron oxide). Dark volcanic rocks are seen at both left and right. (NASA)

(a)   (b)

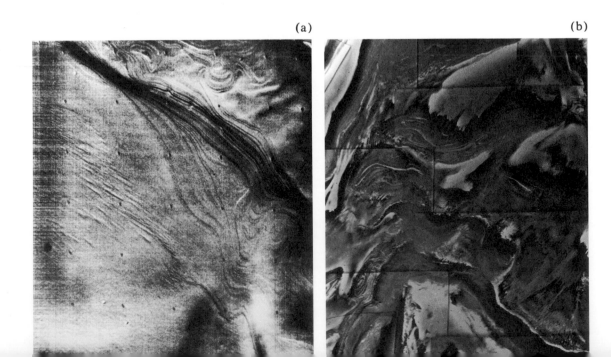

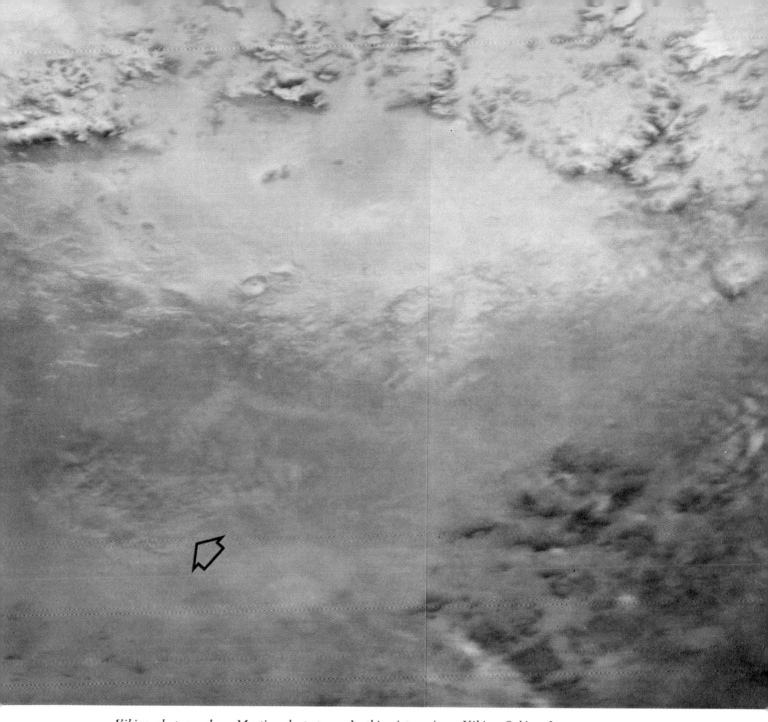

Viking photographs a Martian dust storm. In this picture from Viking Orbiter 2, a tur-
bulent, bright dust cloud (arrow) more than 180 miles (300 kilometers) across can be
seen inside the great Argyre Basin. This is the first color picture of a dust storm taken
from a spacecraft orbiting the planet. (NASA)

Another unexpected discovery about Mars was the presence of layered deposits at the
poles. These imply that there have been climatic changes in which dust and ice may
have alternated, and possibly periods of deposition followed by others of erosion.

(a) opposite page: A picture of part of the south polar region taken by Mariner 9
shows the polar laminated terrain for the first time. The dark lines seem to be ledges
in the terrain. Because there are very few craters on this terrain it is believed to be
very young.

(b) Viking Orbiter spacecraft produced the pictures to make this mosaic of the north
polar cap of Mars. Polar laminations are evident. Again, craters are conspicuous
by their absence. Note also the peculiar striation pattern in the ice, the contrast
of which has been reduced to allow the details to show in reproduction. These are
thought to show evidence of wind erosion of the ice fields. (Jet Propulsion Laboratory/
NASA)

Viking discovered that surrounding the polar cap at the north pole of Mars there is an immense field of dunes. This picture shows part of the dune field. It is a young feature, since there are no craters on it. The dunes probably result from wind erosion of the laminated terrain. (Jet Propulsion Laboratory/NASA)

show a complicated history of magma flows across their surrounds. The largest of these, Olympus Mons, is bigger than any volcano on Earth, much larger than Hawaii even when the part of the big island below the level of the Pacific is included. The third type of volcanic activity is shown in complexes that differ in all respects from the shield volcanoes in shape and form. The volcanic features also differ in age. Some are obviously very old and pock-marked with craters. Others are extremely young, for they have virtually no craters upon them. Even on the shield volcanoes there is evidence of various types of magma flows, indicating a long history of growth.

The Tharsis Plateau is a huge mass of dense rock supported high above the general level of the crust of Mars. How it remains elevated is unknown. It may be that the crust of Mars is extremely thick. Or it may be that there is a continued upwelling of the mantle of Mars holding up the plateau. Tharsis is the highest plateau known, being twice as high as the Tibetan Plateau on Earth.

The areas of chaotic terrain first seen in the pictures returned by Mariners 6 and 7 seem to be explainable by ice collapse. They suggest that

Olympus Mons, the Solar System's biggest volcano, has interesting stories to tell the astute observer. The mosaic of thirteen photos shows part of the summit caldera under high magnification. The smallest features visible on this picture are about 60 feet (18 meters) across, five times smaller than any previous picture of this mountain. The shadows tell us that the craters are about 1.6 miles (2.6 kilometers) deep. Note the almost complete absence of impact craters, which tells us the volcano must have been active until comparatively recently. (Jet Propulsion Laboratory/NASA)

vast areas of the Martian crust may consist of a mixture of rock and ice and that sometimes this melts; the water flows away and leaves behind the jumbled masses of rocky blocks. Generally, however, Mars is a frozen world with perhaps a permafrost extending miles into the crust. While Earth has been dominated by oceans, the "oceans" of Mars never became warm enough to hold water. While periodically large quantities of water flowed across the Martian surface and left its marks on that surface, the water rapidly froze back into the crust. How it was melted from the ice we do not know; possibly it was the internal heat of Mars, possibly it was heat following impact of big meteorites, or perhaps there were periods when the Sun emitted more radiation and heated Mars generally.

The Viking orbiters also provided striking close-up views of the Martian satellites. Phobos was revealed as a remarkable world in that it exhibited a definite layered structure and the typical herringbone pattern of secondary-impact craters. Both moons of Mars are cratered to the point of saturation of their surfaces: complete turnover of the surface by crater overlapping crater. Both satellites are a dirty gray. Both seem to be of low mass. Their origin is still much in doubt, i.e., whether they are natural satellites of Mars condensed from the solar nebula, or captured asteroids. Both satellites are irregularly shaped; Deimos measures $9.3 \times 7.5 \times 6.8$ miles ($15 \times 12 \times 10.9$ kilometers), compared to Phobos' $16.8 \times 13 \times 11.8$ miles ($27 \times 21 \times 19$ kilometers). It is strange that the two moons should be simi-

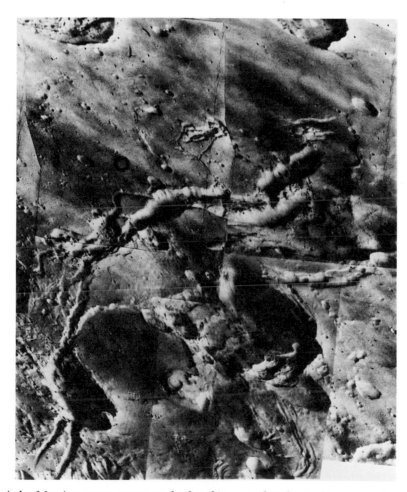

Parts of the Martian crust appear to be breaking up, first by fault zones and then continuing with mass wasting and collapse as though the crust consisted of a mixture of ice and rock and the water melts and flows away along the channels. (Jet Propulsion Laboratory/NASA)

Viking produced this picture of the Plain of Chryse. It is extremely difficult to explain these flow marks except as being created by a great flood of water sweeping over the large Martian plain. Distinct evidence of channeling of the flow between obstacles and flow lines downstream of obstacles are seen. Even the craters have blankets of material ejected from them that appear as though they had been made in a mixture of rock and ice; they look quite different from craters on Mercury and the Moon and more akin to bomb craters in muddy terrestrial soil. (Jet Propulsion Laboratory/NASA)

larly shaped though different in size. Both the satellites revolve around Mars with one face continually toward the planet.

THE QUESTION OF MARTIANS

In addition to finding out more about the physical qualities of the red planet, the two Viking landers had an important task of looking closely at the soil of Mars for any evidence of biological processes in the soil that would indicate the presence of Martian microbes.

The sampling of Martian soil first proved very exciting, then puzzling, then disappointing. The soil was extremely active chemically, but scientists debated as to whether this was a complicated new chemistry or evidence of a Martian biology.

The tests went somewhat as follows.

Three experiments were run at the two landing sites through several cycles. One of the experiments looked for evidence of micro-organisms in the Martian soil that could metabolize simple compounds of carbon. These carbon compounds, which formed a nutrient solution, were labeled with atoms of radioactive carbon 14 so that their presence could easily be detected. A sample of Martian soil was placed in a test chamber and a small amount of the nutrient medium was dripped onto the sample. It was then incubated for several days. This first injection of nutrient was sufficient only to provide a humid condition; a later injection added more nutrient. Any micro-organism in the Martian soil would be expected to consume the nutrient and later respire gases containing the labeled carbon.

All tests with this experiment showed release of radioactive gas, but hardly any release if the sam-

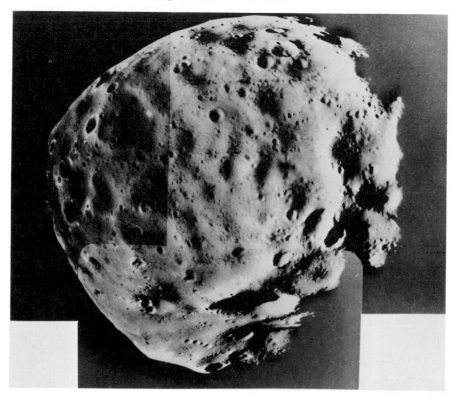

The spacecraft sent to Mars have also thrown new light on the tiny satellites of that planet. This unusual picture of Phobos shows its heavily cratered surface from a distance of 300 miles (480 kilometers) as seen through the cameras of Viking. The spacecraft produced even more detailed close-up pictures later during its mission. The south pole of the satellite is situated within the large crater Hall (named after the discoverer of the Martian satellites), which is at the bottom center, where the pictures overlap. Some features as small as 65 feet (20 meters) across can be seen. There are striations, crater chains, a ridge, and small rocklike objects that seem to be resting on the surface. These hummocky objects are about 165 feet (50 meters) across. They may be debris from impacts that has drifted back to the surface. The gravity of Phobos is insufficient to bring them back at high speed to form secondary craters. (Jet Propulsion Laboratory/NASA)

ple was first sterilized by heating. This experiment seemed to show the presence of micro-organisms in the soil of Mars.

Another experiment took a sample of Martian soil and placed it in a simulated Martian atmosphere containing the radioactively labeled carbon 14. The sample was incubated under simulated Martian sunlight. If there were micro-organisms in the soil they would be expected to incorporate some of the carbon from the atmosphere into their bodily structures. Later, when the soil was heated to break down organic molecules, the labeled carbon should be detected. This experiment also produced positive results for an active cycle and negative results from a control cycle where

the sample was sterilized by heating. At first it seemed to indicate biology, but later tests produced less positive results.

The third experiment placed a sample of Martian soil in a test chamber, providing it first with a humid atmosphere and later with nutrients and observed changes to the atmosphere that might take place because of the presence of micro-organisms. Again there were positive results. The Martian soil caused changes to the surrounding atmosphere when it was humidified. Carbon dioxide and oxygen were released in relatively large amounts and quickly. But the release stopped just as quickly, which suggested that it was due to a chemical reaction rather than micro-organisms in the soil. While the release of carbon dioxide could be explained as resulting from the gas having been absorbed into the soil from the Martian atmosphere, the release of oxygen seemed to require some chemical reaction between the soil and water. When the sample was wetted, some more carbon dioxide was released but the oxygen gradually disappeared. Scientists were sure that these observations resulted from chemical and not biological processes.

The biggest strike against Martian life came from an experiment that was not a life-detection experiment as such. Rather, it was an experiment to determine the composition of the Martian soil; it looked for organic compounds like those found in meteorites. The big surprise from this experiment was that it found no trace of such compounds. There were none at either site; there were none on the surface or in samples dug from beneath rocks. Yet if there were micro-organisms on Mars we would have expected to find an accumulation of their dead bodies and waste materials as we do on Earth. There was no such evidence.

This may be because Martian organisms are efficient scavengers and rapidly consume any organic materials lying about. It may also be because there are no such micro-organisms on Mars because there is no life on the planet. But when one reads NASA Release No. 78–14, dated February 1, 1978, one begins to wonder anew:

Scientists working for NASA and the National Science Foundation have discovered living organisms hidden inside rocks in the frozen deserts of the Antarctic.

The discovery—made in the Dry Valleys, a region whose harsh climate resembles conditions found on Mars—significantly extends the known limits of life on Earth, and also carries important implications for the search for extraterrestrial life.

(a)   (b)

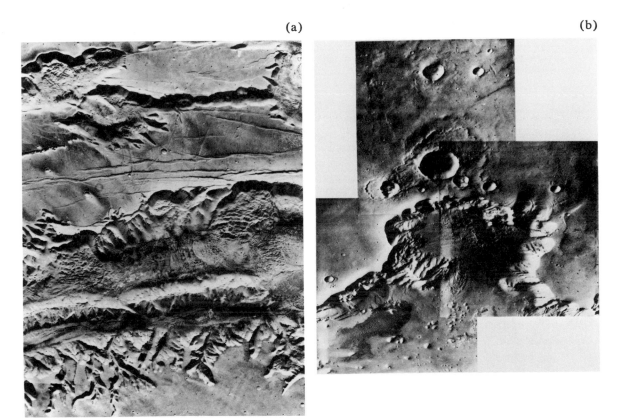

Viking took many thousands of pictures, from orbit and from the surface, that revealed a tremendous variation in the surface features of Mars. (a) *This picture shows the wealth of detail revealed in the floors of the Martian canyons.* (b) *Another type of wider canyon floor with landslides, hummocky hills, and a large tongue of dark dunes (upper right). It will take scientists many years to study all the new details revealed in these pictures.* (Jet Propulsion Laboratory/NASA)

The discovery was made by Drs. E. Imre Friedmann and Roseli Ocampo-Friedmann, of Florida State University, and involved such micro-organisms as bacteria, algae, and fungi. They were found in the air space of porous rocks and in fissures a few millimeters deep.

Back on Mars, all we can say from the Viking mission is that at the two sites in Chryse Planitia and Utopia Planitia, the spacecraft found a chemically reactive soil but no evidence of organic compounds in it. However, these sites were chosen to provide the best chance of landing safely. They were not ideal from the standpoint of biology if life on Mars is not ubiquitous, as on Earth, but is confined to parts of the planet only. The right places to search may be in the fissures of rocks in a volcanic crater or a deep canyon or around the polar cap, places where there is more water or warmer ground. The question of life on Mars was by no means settled by Viking, whose biology instruments were turned off late in May 1977 after some six months of operation. The search needs to go on.

A MANNED MISSION TO THE RED PLANET

While other unmanned missions will undoubtedly follow the Vikings, possibly exploring Mars with automated roving vehicles and arranging for a return of Martian soil to the Earth, the twenty-first century will most likely witness a manned mission to the red planet. And later still there will probably be bases established on Mars, as they once were placed in Antarctica, in which people can stay for years to explore that fascinating planet.

A typical landing operation might begin with the assembly of spaceship modules brought up into Earth orbit by space shuttles. Although the trip to Mars could be made with a single spaceship, two are preferable, since each would be capable of accommodating the other's astronauts in the event of a major malfunction during the mission. If both should arrive safely, maximum advantage would be taken of the full complement of exploration gear carried aboard the two ships.

Each spaceship would weigh about 750 tons and be made up of three nuclear propulsion modules (PMs) plus a crew or mission module (MM). The outer two PMs would commence firing in Earth orbit to place the spaceship on its elliptical path from the orbit of Earth to the orbit of Mars. They would then separate and could be returned to Earth for reuse in another mission. Meanwhile, the main spaceship would progress toward the red planet. The length of the trip would depend on the type of trajectory followed, which in turn depends upon when the mission is actually flown.

En route, the astronauts would conduct many scientific experiments and make observations in readiness for the arrival maneuvers. By the time they enter into orbit around Mars, the weight of their spaceship would be down to about 300 tons because of the earlier separation of the two PMs and the consumption of supplies along the way. The orbiting crews would remain in their ships for several days at least as they prepare their surface exploration craft (the Mars Excursion Modules, or MEMs) for the landing, observe conditions at the landing area, and launch remote-controlled exploratory probes down to the Martian surface. These probes would check atmospheric composition, observe the weather situation, and when they reached the surface would collect soil samples and return them to the orbiting spaceships. In the laboratories of these spaceships the soil samples would be carefully analyzed to guard against there being harmful organisms or chemical substances that might prove hazardous to the landing parties.

When all would have been checked, the MEMs would be released from their parent spaceships and would descend to the ground, somewhat as the earlier Viking landers had. Each MEM would, however, be much larger than Viking. Each would weigh about fifty tons and would provide living quarters for its three-man crew together with a scientific laboratory. While on the surface the astronauts would conduct experiments similar to those of the Apollo crews on the Moon. But there would be at least one significant difference: The Martian astronauts would be looking for remains of early life on Mars if no present life had been discovered to that time. And the crew would check whether terrestrial life could survive on Mars. While at that stage they would take all precautions against contaminating Mars with terrestrial life forms, it may be decided later (if there is no evidence of Martian life) that we should attempt to seed the red planet with life from Earth.

Once their thirty-day exploration period is

over, the crews would ascend to orbit in the upper stages of their MEMs, the lower stages remaining on Mars as testimony of the visit by man. Rock and soil samples, photographs, and other items now aboard the main spaceships, the crews would prepare for departure to their home planet. The single remaining PM on each ship would be fired. The trip home would begin. Months later the spaceships would enter into Earth orbit and rendezvous and dock with a space station. There, the crew would go through quarantine, later to return to Earth's surface on a space shuttle.

This particular approach to manned travel to Mars was first proposed over a decade ago and was based, to a considerable extent, on a concept advanced even before Sputnik 1 ushered in the space age. The technology is available and the mission is neither more hazardous nor more extraordinary than were the voyages of Captain Cook two centuries ago.

When will a manned Mars expedition become a reality? At least a decade and perhaps two after man decides to go and establishes an Apollo-type endeavor to carry it out. Though much technical development would be required, especially in the selection of the propulsion system, the big stumbling block is financial. It may well be that the first trip will have to be international in scope, building for example on the U.S.-U.S.S.R. Apollo-Soyuz experience and that being generated by America's space shuttle and its European Spacelab payload.

The Mars mission may signal the end of the one-flag approach to space exploration.

9
TITIUS 28: A MISSING PLANET?

Before he developed his famous relationships of elliptical planetary orbits, Johann Kepler wondered about the large gap between the orbits of Mars and Jupiter and made a point of writing in one of his notebooks that between those planets there should be another.

In 1772 Johann Daniel Titius developed an empirical rule for the distance of the then-known planets. Start with the series of numbers, he said, in which each number after the first is doubled to give the next number; namely, 0, 3, 6, 12, 24, 48, and 96. Then add 4 to each. This produces the numbers 4, 7, 10, 16, 28, 52, and 100. And if the Earth's distance from the Sun is taken as being 10, then the other planets fit the series remarkably; namely, Mercury 4, Venus 7, Earth 10, Mars 16, Jupiter 48, and Saturn 96. But there is no planet to fit the Titius 28.

A short while later, Johann Elert Bode, who became director of the Berlin Observatory in 1786, publicized this relationship, and in due course it became widely known as Bode's law. The series remained somewhat of a mathematical curiosity until Sir William Herschel discovered the planet Uranus, in 1781. A flurry of excitement ensued because Uranus fitted the next number in the series, namely 196. Astronomers began thinking if one new planet could be discovered by an amateur such as Herschel, maybe there were other planets to be found.

It was not long before Bode had organized a group of German astronomers to look for the missing planet between Mars and Jupiter. The group included such well-known figures as Johann Hieronymus Schröter and Wilhelm Matthäus Olbers. Aided by Baron Franz Xavier von Zach, Bode called a conference of these astronomers, which met at Schröter's Lilienthal observatory. The zodiac was divided into twenty-four zones, each of which was allocated to an individual astronomer. One of the zones was allocated to a professor of mathematics and director of the observatory at Palermo, Giuseppe Piazzi. The exact details of how Piazzi discovered the missing Titius 28 are not clear, but discover it he did. When most other people, including astronomers, were celebrating New Year's Eve 1800, Piazzi, being a monk, decided to check some work he had been doing on a star catalogue of part of the constellation of Taurus. In doing so he saw a faint star that had not been catalogued, a star that was barely visible without telescopic aid. Was it a comet? He was not sure. So he wrote to Bode. Piazzi, it seems, had not attended the conference at Lilienthal and a letter was on the way asking him to look for Titius 28. But well before he could have received the letter he had already discovered the object in Taurus and had written to Bode about it.

Bode was excited when he received Piazzi's letter and immediately contacted Von Zach. They were certain it was not a comet. The great mathematician Carl Friedrich Gauss was consulted, and through this discovery directed his mathematical talent to astronomy. From the few observations of Piazzi he computed the orbit of the object and showed clearly that it was moving around the Sun between the orbits of Mars and Jupiter, at a distance that fitted the Titius 28 position. Everyone was elated. A second new planet had been added to the Solar System. Piazzi named it Ceres.

The new planet was lost for a time, but the work of Gauss allowed its future positions to be predicted, and exactly one year after its first sighting it was again seen, by Olbers.

Olbers continued to observe the planet to provide Gauss with a better set of co-ordinates for his orbital computations. Then came a big surprise. On March 28, 1802, Olbers saw another planetary object not far from Ceres. It, too, was faint. It was named Pallas by Olbers. Astronomers started to question the faintness and small size of these planets and whether they represented the debris from the catastrophic breakup of a larger body. And were there still other small bodies to be found which might properly be called planetoids?

The questions were intriguing and helped to keep the German group together and motivated

The discovery of minor planets between the orbits of Mars and Jupiter led to questions of how many there might be and the hazards that might be presented to a spacecraft such as that shown here passing through the belt. If the asteroids were undergoing mutual collisions, there might be very large numbers of small particles in addition to those larger asteroids that we see from Earth. There was no way of finding out the true picture in the asteroid belt without sending a spacecraft through it. (NASA)

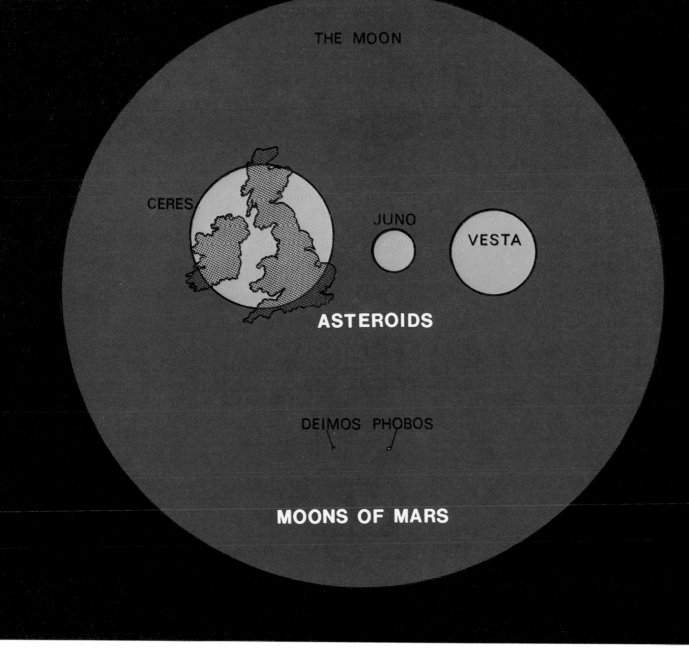

THE MOON

CERES

JUNO

VESTA

ASTEROIDS

DEIMOS PHOBOS

MOONS OF MARS

Even the larger asteroids such as Ceres, Vesta, and Juno are small when compared to our Moon. Most of the minor planets do not even approach their sizes, being more comparable to the two tiny moons of Mars, Deimos and Phobos. (Arfor Picture Library)

for a continuing search. Sure enough, Karl Ludwig Harding found a third small body on September 1, 1804, and named it Juno. And Olbers discovered Vesta on March 29, 1807.

All four were very small bodies relative to the other planets, so small that their discs were difficult to measure. It was not until recent years

that their diameters have become known with any reasonable accuracy, and that as a result of observing the asteroids, as they were more often being called, at wavelengths other than those of visible light. The best estimates today give these four bodies the following diameters: Ceres 635 miles (1,022 kilometers), Pallas 348 miles (560 kilometers), Vesta 312 miles (502 kilometers), and Juno 140 miles (225 kilometers).

After 1807, no further asteroids were discovered for quite some time. The German group ended its search in about 1815, and a full thirty years passed before an amateur astronomer, Karl Ludwig Hencke, discovered another asteroid, Astraea, on December 8, 1845. Two years later, he found Hebe. Many astronomers then joined in the search, and by the end of the nineteenth century hundreds of asteroids had been found— all very small objects. Professor Max Wolfe, of Heidelberg, used photography of star fields to search for asteroids and discovered more than 230 of them. In the 1890s one German astronomer referred to the number of these bodies as the plague of the minor planets. Today we know of about eighteen hundred named bodies plus another fifteen hundred recorded in ephemerides and many more seen once or twice and then lost. To cope with this "plague" a Minor Planet Center was established at Cincinnati Observatory to co-ordinate asteroid studies and act as a clearinghouse for information about them. It publishes circulars, keeps files on world-wide observations, and computes orbital elements and ephemerides.

But despite their tremendous number, the asteroids do not even add up to a mass as large as a planet. In fact, astronomers suspect that the total mass is much less than that of our Moon.

A PLANET THAT NEVER WAS?

Do these little worlds represent debris from the breakup of a planet? Or are they the parts of a planet that never was, a planet that for some reason did not form?

Michael W. Ovenden, a Canadian astronomer in Vancouver, British Columbia, has been investigating the long-term stability of the orbits of the major planets. He finds that any serious disturbance to the gravitational field of the system ceases to show its effect after about 200 million years— the time it takes the system to relax. Ovenden finds that the Solar System is not relaxed now and that it shows evidence of a major disruption as recently as 16 million years ago. At that time, he says, a body at the distance of the asteroid belt was broken into fragments and dispersed. That body was ninety times more massive than the Earth—about the same size as Saturn.

This theory has been invoked by Thomas C. Van Flandern, of the U. S. Naval Observatory, to explain the origin of a peculiar class of comets that he found. Van Flandern studied over five hundred comets that have orbits so large that their periods of return to the Sun must be measured in millions of years. He found that sixty of these comets pass through the plane of the planetary system at the distance of the asteroid belt, at Titius 28. Further, if one could look at them from the Sun as they passed northward or southward through the plane of the planetary orbits, these crossover points (or nodes) seemed to cluster around the same part of the sky. Now, if a comet moves from a great distance into the heart of the planetary system, it is tugged and pulled by the planets in ways that depend upon their positions in the Solar System at the time of the comet's return. As a consequence, the comet's path is disturbed and it does not follow the same orbit to the limits of the Solar System as it did on the way to the Sun. And each time a comet returns, the planets are in different configurations and the comet's orbit is affected differently.

Van Flandern believes that his class of sixty comets have not had their orbits disturbed in this way because they have made only this one return to the inner Solar System; it is their first time around their orbits. He traced each comet's orbit back into space taking into account known perturbations and was surprised to discover that all the comets seemed to have left one point in space in a great hurry and at one specific time. They seem to have originated in a common explosive event that took place 16 million years ago. This coincidence has piqued the curiosity of many astronomers. If there was such an explosion violent enough to produce these comets, what could have happened to the rest of the material that made up the ninety Earth masses of the original body? Surely the amount of energy released would have evaporated the icy material of the comets rather than pushing it out intact along cometary orbits?

We know that the inner planets were indeed bombarded from space sometime in their past. But most other evidence points to this bombardment having taken place billions, rather than millions, of years ago. It is difficult to answer these questions until we have sampled more thoroughly the surface of the other worlds of the Solar System. But if we do find out that the theory is valid and a planet indeed blew apart some 16 million years ago, we are then left with an even bigger question: What caused the explosion?

The alternative is that the asteroids did not originate from a broken-up planet but, rather, represent material that never made the planetary grade. To this end scientists have compared the rotation periods of the planets and the larger asteroids. Except for the Sun-dominated inner planets, Mercury and Venus, the periods of the planets range from about ten to twenty-five hours. Pluto, which may be a special case of a satellite dislodged from another planet, has an unusual rotation period: six days nine hours. The periods of asteroids for which we have reasonable data range from approximately three to eighteen hours, the average being eight hours. There are a few exceptions that rotate faster (Icarus is 2.27 hours) and slower (an asteroid numbered 532 has a rotation of 18.8 hours).

This relatively narrow range of periods seems unusual when the asteroids cover such a large range of sizes. Perhaps even stranger is the fact that the rotation rates of the asteroids are not very different from those of the planets. If the asteroids were indeed products of a disintegrated world, their rotation rates would not likely be similar to those of the planets but more randomly spread over a wider range. So astronomers have concluded that the asteroids may be planetesimals —tiny, embryonic planets—created at the same time as the larger planetesimals that fell together to form the planets. Why they never became a full planet in unknown, but it is possible that the massive Jupiter may have had a disruptive effect and prevented the accretion of planetesimals that would have led to a planet between Mars and Jupiter. We know, for example, that even today Jupiter affects the distribution of the asteroids, and that small groups of them move ahead and behind Jupiter in its orbit around the Sun.

THE EFFECTS OF JUPITER

Whatever the detailed history of the asteroids, their total mass today is estimated at a thousandth or less that of Earth. If the smaller asteroids are continuously fragmenting, as many investigators believe, more and smaller bodies are ever being formed. However, velocity changes resulting from collisions cause small asteroids and meteoroids to be continuously ejected from the asteroid belt, so that over the long period it must be losing mass. But as the mass in the belt is reduced, the number of collisions may be dropping too, so that the rate of mass loss will be reduced and the belt will never be clear of asteroids. The ejected material is swept up by the planets and their satellites and some falls into the Sun.

Other than the processes of collision, perturbations by the planets—Jupiter in particular— affects the stability of the asteroid orbits. One of the results of the giant planet's influence is the presence of gaps in the belt of asteroids, which have been named the Kirkwood gaps, for Daniel Kirkwood, mathematics professor of Indiana University who in 1858 calculated that asteroids could not revolve in orbits around the Sun that had periods equal to simple fractions of that of Jupiter. Such gaps have been found. There are few asteroids in orbits with periods of one third, two fifths, and one half that of Jupiter, namely, 4, 4.8, and 5.9 years. Such orbits would have a radius of approximately 2.5, 2.95 and 3.3 times that of the Earth's orbit. Not all orbits that are simple fractions of Jupiter's lead to gaps. Some actually lead to a stability. Thus, the period of a group of thirty-three asteroids (called the Hilda group after the name of one of its members) is two thirds that of the Jovian period. For the Hungarian group of twenty-four asteroids, the orbital period ratio is two ninths. There is even one asteroid, Thule, that has a period three quarters that of Jupiter.

The dynamic effect of Jupiter is revealed in two other groups of asteroids, known as the Trojans; they orbit in the same period as Jupiter, one group leading and the other following the giant planet in its orbit around the Sun. These asteroids keep a position in space that is always ahead or behind Jupiter by a distance equal to their distance (and that of Jupiter) from the Sun. These positions in space are known as Lagrangian equilibrium points and have come into prominent notice in recent years because of the suggestion of placing space colonies at similar points relating to Earth and the Moon.

They have, however, been known since 1772 as a mathematical curiosity. Their name derives from that of Joseph Louis Lagrange, the famous French mathematician who became director of the Berlin Academy of Sciences. In looking at the problem of how three astronomical bodies might move in space under the influence of their individual gravitational fields, he found a solution if one of the bodies could be assumed to have negligible mass and therefore negligible gravity of its own. And he identified positions in space that possessed unusual attributes. He termed them libration points. There was one position inside Jupiter's orbit at which a body might revolve around the Sun in the same period as Jupiter, and one similar point outside of the orbit of Jupiter. These he termed colinear libration points. A third, small body at either of two points would, however tend to drift away from them. In contrast, there are two points ahead and behind Jupiter, the corners of equilateral triangles with Sun and Jupiter at the other two corners, where a body would not tend to drift away. These he called triangular libration points, or equilaterals. The first two he called L_2 and L_3; the equilaterals he called L_4 and L_5, L_4 being ahead of and L_5 behind Jupiter in its motion around the Sun.

In 1906, C. V. L. Charlier, of the Lund Observatory, discovered that the asteroid Achilles traveled in the same orbit as Jupiter, at L_4. And soon afterward the asteroid Patroclus was discovered at L_5.

These new asteroids were named differently from those discovered earlier: a precedent was established when they were given Trojan War hero designations. Earlier asteroids had been named with classical female names, and when these had become scarce, female names that sounded classical were added. Actually these female names had started to run out, and when an asteroid was discovered that orbited between Earth and Mars it was given the name Eros by its discoverer.

The Trojan names led to the asteroids at the Lagrangian points often being referred to as the Trojans; actually they carried both Trojan and Greek male names. While only fifteen are officially catalogued, many more have been identified. In fact, C. J. Van Houten suspects that up to seven hundred asteroids brighter than magnitude 20.5 are situated in or around the L_5 position alone. They do not, of course, occupy exactly the same position but follow small orbits centered on the Langrangian point. The Trojans that are catalogued are relatively large asteroids, and many are irregularly shaped as determined by the way their brightness varies as they rotate. For example, Hektor is about 70 miles (113 kilo-

meters) long and 26 miles (42 kilometers) wide, while Patroclus has a diameter of about 170 miles (273 kilometers).

THE WANDERERS

The majority of asteroids move in fairly circular orbits around the Sun, but their orbits are a little more elliptical than those of the planets. And the inclinations of their orbits to the plane of the ecliptic, which is the plane of Earth's orbit, is on the average much greater than those of the planets. Some asteroids have orbits that are inclined as much as 67°; the greatest inclination of the inner planetary orbits is that of Mercury, which is only 7°. Pluto, in this, too, is something of an oddball, with an inclination of 17°.

While most asteroids orbit in a belt between 2.1 and 3.3 times Earth's distance from the Sun, there are exceptions in addition to the Trojans. Hidalgo reaches out almost to the orbit of Saturn at its aphelion and plunges almost to the orbit of Mars at perihelion. Its average distance is 5.8 times Earth's distance from the Sun. By contrast, Icarus has an average distance of only 1.077 times Earth's distance from the Sun.

These erratic asteroids, wanderers from the general belt, have been known for over a century. James C. Watson, of the Ann Arbor Observatory, discovered Aethra on June 14, 1873, and it was soon apparent that its perihelion was inside the orbit of Mars. By the mid-1970s, over thirty asteroids that crossed the orbit of Mars had been catalogued. Probably these errant bodies originated from orbits being distorted by planetary perturbations and collisions between asteroids.

A most exciting discovery came in 1932. A Belgian astronomer named Delporte reported he had discovered an object moving toward the Earth. Astronomically it was close. In fact, it came closer than 10 million miles (16 million kilometers) by March 21 of that year. A month later, Karl Reinmuth, of Heidelberg, discovered a 1-mile (1.6-kilometer)-diameter asteroid that came to within 6 million miles (10 million kilometers) of Earth. It was named Apollo, which in turn became the group name for asteroids that cross Earth's orbit. Then it was lost for another forty-one years, until rediscovered by Richard E. McCrosky and Cheng-Yuan Shao, of Harvard

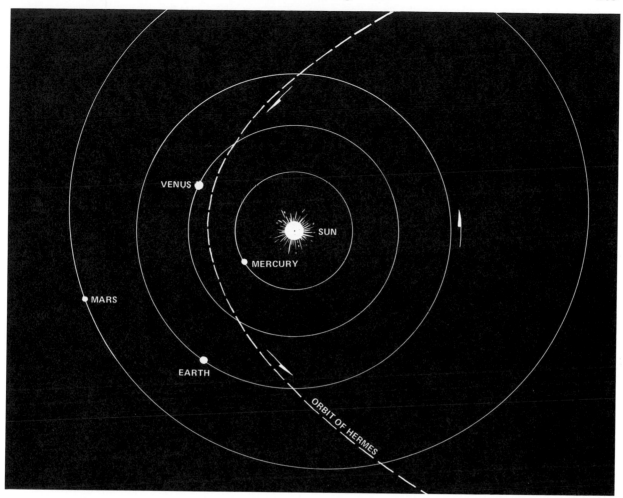

All asteroids are not confined to the belt between Mars and Jupiter. Tiny Hermes, for example, less than 1.25 miles (2 kilometers) in diameter, can pass within 600,000 miles (1 million kilometers) of Earth. It has a short period relative to the other asteroids, orbiting the Sun in 545 days as it follows a path that carries it outside the orbit of Mars and inside that of Venus.

Observatory, on March 28, 1973. This small body has its perihelion inside the orbit of Mercury.

By the mid-1970s, some fourteen Apollo asteroids larger than 0.6 mile (1 kilometer) in diameter had been discovered. Astronomers suspect that there may be fifty or even one hundred of these Earth-crossing asteroids, and hundreds, perhaps thousands, of smaller, meteoroid-size bodies as well. In 1970, for example, the Lost City meteorite, which hit the ground in Oklahoma, was following an Apollo orbit. So was a 10-foot (3-meter) -diameter body that smashed into the surface of the Moon in May 1972 with an impact that shook the seismometers at four of the sites established by Apollo spacecraft astronauts. The vibrations continued for about an hour.

There has been renewed interest in trying to track these bodies in recent years, because one of them may someday impact on the Earth. While there is very little that could be done to avoid the resultant catastrophe—a ½-mile (1-kilometer) -diameter object might devastate a whole state—the poised nuclear weapons of world powers make it imperative that the nations should know in advance of the asteroid danger and not think that it is the beginning of an atomic attack. When some astronomers started such a search, they rapidly located new members of the group.

The characteristics of these wanderers suggest that many of them may not be true asteroids but, rather, extinct comets whose nuclei travel along these elliptical orbits. Audouin Dollfus, who has specialized in obtaining information about the surfaces of planetary bodies by the way in which they reflect light, suggests that Icarus has a surface that appears to be a fluffy, loosely bound together mass of dust. But his data would also fit a body consisting of stones embedded in ice, like a cometary nucleus. Moreover, Icarus has an almost spherical shape, which might also suggest that it is a dissipating cometary nucleus, analogous to a melting dirty snowball. Those Apollo asteroids, however, that have irregular shapes may be true asteroids that have been dislodged from the belt by collisions.

AND IF A WANDERER HITS THE EARTH?

About thirty thousand years ago in the desert near what is now Winslow, Arizona, not far from the Grand Canyon, groups of primitive people who had just begun to spread across the North American continent after their hazardous crossing of the ice bridging the Bering Strait, witnessed an awesome spectacle. Without warning, a whole section of sky erupted in brilliant light. The desert exploded. Molten rock bombs shot high into the air. Curtains of debris swept across the desert for miles, raising immense clouds of dust and punctuated by falling rocks that dug craters from the size of little pits to great gaping holes. The ground shuddered as primitives quaked in wonder.

The dust gradually settled after a series of spectacular sunsets. And when it did, those frightened people who remained in the area saw a strange, flat-topped mountain from which eerie glows reddened the night sky. It may have been years or decades before a few brave and adventurous individuals climbed the mount of the fire god and looked down into an immense crater, 1 mile (1.6 kilometers) across and almost half a mile deep.

These people had witnessed the impact of an Apollo asteroid on the Earth. Still there in the desert today, the crater is a mere shadow of its original grandeur, because it is now half filled with silt and its walls have been eroded. Many other craters have now been discovered elsewhere on the Earth, but most have been weathered so much that they could not have been recognized until aerial searches were made. And strangely, the search was triggered by the space program, because scientists wanted to know as much as possible about big impact craters before we went to the Moon.

We may even have experienced one of these impacts in recent times. On the morning of June 30, 1908, there was a tremendous disturbance in a remote region of central Siberia. It centered near the small village of Vanovara and devastated nearly 800 square miles (2,000 square kilometers) of forest. People and animals were knocked over and houses shaken over 30 miles (50 kilometers) from the village. Peasants clapped hands over their ears to shut out the deafening explosions in villages 400 miles (650 kilometers) from Vanovara. But the area was so remote and political conditions in Russia so volatile that no thorough scientific investigation of what happened could be made until half a century later; then much of the important evidence had been lost and memories dimmed by the years. Kirill P. Florensky led three well-organized expeditions, in 1958, 1961, and 1962. He concluded that a comet nucleus impacted the Earth and caused the 1908 explosion. Among the reasons for his conclusion he cited ". . . the unusually loose structure (of the nucleus), which led to breakup in the atmosphere; the dust trail pouring away from the Sun, that caused unusual sunsets over nearly all Europe; the nature of the orbit; and the lack of . . . fragments." The loose structure of the nucleus was inferred because the object exploded high in the atmosphere and did not dig a crater. He estimated that the cometary nucleus was moving at about 3 miles (5 kilometers) per second when frozen gases in its head were melted by atmospheric friction as it plunged toward the Earth's surface. Had it been an Apollo asteroid like the one that blasted the meteor crater in Arizona the effects would have been far more serious. It has been estimated that a body having a diameter of 0.6 mile (1 kilometer) of the metallic rocky material of a true asteroid would blast a crater from 6 to 10 miles (10 to 15 kilometers) across, and would obliterate an area up to 5,000 miles (8,000 kilometers) square.

The Vanovara incident has, however, a number of mysteries connected with it. Some astronomers have asked why the comet was not seen before it impacted. Other scientists have asked why the object approached Earth so slowly. Eyewitness accounts given to members of earlier expeditions told of an explosion and column of fire and smoke very similar to those associated with nuclear weapons. Speculation has run rife; the explosion over Siberia has been explained by encounter with Earth of a small piece of antimatter —matter in which all subatomic particles have electrical charges opposite to those in our matter —or with a mini black hole, in which normal matter has supposedly been concentrated almost into a mathematical point and its intense gravity effectively pulls surrounding matter apart as it falls into the black hole. There have even been speculations that the Vanovara incident represents the failure of an extraterrestrial spaceship trying to make an emergency landing on our planet. But all the explanations are highly speculative and none have been proved. The best evidence we have are photographs obtained by an early explorer of the region who made several trips during the period before World War II. These pictures show great areas of devastation in which trees have been pushed over radially from a site some 40 miles (60 kilometers) from Vanovara. But no crater such as that in Arizona was found.

Undoubtedly there is always the chance that a mile-sized object from space will impact the Earth. However, in addition to the Apollos there is another class of asteroidal bodies that approach us, though they do not actually cross our orbit. The first of these bodies, called Amor, was discovered by Eugene Delporte, of Belgium, on March 12, 1932. It approaches within 9.3 million miles (15 million kilometers) of Earth and appears to be about half a mile in diameter. A more recent member of this group approached the Earth to within 10 million miles (16 million kilometers) in 1973.

In addition to Trojans, Mars-orbit crossers, Apollos, and Amors, there are other groups of asteroids. Kityotsugu Hirayam, the director of Tokyo Observatory, conducted detailed studies of asteroids during much of the first three decades of this century. In 1918 he identified a number of main-belt "families" by their orbital elements. Naming them after the brightest asteroid in each group, he classified the families according to their distance from the Sun. He discovered fifty-seven asteroids moving in orbits at 2.2 times Earth's distance. He called them the Flora family. The

Maria family has thirteen asteroids at 2.5 times Earth's distance. The Koronis family, fifteen at 2.9 times Earth's distance. There are also an Eos family, of twenty-three members, and a Themis family, of twenty-five members. These are at distances of 3.0 and 3.1 that of Earth, respectively.

Many astronomers believe that the members of these families are the remains of large masses that broke up as a result of collisions between asteroids. One possibility is that a fairly large asteroid—now the brightest member of the group— was crashed into by smaller bodies. Whereas on a big world such as the Moon, impacts produce craters and the debris falls back to the surface, the gravity of an asteroid is insufficient to hold on to the cratering debris. Thus, it may now make up the other members of the family group. Families that do not have one giant member might result from the total disruption of an asteroid that would have been caused if the impacting body was a large body.

More recently, other astronomers have identified more of these asteroid families. Dirk Brouwer and J. R. Arnold found six, N. M. Shtaude found fifteen, and many more were discovered as a result of a special survey of minor planets made in 1970.

An important new tool for astronomers has been the powerful digital electronic computers that have been developed during the past decade. With such instruments astronomers are able to calculate backward and forward in time to show how the paths of astronomical bodies are affected by others. One of these activities has been applied to calculations of the orbits of asteroids under the influence of the perturbations of the major planets. While such calculations have revealed some interesting characteristics of orbits of the asteroids, such a long-term stability of the perihelion but considerable changes in the inclination of the orbit, no consistent picture has been obtained to throw light on the origin of these bodies. The big question still remains: Did the asteroids originate from the breakup of a planet, or are they particles that should have formed a planet but were prevented in some way from doing so? And can asteroids have satellites? Edward Bowell and Michael A'Hearn of Lowell Observatory, Keith Horne of Caltech, and James McMahon, an amateur astronomer, believe that they have discovered a tiny 30-mile (45-kilometer) -diameter moonlet revolving around the asteroid Herculina, at a distance of slightly less than 1,000 kilometers. It may take a spacecraft to confirm this unexpected "discovery."

If we could answer this question—and we might do so by sending instrumented spacecraft to these tiny worlds to photograph their surfaces and perhaps obtain samples of their material—we might also come close to understanding how the planets formed. We are still in doubt as to the sequence in which the planets formed, whether they fell together in layers of various materials that condensed from the ancient nebula at different times, or whether they formed from a well-mixed material that subsequently separated into light and heavy materials when the bulk of the planet heated after its formation. Another important question is how the accretion, or falling together, took place—whether the material condensed from the nebula around a few large centers, or whether it formed all over the Solar System around smaller centers that later formed into streams and later still coalesced into the large bodies.

ASTEROID MISSIONS

Many tasks await the first spacecraft to be dispatched to the asteroids. A mission to Eros, for example, might clarify part of the history and evolution of our Moon. If the Moon were once a separate planet before it became a satellite of Earth, it may have followed an Eros-type, Earth-crossing orbit. Eros and the Moon might thus belong to the same family.

Because asteroids are thought to be too small to have melted since their formation, an examination of their material might throw light on the composition of the Earth and other planets before they heated and melted because of their stores of radioactive elements. Probes to the asteroids would also provide firm information about sizes, rotation rates, masses, magnetic fields, densities, and isotopic composition. They could also check for evidence of prebiological organic molecules of the types discovered in meteorites, thus providing other links in the chain of events that might explain the origin of life in our Solar System.

While some astronomers would be excited to see a mission to a larger asteroid such as Ceres, off of which radio astronomers were successful in bouncing radar echoes for the first time in 1977, others would prefer to send the first missions to smaller bodies. Selecting a large body would be important because it might represent an original asteroid that had not been fragmented by collisions. However, other astronomers say that the likelihood of an asteroid having passed through 3.5 billion years or so without a collision is not very great. Moreover, an expedition to a broken body might be even more important because its surface may provide a historical record analogous to the geological record on the walls of the Grand Canyon. If we could find a body that has been broken cleanly in two, we might be able to view the layering sequence of formation of the original body. But most astronomers are agreed that a close inspection of almost any asteroid should lead to exciting if not surprising results.

Whatever the targets, and whenever the decision may be to dispatch spacecraft, low-mass asteroids can best be approached by low-thrust advanced propulsion systems based on ion, plasma, or arc-jet systems. These are essentially rocket propulsion devices that eject a stream of ionized particles instead of the chemical molecules of the conventional rocket engine. Ernst Stuhlinger, a pioneer in electrical propulsion systems, has described some of the advantages as ". . . a high incremental velocity, an almost unlimited re-ignition capability, a large payload fraction, a long operating lifetime, and a sizable electric power source available for data transmission after the target has been reached." Basically, the high velocity of the electrically produced jet means that much less propellant has to be consumed for a given journey than when a conventional chemical rocket engine is used. And the electrical engine is much easier to start in space; it does not require reignition devices. The electrical generators needed to produce the jet can later be used to send large quantities of data to Earth.

The total mass of such a spacecraft when it leaves Earth for an asteroid—actually from an orbit around the Earth to which it will have been carried by a space shuttle—is two to three times smaller than that of a chemical rocket that would be required for the same mission.

A particularly enticing target for such a mission is Eros, which is thought to be a 14×4-mile (22×6-kilometer) fragment of a larger asteroid. Mission planners have calculated that the spacecraft would take about three years to journey to the planetoid, land on it, conduct experiments and gather samples, and return it to Earth orbit. Since the rotation period of Eros is only 5.5 hours, its extremities must be moving at about 11.5 feet per second (3.5 meters per second), adding some complexity to the landing maneuver, but those who have studied such missions in detail have no doubt that a landing could be made.

Another exciting mission would be to Icarus, a small world that follows such an elongated orbit that its surface temperature varies by 900° F (500° K) between aphelion and perihelion. Its small size and high velocity would make this a more difficult mission than that to Eros. Somewhat easier would be a mission to one of the big four, Vesta. Opportunities to send missions to the regular asteroids are available fairly frequently. The more elliptical the orbit of the asteroid, the less frequently opportunities to visit it from Earth come around. A good time to launch a mission to Vesta would be in 1985. Missions to Juno, Pallas, and Ceres had good opportunities in 1976, 1979, and 1984, respectively. As these dates pass and missions are not prepared, we have then to wait for the next suitable times.

Instead of aiming for one single asteroid, spacecraft might also be sent on a mission during which a single spacecraft would pass fairly close to several without stopping at any of them to collect samples. Such a mission would be like those of the early planetary fly-bys on which we

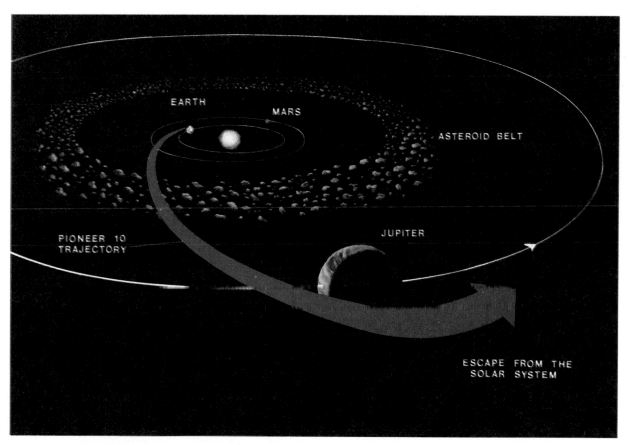

Pioneer 10, man's first Jupiter probe, passed through the asteroid belt during a 200+-day flight that began in mid-July 1972 and ended in mid-February 1973. Surprisingly, the belt was discovered to be so free of micrometeoroids that one experimenter concluded it would present "little hazard to future spacecraft going through to explore the outer planets." Months later, on August 18, 1973, the companion Pioneer 11 Jupiter probe began its seven-month journey through the asteroid belt, exiting on March 12, 1974. (Arfor Picture Library)

would be photographing their surfaces and measuring their physical characteristics, such as size and mass. The spacecraft would fly by each asteroid at about 3 miles (5 kilometers) per second at a distance of about 9,300 miles (15,000 kilometers), thereby providing about one and three quarters hours of good observations at each asteroid.

HOW MUCH DEBRIS?

Astronomers and deep-space-mission planners are extremely interested in the density of the shattered remains of asteroids that may be present in the asteroid belt. If there is a tremendous number of much smaller, invisible bodies, the belt could present a real hazard to spacecraft passing through it. There is no way of finding this out from Earth. About the middle of the belt, where particles orbit the Sun at a speed of 38,000 miles (61,200 kilometers) per hour, a particle with a mass of only one thousandth of a gram—a mere speck of dust—would penetrate a sheet of aluminum one centimeter thick.

The first spacecraft to venture into this region was Pioneer 10, on its way to Jupiter. About mid-July 1972 the spacecraft entered the belt and everyone waited expectantly. The particle counters on the spacecraft did not show any major changes. The belt did not seem to be permeated by debris. By mid-February 1973 Pioneer 10 emerged unscathed from the 155-million-mile-wide (250-million-kilometer-wide) belt. Spacecraft designers, already thinking about plans for sending much larger spacecraft into the outer Solar System, continued their planning with renewed enthusiasm. But perhaps Pioneer 10 had

been lucky. In mid-August 1973 Pioneer 11 started into the belt. It, too, emerged safely, in mid-March 1974. The belt did not seem to present any major hazard to spacecraft.

The number of small particles in the belt did not depart significantly from elsewhere in the Solar System. Now a new question arose, as to what happens to the particles if the asteroids are, indeed, continually colliding with one another. If such collisions are taking place, then the small particles must somehow be swept out from the Solar System. A possibility is that solar radiation is doing this. However, this could apply only to a limited range of particle sizes. Another possibility is that heating of one side of the particle coupled with radiation into space from its shadowed side would cause the particle to spiral in toward the Sun—what is known as the Poynting-Robertson effect.

Whatever the mechanism that is responsible, the experiments carried out by the two Pioneer spacecraft showed that the asteroid belt does not present a hazard to spacecraft, except from collision with the larger bodies, and if collisions between the asteroids occur, the debris is continuously being cleaned up from the belt. But the question is crucial, because we really want to know whether or not the collisions are taking place and whether or not the belt is a breaking down of larger bodies or a delayed attempt to form larger bodies from smaller bodies. As yet, we have no firm answer either way.

But the path through the belt had been opened by the Pioneers, and soon other, larger spacecraft were following. By the twentieth anniversary of the orbiting of the first Earth satellite, the exploration of the outer Solar System had started.

10
THE SYSTEM WITHIN THE SYSTEM

GAS AND ICE SPHERES OF BROBDINGNAG

An alien starship visiting our Solar System would see things very differently from our Earth-centered view. As we expanded our awareness of other celestial bodies we paid most attention to the worlds closest to us: the Moon, Mars, Venus. This was understandable, because we could observe them most easily. And even with the advent of the space age, we again concentrated on the small, inner worlds of our Solar System, because they were the easiest for our spacecraft to reach.

The occupants of a spaceship approaching the Solar System from outside would see things from another perspective. For they would first be aware of a group of large planets orbiting the Sun at a distance of between some 460 million miles (740 million kilometers) and 2,800 million miles (4,480 million kilometers)—two gas giants and two giant spheres of ices. As they inspected these planets, they would doubtless be intrigued by their satellites. Even the outermost of the giant planets, with only two satellites, presents unusual features. But the inner gas giants have retinues of many worlds. And if the aliens looked into the inner Solar System they would see some small worlds that might appear to be escaped satellites of the giant planets, rather than planets in their own right.

Beyond the asteroid belt is the remote outer domain of the solar empire, a region in which, after the Sun, most of the matter of the Solar System is concentrated. There, instead of the rela-

tively small, hard-surfaced worlds that we inspected during the first two decades of space exploration, we find the four giants, with their thick and—to us—noxious atmospheres, and a small, dense, and incredibly remote planet about which little is known. The environments of the giant planets—Jupiter, Saturn, Uranus, and Neptune—seem so inhospitable that they will likely remain beyond the direct exploration of man forever. But their families of satellites are more like the inner planets of the Solar System and offer potential for direct human exploration by future space expeditions.

We have learned much about the atmospheres of the giants, especially Jupiter, which is nearest to us, and we have reasonable theories of what the interiors are like. The thirty-one known satellites are mainly airless bodies, though some do possess atmospheres. We know that there are details on the surfaces of these worlds, but such details are as elusive as the features of Mercury had been to Earthbound telescopes. This is because the satellites are so far away from us compared with the distances of the terrestrial planets. Jupiter and its family of smaller worlds never comes closer to Earth than 365.9 million miles (588.7 million kilometers), compared with Mars's *greatest* distance from Earth of 249.3 million miles (401.1 million kilometers). Saturn is never closer than 742 million miles (1,194 million kilometers); Uranus, 1,605 million miles (2,582 million kilometers); and Neptune, 2,674 million miles (4,303 million kilometers).

THE JOVIAN SYSTEM

As its name suggests, the planet Jupiter is the most important object in the Solar System (except, of course, for the Sun). It has over 1,300 times the volume of Earth and is nearly 320 times more massive. In fact, if placed on one side of an imaginary balance, we would have to have all the other planets twice over on the other pan to approach a balance. After Venus and Mars, it is the brightest object in the night sky. It is also large enough that, despite its distance, it is a rewarding object for study with even relatively small telescopes.

There are three things that an observer sees almost immediately when looking at Jupiter through a telescope: the bright disc of the planet is crossed by darker belts paralleling the planet's equator, the disc is decidedly flattened at the poles and bulging toward the equator, and the planet is accompanied by several starlike companions. If you continue to watch for several days, you will see that these starlike companions move from side to side of Jupiter, and you will ascertain that there are four of them. And during this period, if your telescope is large enough to show details on the dark belts, you will observe that the planet rotates rapidly on its axis, completing one revolution in just under ten hours.

Because it is such a large world and appears so bright in terrestrial skies, Jupiter, like the Moon and the inner planets, has been known from antiquity. Since it revolves around the Sun beyond the orbit of the Earth, it is an exterior planet that,

The planet Earth is quite small compared with the enormous disc of Jupiter; eleven Earths would fit across Jupiter's disc. Jupiter has 1,300 times the volume of Earth and is nearly 320 times more massive.

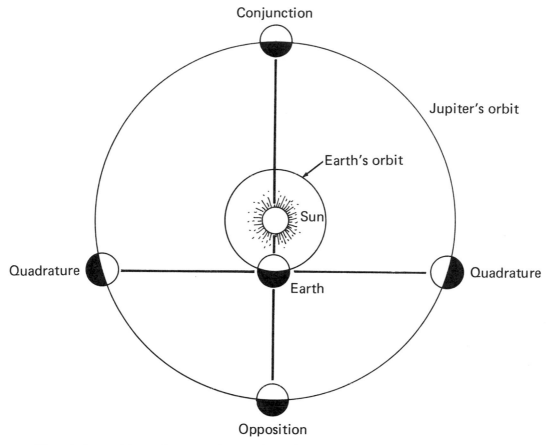

The relative positions of Jupiter, Sun, and Earth are shown when Jupiter is at opposition, conjunction, and quadrature.

like Mars, has conjunctions and oppositions and exhibits a slightly gibbous phase at the quadratures; i.e., it is slightly less than fully illuminated.

In 1610, Galileo was the first to look at Jupiter through the newly invented telescope, and he discovered the four large satellites. This discovery of four bodies orbiting another world had a profound influence on our view of the universe. The fact that four bodies were orbiting a world other than the Earth was a major influence toward a more general acceptance of the Copernican system, which said, contrary to the then prevalent view, that the Earth and the other planets revolved around the Sun. The traditional view was that everything revolved around the Earth as the center of the universe.

Although there is some doubt that Galileo was actually the first to record an observation of the four big satellites (Simon Marius, of Ausbach, Germany, may have actually observed them about ten days before him), they are often referred to as the Galilean satellites of Jupiter. But their names are those suggested by Marius from popular mythology: Io, Europa, Ganymede, and Callisto.

The bands of Jupiter were first mentioned by G. Campani in 1664, but there is no clear evidence of who first noticed the polar flattening.

The banded atmosphere is usually classified by astronomers in terms of a series of zones and belts. Nowadays believed to be flow features of all large, rapidly rotating, semigaseous planets, the zones on Jupiter range from white and pale yellow to varying shades of red, while the belts tend to be slate gray and brown. Giovanni D. Cassini discovered that the banded atmosphere rotates rapidly as a whole. In 1665 he reported that part of the atmosphere travels once around in 9 hours and 50 minutes and another part does it in 9 hours and 56 minutes. Following more than one hundred years of sporadic observations, William Herschel confirmed the two rotational

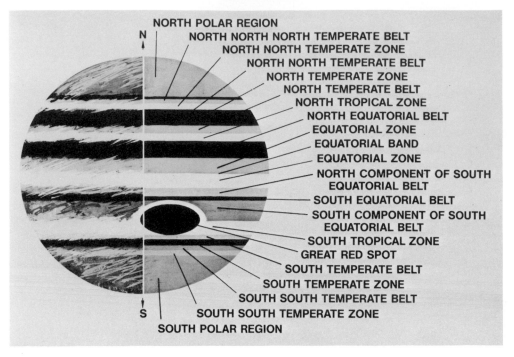

NORTH POLAR REGION
NORTH NORTH NORTH TEMPERATE BELT
NORTH NORTH TEMPERATE ZONE
NORTH NORTH TEMPERATE BELT
NORTH TEMPERATE ZONE
NORTH TEMPERATE BELT
NORTH TROPICAL ZONE
NORTH EQUATORIAL BELT
EQUATORIAL ZONE
EQUATORIAL BAND
EQUATORIAL ZONE
NORTH COMPONENT OF SOUTH
EQUATORIAL BELT
SOUTH EQUATORIAL BELT
SOUTH COMPONENT OF SOUTH
EQUATORIAL BELT
SOUTH TROPICAL ZONE
GREAT RED SPOT
SOUTH TEMPERATE BELT
SOUTH TEMPERATE ZONE
SOUTH SOUTH TEMPERATE BELT
SOUTH SOUTH TEMPERATE ZONE
SOUTH POLAR REGION

Conventional nomenclature of Jupiter's visible surface as used by astronomers. (NASA)

rates, which are now termed systems I and II, for the shorter and longer periods, respectively. The shorter period is confined to the equatorial region of the planet and seems to be caused by a rapidly flowing wind around the equator, moving clouds there some 250 miles per hour (400 kilometers per hour) faster than clouds about 10° north and south of the equator.

The four Galilean satellites form only a part of the enormous Jovian system. We now know that Jupiter has at least fourteen satellites. The fifth satellite was discovered in 1892, the others in later years, with discoveries being made even as spacecraft were flying to the planet. While the Galilean satellites follow fairly regular orbits around Jupiter, the smaller, outer moons trace an extremely complex orbital pattern. The mean distance of the outermost satellite from Jupiter is 14.7 million miles (23.7 million kilometers). The innermost five satellites and three small satellites in an outer group all orbit Jupiter in the normal, prograde way, i.e., in a counterclockwise direction as seen from above the north pole of Jupiter. The outermost group of satellites consists of small worlds in highly elliptical and inclined orbits that move in the opposite (retrograde) direction.

There is another aspect of the Jovian system not visible to the eye: it is the highly complex environment of particles and fields. Early in 1955, two radiophysicists were testing a telescope for a radio survey of the heavens by the Carnegie Institution, of Washington. These two men, Bernard F. Burke and Kenneth L. Franklin, were surprised to find radio waves coming from the area of the sky where Jupiter was situated at that time. Subsequent observations proved that these waves came from Jupiter. The discovery ushered in a new science, of planetary radio astronomy. Further research showed that the radio bursts from Jupiter were associated with a region of space surrounding the planet; Jupiter obviously possessed radio-emitting belts similar to those of the Earth. In turn, this suggested that the big planet must also have a magnetic field. Subsequently it was found that the magnetosphere of Jupiter was in keeping with the size of the planet and many times larger and stronger than that of the Earth.

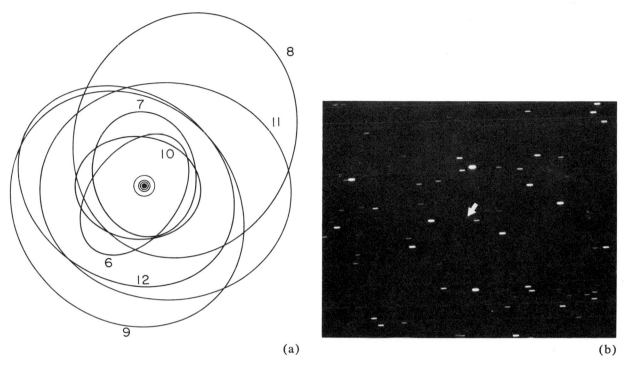

(a)　　　　　　　　　　　　　　　　　　(b)

The outermost satellites of Jupiter follow interlacing orbits to great distances from the planet. (a) The Galilean satellite orbits are the small circles close to Jupiter. (b) A photograph of one of the smallest of Jupiter's moon, tiny Andrastea (J12), caught on a plate made at the U. S. Naval Observatory's Flagstaff Station on April 14, 1958. (U. S. Navy)

The innermost five moons (Amalthea and the Galilean satellites) and the three small satellites in the first outer group all orbit in a prograde direction. The moons of the second outer group orbit in a retrograde fashion. The orbit of Poseidon is highly inclined to the equator of Jupiter so that the satellite is difficult to locate.

A GREAT BALL OF LIQUID

When astronomers calculated the mass of Jupiter, they found out how very different the planet must be from ours. In 1842, Friedrich W. Bessel developed a way to estimate the mass of Jupiter from the manner in which the planet affects the motions of its satellites. He showed that, though the giant planet is 1,250 times the volume of Earth, its mass is only 388 times that of Earth. So it must be made of very lightweight material. Its density was calculated by Bessel as only one and a third times that of water. By contrast, the average density of the Earth is 5.5 times that of water. Bessel's findings were very close to the values accepted today.

Models of the interior of Jupiter must take into account this extremely low density. After years of study, astronomers concluded that Jupiter probably has a dense core of perhaps ten times the mass of the whole Earth. This core consists of rocks and metals and is extremely hot, though not hot enough for hydrogen surrounding it to be triggered into a thermonuclear reaction, as takes place in the hot center of the Sun. The core is surrounded by a deep shell of hydrogen, but it is a very special kind of hydrogen, like none that we find on Earth. It is special because high pressures and temperatures have changed the characteristics of the normally gaseous element. Hydrogen when sufficiently cooled in a laboratory changes from a gas into a liquid. If it is cooled still further, to —434°F (14°K), the liquid freezes into a solid, even at the relatively low pressure of Earth's sea-level atmosphere.

Physicists at the Soviet High-Pressure Research Institute claim to have observed abrupt changes in the density of hydrogen under extremely high pressures. In one test, they reported that the density increased from 1.03 to 1.3 times that of water and explained that it was caused by hydrogen passing into the solid state. Scientists outside the U.S.S.R. are not so certain that metallic hydrogen has been created, however, since they have not been able to duplicate the experiment. Chemists say that a material is metallic when all its atoms are uniformly distributed within the material. At high pressure, the atoms of hydrogen are thrust close together so that they do not pair into molecules, and individual atoms are close to being equidistant from each other with their electrons uniformly distributed throughout the material.

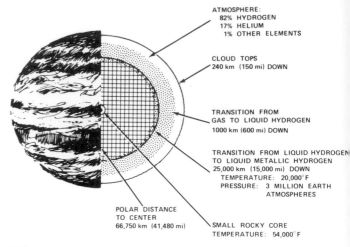

This cross section of Jupiter illustrates our best concepts of what the interior of the planet may be like. Clearly, it is a world made up principally of hydrogen. (NASA)

The gravity of Jupiter is so powerful that the weight of all the material of the planet is sufficient to pressurize an inner shell of hydrogen into the metallic state. This shell of metallic hydrogen may reach almost 80 per cent of the way to the visible cloud tops of the planet. And because the metallic hydrogen is such a good electrical conductor, it makes possible the circulating electric currents that probably generate the strong magnetic field of the planet.

Above the metallic-hydrogen shell is a shell of liquid hydrogen, a great ocean that reaches almost to the visible surface. But as the surface is approached, pressures fall and the ocean of hydrogen gradually merges into an atmosphere of hydrogen gas. At the outer limits of a relatively thin shell of atmosphere are the clouds that make up the belts and zones seen from Earth as the visible surface of the mighty planet.

Even before spacecraft visited the Jovian system, astronomers had a good idea of how pressure and density varied with distance from the visible clouds down to the center of the planet. They had a more difficult time estimating temperature, because it depends on several unknown factors, including the amount of helium mixed with the hydrogen of the planet. Estimates of core temperature varied from a few thousand to tens of thousands of degrees. They also had difficulty understanding why Jupiter appeared to be sending into space two and a half times more heat than it was receiving from the Sun. Some researchers thought that the planet was simply giving off primordial energy from its formation, like the cooling of a hot piece of coal. Others believed that the planet might be contracting slowly, for a shrinkage of only a millimeter each year in the diameter of Jupiter would supply the heat energy now flowing from the planet. Still others speculated that helium, which is heavier than hydrogen, was slowly settling toward the core of the planet and this was releasing energy. Those who wanted a better chance of being right, suggested that all three processes might be taking place concurrently.

Essentially, Jupiter is seen as a great ball of liquid spinning rapidly on its axis so that its polar diameter is shortened to 85,750 miles (138,000 kilometers) compared with its equatorial diameter of 88,760 miles (142,820 kilometers).

BELTS, ZONES, AND SPOTS

The banded atmosphere of Jupiter has long been the subject of special attention. Astronomers carefully recorded the intricate details of the motions of and changes in these cloud formations, generation after generation. And with the application of photography to astronomy, many picture records were produced of the turbulent atmosphere of the giant of the Solar System.

By analyzing the spectrum of sunlight reflected from the clouds of Jupiter, astronomers discovered the presence of hydrogen, methane, and ammonia. They also suspected the presence of helium and water vapor. Most astronomers assumed that the cloud tops were made up of ammonia vapor and crystals, and that the grayish polar regions consisted of condensed methane. They thought that above the cloud tops a transparent atmosphere extended to a height of some 40 miles (60 kilometers) and consisted predominantly of hydrogen. They have, however, been steadily increasing their estimates of the amount of hydrogen in the atmosphere as a whole, and by 1970 they believed it accounted for 60 per cent of the mass of the Jovian atmosphere. In 1974, S. T. Ridgeway, of Kitt Peak National Observatory, Arizona, added ethane to the constituents of the Jovian atmosphere, as well as acetylene.

The presence of methane, ammonia, and presumably water vapor in Jupiter's atmosphere aroused great interest among biochemists, who believed these materials were the very substances that gave rise to, or at least preceded, the appearance of life on Earth. They speculated that simple forms of living organisms may exist today in the Jovian atmosphere, suspended at regions where temperature and other conditions are suitable, just as algae and plankton float freely at various levels in Earth's oceans. Even before the first space flights to Jupiter, Cyril Ponnamperuma, of the University of Maryland, looked upon reddish areas of the atmosphere of Jupiter as being colored by life-precursor molecules that were being created by the conditions within the Jovian atmosphere.

Explanations for the colors of the banded atmosphere have varied. One suggestion is that they are produced by solutions of metallic sodium in ammonia. Another suggests that free radicals —groups of atoms that behave like single atoms and can readily link with or separate from other atoms or groups of atoms while remaining unchanged as a group—are in the ammonia clouds, possibly trapped there. These might account for the color. Other suggestions have been that conditions within the clouds of Jupiter could lead to molecules bonding together into giant molecules known as organic polymers. Also, solar ultraviolet radiation penetrating the upper layers of the atmosphere could break down hydrogen sulfide into chemical groups that produce color.

Whatever the cause, and we are far from solving the problem, colors in the Jovian atmosphere vary considerably with time. Sometimes they appear quite intense; at other times the planet is almost colorless. In recent years the colors have been strengthening, particularly in regions that astronomers call the North Equatorial Belt and the Equatorial Zone.

Jupiter's face also suffers from spots, including a giant red one that has persisted for centuries. But there are small red and white spots which last for varying lengths of time. While the smaller spots seem connected in some way to the general turbulent nature of the Jovian cloud system, the big spot, called the Great Red Spot, seems to be a unique phenomenon. We have seen nothing like it elsewhere on Jupiter (other red spots are much smaller and short-lived) or on the other giant planets.

The Great Red Spot was discovered by Robert Hooke in 1664. The following year, D. Cassini

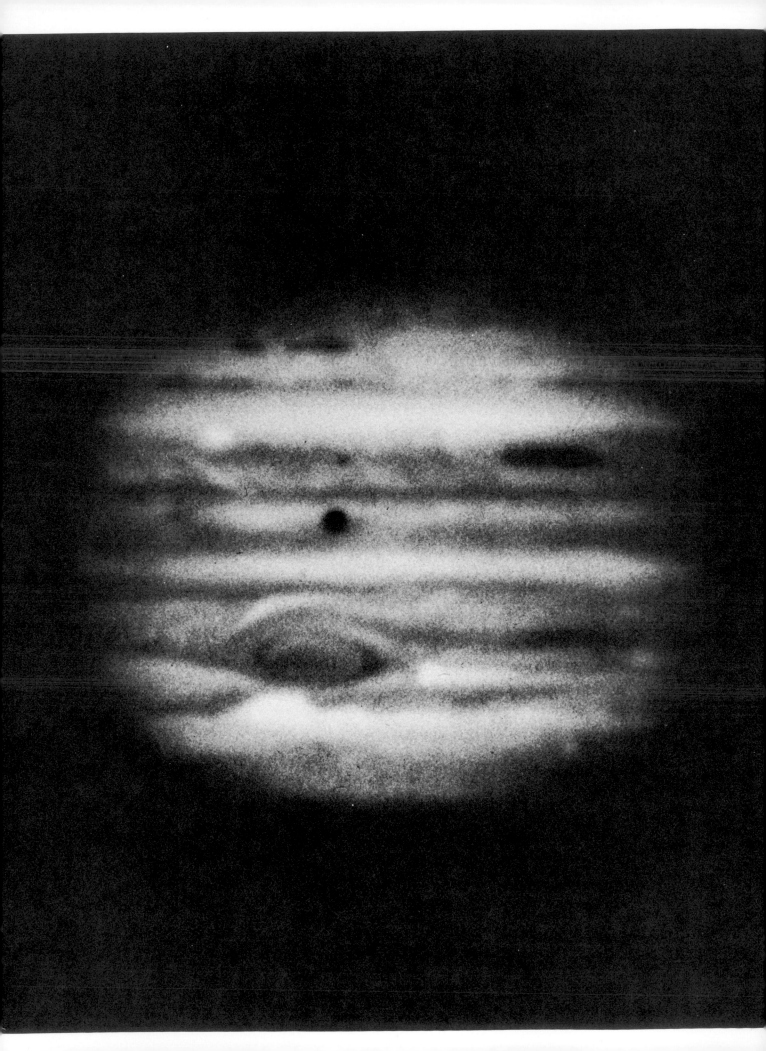

Jupiter photographed from the Pioneer 10 spacecraft in 1973. The picture shows part of the disc only, as the spacecraft made its close pass by the planet. Note the wealth of details and colors in the belt and zone structure of the planet. (NASA)

referred to the spot as the "eye" of Jupiter. At present, the spot is about 15,000 miles long and 4,000 miles wide (24,000 and 6,500 kilometers), but it has been known at times to expand to almost 30,000 miles (48,000 kilometers). The Great Red Spot lies in the South Tropical Zone and rotates around the planet with the other cloud systems but in a different period. In 1872 its period was 9 hours, 55 minutes, and 31 seconds, compared with 9 hours, 55 minutes, and 38 seconds in 1920–21, and 9 hours, 55 minutes, and 43 seconds in 1948. The spot seems to drift gradually in longitude, so that other markings in the South Tropical Zone overtake and pass it. A part of the South Tropical Zone developed bright spots at both ends and heavy shading between

them. It was referred to as the South Tropical Disturbance when it formed, in 1901. In 1902 it caught up with the Great Red Spot and leaped across it, the bright spot at the front end suddenly appearing ahead of the Great Red Spot. This process was repeated some nine times until the disturbance faded, in 1940.

Astronomers concluded that the spot must be high in the Jovian atmosphere for other features to pass beneath it. But, just as it wanders in longitude, the spot moves up and down in the atmosphere. When high up it becomes a vivid reddish brown. At lower elevations it loses its bright color and becomes pinkish white, occasionally fading completely. The spot seems to brighten and darken in a period of about thirty years.

Jupiter photographed in blue light shows the banded structure of the giant planet's atmosphere and the shadow cast by the moon Io. (Lunar and Planetary Laboratory, University of Arizona)

The Great Red Spot of Jupiter as it might be seen from a satellite such as Io. (NASA)

The cause of the Great Red Spot has puzzled generations of astronomers. Agnes M. Clerke, astronomical historian, wrote at the beginning of the twentieth century: "It certainly does not represent the outpourings of a Jovian volcano; it was in no sense attached to the Jovian soil—if the phrase have any application to the planet; it was not a mere disclosure of a glowing mass elsewhere seethed over by rolling vapours." In 1967, as plans were being prepared for a spacecraft mission to Jupiter, the NASA *Handbook of the Physical Properties of the Planet Jupiter* divided the theories for the origin of the Great Red Spot into two classes: those that suggested it was somehow connected with a feature on an underlying solid surface, and those that suggested it was some sort of floating body. The comparison of various theories concluded with the statement

that "it has not yet been shown conclusively that the Great Red Spot is a feature of atmospheric circulation, rather than an actual body."

Prior to spacecraft visiting Jupiter, it seemed impossible to decide whether the spot was an enormous region of turbulence caused by a geographic structure on a Jovian surface, what was referred to as a Taylor column, or a "raft" of hydrogen ice floating on the sea of liquid hydrogen.

POWERFUL RADIO TRANSMITTER

In addition to radiating heat from its interior into space, Jupiter was discovered to be an active emitter of energy at radio frequencies. This radio emission is of three distinct types, and its intensity is greater than that from any other extraterrestrial source except the Sun. Thermal radio waves at wavelengths of less than a few centimeters are produced by molecules moving about in the atmosphere of Jupiter. At longer wavelengths, from a few centimeters to tens of centimeters, decimetric radio waves are produced by electrons moving above the atmosphere. The third type of radiation is more energetic and is at wavelengths of tens of meters. This decimetric radiation is thought to be produced by electrical discharges in the Jovian atmosphere similar to lightning flashes.

After the discovery of the most powerful decimetric radiation, in 1955, C. A. Shain, of the Division of Radiophysics of Australia's Commonwealth Scientific and Industrial Research Organisation (CSIRO), in Sydney, began to search early records. He found that in 1950 and 1951 a series of radio bursts had been recorded and accepted as terrestrial interference. In 1956 he said these confirmed the discovery of Burke and Franklin and showed that the radio signals from Jupiter were associated in some way with the rotation of that planet. This was confirmed by later observations over a wider band of frequencies. It was not long before radio astronomers were routinely using emissions as a way of establishing the spin rate of Jupiter. They were mystified to discover that, at 10.3 centimeters wavelength, Jupiter's period of rotation was slightly longer than when it was measured by optical viewing of the cloud features. Another interesting discovery was that the radio bursts were somehow linked to the position of the Galilean satellite Io. The bursts of energy that cause the radio waves when released by Io were found to be equivalent to billions of lightning flashes on Earth occurring all at once.

In 1964, E. K. Bigg, of CSIRO, offered the then astonishing suggestion that the innermost Galilean satellite, Io, was influencing the Jovian decimetric bursts. He theorized that a single principle source region for decimetric radio waves exists on Jupiter, and the orientation of a line joining the source to Io governed whether or not we received the signals at Earth. Thus, the pulsations of radiation would be caused by alignments of Jupiter, Io, and Earth.

A couple of years later, CSIRO's J. H. Piddington, together with J. F. Drake, of the University of Iowa, noticed that the Jovian magnetosphere rotates with the planet. They also observed that Io travels westward through the magnetosphere as it moves along its orbit. They

suggested in 1968 that "this motion may cause an electromagnetic disturbance which contributes to the radio emission." The following year, Peter Goldreich, of the California Institute of Technology, and Donald Lynden-Bell, of the Royal Greenwich Observatory, England, examined many specific characteristics of the decimetric radio bursts that seemed to be controlled by Io. As a result, they suggested that the bursts were related to instabilities in a circulating current produced by the interaction of Jupiter's ionosphere and magnetic field with Io. But, for this process to work, Io had to be conductive enough to carry enormous electric currents.

Measurements and theory alike had made Io a very cold world, with an average surface temperature of about −279° F (100° K). The surface itself seemed to consist of rocks and ice. Therefore it appeared that its conductivity would be much too low to satisfy the Goldreich/Lynden-Bell theory. David L. Webster and some associates at NASA's Ames Research Center sought ways to overcome the difficulty. The low temperature of Io might be just enough to hold some gases to its surface and thus allow it to have an ionosphere, a tenuous upper atmosphere of electrically charged atoms and molecules. Such an atmosphere might provide a path for the enormous currents needed.

Assuming that it could, Webster then attempted to find out what gases were likely to be present in an atmosphere of Io and whether these were gases that would ionize under the influence of incoming solar radiation. They thought that methane might be the most likely constituent, with perhaps some argon and nitrogen, and concluded that such an atmosphere might have a sufficient density to produce an ionosphere. This is where the matter rested until more definitive information could be obtained either from better Earth-based observations or from space probes.

Meanwhile other radio observations of Jupiter showed that the decimeter-wavelength radiation originated well above the Jovian surface in either an ionosphere or the equivalent of Jovian Van Allen radiation belts. The existence of such belts would also imply the presence of a magnetic field of Jupiter, as mentioned earlier. A magnetic field was required to keep the charged particles spiraling along magnetic field lines from pole to pole to generate the radio emissions. This was good news in another way, because it suggested that theories for the origin of the Earth's magnetic field might be correct. These theories suggested that a planet generates a magnetic field like a huge dynamo due to the flow of conductive metallic material within its bulk acting like the current-carrying coils of a dynamo, while the rotation of the planet acted as the motor driving the coils. The presence of a field at Jupiter also reinforced the theory of an inner shell of electrically conducting, liquid, metallic hydrogen.

THE FASCINATING SATELLITES

A fifth satellite of Jupiter was not discovered until almost three centuries after the discovery of the four Galilean satellites. The latter have numbers as well as names, J1, J2, J3, J4, in order of distance from Jupiter. But starting with the new discovery, the satellites were given numbers in sequence. J5 was discovered by E. E. Barnard in 1892. It was also named: Amalthea. Only 100 miles (160 kilometers) in diameter, it orbits 66,000 miles (106,000 kilometers) above the Jovian cloud tops at a rapid 16.2 miles per second (26 kilometers per second), so that it revolves once around the planet every twelve hours, only a little longer than it takes Jupiter to rotate on its axis. Its orbit is almost as regular as those of the big satellites.

The Galilean satellites are major worlds in their own right. Two of them are as big as Mercury, and the others rival Earth's Moon in size. Because they are large, hard-surfaced worlds relatively close to their primary, the four satellites attracted much attention in the years immediately preceding the first spacecraft mission to Jupiter. The spacecraft would be expected to obtain some pictures of them. The 1973–74 period was particularly interesting to observe the satellites, because the plane of their orbits was almost aligned with Earth and Sun. This meant that the moons passed behind Jupiter in occultations, cast shadows on the giant planet as they crossed in front of it, and went through eclipses as they entered its shadow. These conditions are met every six Earth years, but the periods in 1961–62 and 1967 were not suitable for observations, because Jupiter appeared close to the Sun in Earth's skies. But in 1973–74 Jupiter was in opposition and excellently placed for observation at the time the satellite orbits lined up with Earth and Sun. Astronomers observed the eclipses, transits, and occultations to make better measurements of the satellites' sizes, colors, possible atmospheres, and albedos (the reflective properties of their surfaces). Robert E. Murphy, of the University of

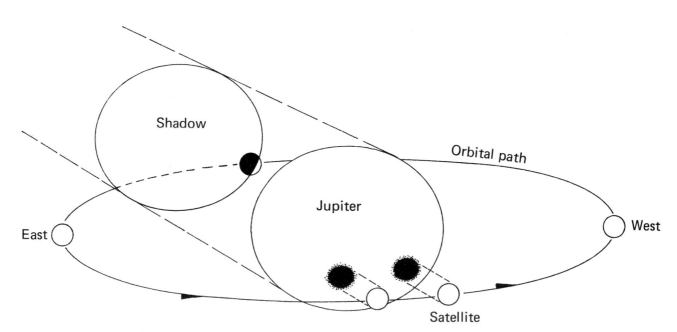

The behavior of a Galilean satellite in orbit around Jupiter, showing how it casts a shadow on the planet and how it disappears in Jupiter's shadow.

Hawaii, and Kaare Aksnes, of the Smithsonian Astrophysical Observatory, observed two occultations of Europa by Io. The observations were made with a very sensitive instrument that measures the amount of light received from the satellite and records the changes in this light with great precision. From these changes during the occultations the scientists deduced that Europa must possess a bright polar cap that extends some 30° from the geometrical north pole of the satellite.

But in the years before the first spacecraft went to Jupiter, perhaps most interest centered on Io. As early as 1964, Alan B. Binder and Dale P. Cruikshank, of the Department of Geology and Lunar and Planetary Laboratory of the University of Arizona, observed Io brightening as it emerged from the shadow of Jupiter. They attributed this effect to the presence of an atmosphere on Io. They supposed that some of this atmosphere condensed on the Ionian surface when it was cooled by passing through the shadow of Jupiter. As a result, the surface would be brightened by this condensate as it emerged from the shadow. But the returning solar radiation would rapidly evaporate the material back into the atmosphere. There was, however, a possibility that the brightening might be due to a fluorescence stimulated by a warming of the surface of Io following the eclipse.

An opportunity to test the density of the supposed atmosphere of Io came in 1971. On May 14 of that year a rare astronomical event took place: The satellite Io passed directly in front of a fairly bright star as seen from Earth. Referred to as a stellar occultation, it was important because the distant star is a point source of light. If occulted by an airless world its light should be cut off abruptly. But if the world had an atmosphere there would be a slight diminution of the light by the atmosphere before the cutoff. The light source in question was Beta Scorpii C, a fifth-magnitude star, visible to the unaided eye. Bradford A. Smith, of New Mexico State University, and Susan A. Smith, of the California Institute of Technology, traveled to Jamaica, which was situated close to the center of the path of the shadow of Io cast by the star on the surface of the Earth. The results were disappointing. The star was occulted so abruptly that it seemed that Io's atmosphere probably does not possess sufficient body for condensations to occur during the shadowing by Jupiter as observed by Binder and Cruikshank.

Yet there were still astronomers who contended that all the Galilean satellites might possess slight atmospheres, possibly less than one thousandth the pressure of Earth's atmosphere at sea level. This opinion was not confirmed by subsequent observations. In 1974, R. L. Millis and D. I. Thompson, of the Lowell Observatory, in Arizona, and B. J. Harris and other astronomers at Perth Observatory, Australia, were not even able to find evidence of the brightening of Io after an eclipse. Meanwhile, other observations suggested that there might be a very thin ionosphere some 60 miles (100 kilometers) above the surface, implying that Io possessed an atmosphere.

Investigations of Io and the other Galilean satellites focused on temperatures, because they give important clues about surface composition. Mauna Kea Observatory, in Hawaii, is over 13,000 feet (4,000 meters) above the Pacific Ocean. The barren cone of the ancient volcano pokes high above the main moisture-laden layers of Earth's atmosphere and allows astronomers to detect infrared radiation from celestial objects. Using the 88-inch (224-centimeter) telescope at the observatory, astronomers measured the temperatures of the Galilean satellites' surfaces. These were: Io, −231°F (127°K); Europa, −245°F (119°K); Ganymede, −218°F (134°K); and Callisto, −191°F (149°K). The surface temperature of Ganymede during one eclipse dropped 48.6°F (27°K) thereby indicating that the materials on the surface must be very poor conductors of heat. Scientists guessed that the satellite might have a crust of ice covering a satellite-wide ocean of water and ammonia, and below this deep ocean a core of rocks and iron oxide.

For some time, astronomers had believed that Europa was an ice-covered world because of the way the surface material reflected light from the Sun. However, similar observations of Callisto produced an anomaly. Despite the low density of this satellite, which would suggest that it consisted almost entirely of water, scientists could detect no ice on its surface, although they speculated that the ice might be covered by a layer of dust.

The absence of ice on Io was readily explained: it possessed dark, rather than light, polar caps, so it was unlikely that there was any ice on the satellite. An emerging theory of planetary and satellite formation suggested that satellites formed near their primaries might have been unable to condense water from the primordial nebula, because temperatures were too high.

Many new theories began to emerge about the composition of the surface of Io. Godfrey T. Sill, of the Lunar and Planetary Laboratory, speculated that it might consist of sodium and its compounds, the sodium originating from interplanetary dust swept up by the Jovian system in its orbit around the Sun. Energetic particles—charged particles moving at high velocities—trapped in the radiation belts of Jupiter, through which Io travels, might decompose the sodium compounds so that elemental sodium could be deposited on the surface of Io. T. V. Johnson, of the Jet Propulsion Laboratory, suggested that the surface of Io might be salt-encrusted, the salt having been brought to the surface by water seeping upward from the interior and evaporating into space, leaving its salts behind. It appeared that Io and Europa were silicate bodies, while the outer two satellites were icy bodies. As a result, if this is true, the surfaces of Io and Europa might be cratered like those of the inner planets of the Solar System, whereas the surfaces of Ganymede and Callisto would be comparatively smooth, because ice would be expected to "heal" itself following a meteorite impact.

The Galilean satellites and Amalthea are the inner moons of Jupiter. Revolving far outside them are two more groups of satellites. The first comprises four satellites that range in size from about 12 to 37 miles (20 to 60 kilometers) and follow normal counterclockwise orbits between 7.15 and 7.27 million miles (11.5 and 11.7 million kilometers) from Jupiter. Their orbits are all inclined 28° to the plane of Jupiter's equator, in sharp contrast to the inner satellites, which are all almost in the plane of the equator. One of these satellites was discovered as recently as 1974. They are Himalia (J6), Elara (J7), Lysithia (J10), and Leda (J13).

The outermost group is made up of Pasiphae (J8), Sinope (J9), Carme (J11), and Ananke (J12). These outer satellites take about 700 days to orbit Jupiter, at a distance of between 12.86 and 14.73 million miles (20.69 and 23.7 million kilometers). Oddly, they follow reverse, or clockwise, orbits, indicating that they may be captured

asteroids. All are so far from Jupiter that the gravitational field of the Sun exerts the principal perturbing force on their orbits, which are inclined about 25° to the equatorial plane of Jupiter. These outer moons are very small bodies, probably between 9 and 18 miles (15 and 30 kilometers) in diameter. Their small sizes and great distance from Earth make it very difficult to measure their diameters with precision. It has been speculated that the two outer groups may also represent the fragments of larger moons that broke up, or even planetesimals that were not able to form into a larger satellite.

PIONEER ODYSSEY TO THE GIANT PLANET

A major leap forward of our understanding came in the early 1970s when Pioneer 10 and Pioneer 11 were dispatched to the Jovian system. Both unqualified successes, the two craft confirmed much that was already known about the Jovian system, clarified some of the speculation, and filled in many missing details. Additionally, because of the impetus of the space experiments, ground-based activities stepped up. This allowed information gathered by the two probes to be interpreted on the basis of the latest observations made from Earth.

After a delay of several days because of high winds and thunderstorms, Pioneer 10 took off on March 2, 1972, from the John F. Kennedy Space Center, Cape Canaveral, Florida. As the flame of the Atlas' three rocket engines speared through the night, distant lightning played on the cloud tops. But all went well, and the spacecraft later separated from its launch vehicle and flew toward Jupiter at a record speed, of 32,129 miles per hour (51,695 kilometers per hour), 7,584 miles per hour (12,202 kilometers per hour) faster than the Apollo spacecraft had traveled to reach the Moon. In fact, Pioneer 10 passed the distance of the Moon's orbit in about eleven hours, a journey that took the Apollos four days.

After traveling for twenty months, the spacecraft made its closest approach to the planet, on December 3, 1973, flying 81,652 miles (131,378 kilometers) above the turbulent cloud tops of the planet. During approach to the planet, fly-by, and departure, Pioneer 10's many scientific instruments collected an enormous amount of data. A

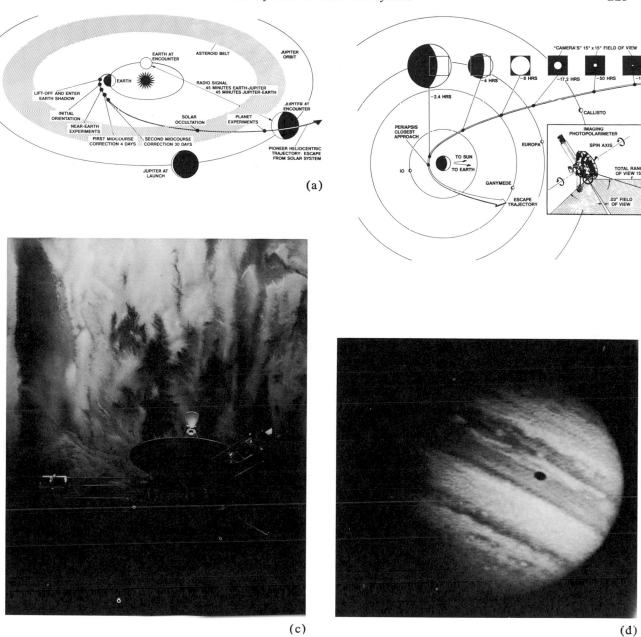

(a) *Relative positions of Earth and Jupiter at launch and at encounter of Pioneer 10 with the giant planet.* (b) *The trajectory of Pioneer 10 passed Jupiter, showing how for the first time it would be possible to photograph Jupiter at the crescent phase after the encounter.* (c) *Artist's concept of the spacecraft zooming past Jupiter, almost skimming the cloud tops.* (d) *View of Jupiter at a distance of 2.5 million kilometers, taken by the fast-approaching Pioneer 10 spacecraft. Notice the Great Red Spot and the shadow of the moon Io. A seventy-hour time lag is involved in the image processing.* (NASA)

special light sensor attached to the spinning spacecraft enabled it to obtain images of the planet and several satellites, which were as-

sembled by computers on the ground to provide photographs.

A combination of Jupiter's powerful gravitational field and the planet's orbital motion acted as a slingshot to fling the spacecraft at 82,000 miles per hour (132,000 kilometers per hour) right out of the Solar System. Pioneer 10's Jupiter-influenced trajectory set the spacecraft by the orbit of Saturn in February 1976, that of Uranus in September 1979, Neptune in July 1983, and Pluto in January 1987. Slowed down somewhat by the powerful pull of the Sun, the spacecraft's escape velocity from the Solar System into interstellar space was about 25,000 miles per hour (40,000 kilometers per hour). Pioneer 10 is expected to become completely free of the solar gravitational field when it reaches a distance of about five light-years, some 126,000 years hence. It will then orbit the Milky Way Galaxy like the stars. In approximately 100 million years Pioneer 10 will be about 2,000 light-years from Earth, but in about 200 million years it may approach once again to within 260 light-years.

Pioneer 10 is thus mankind's first emissary to the stars. Realizing this as he watched the space-

craft being tested prior to launch, Eric Burgess suggested to Carl Sagan that a message to extraterrestrials should be designed and carried on the spacecraft. Sagan and Frank Drake designed a plaque, which was carried on the spacecraft—an interstellar cave painting that has a chance of surviving all the marks of man on Earth even if it is never viewed by other intelligences on distant star systems.

Pioneer 10 carried 66 pounds (30 kilograms) of scientific instruments, about 11.5 per cent of the total weight of the spacecraft. It had a big, dish-shaped antenna, 9 feet (2.75 meters) across, which beamed radio communications toward Earth and received commands from the spacecraft's controllers. Because it would go so far from the Sun—at Jupiter the solar radiation's intensity is only one twenty-fifth that at Earth—Pioneer 10 carried an electrical generator that changed the heat derived from radioactive decay of plutonium 238 into electricity. Solar cells, as

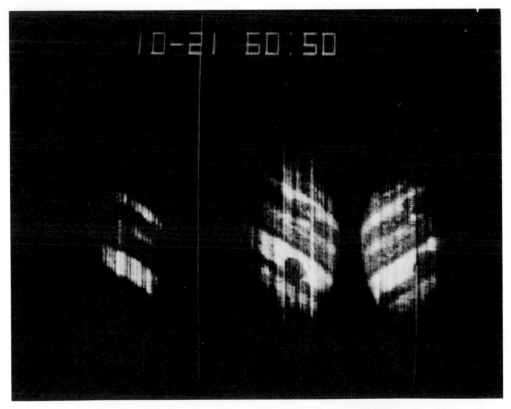

To assist scientists evaluate the progress of the Pioneer mission during the fly-by, data transmitted from the spacecraft were assembled by a computer into rough images as quickly as they were received at Earth. The images were then displayed on TV-type viewing screens. These are typical images showing the Great Red Spot. (NASA)

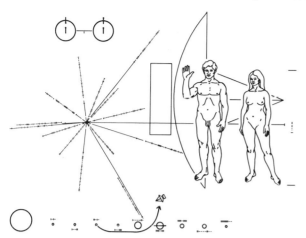

The Pioneer spacecraft escaped from the Solar System after flying by Jupiter. Each carried a plaque designed to show where they came from and who sent them. There is a remote possibility that millions or even billions of years hence the spacecraft might be intercepted by intelligences of a planet circling a distant star system. (NASA)

used on spacecraft flying about the inner Solar System, would not be able to generate sufficient electrical energy in the outer Solar System, because of the reduced intensity of solar radiation there.

The twelve scientific experiments of Pioneer 10 gathered information about interplanetary space and the environment, the asteroid belt, as well as the Jovian system, though most of the experiments were directed toward Jupiter and its environment.

Pioneer 10 was the first of two identical spacecraft. Pioneer 11 blasted off from its launch pad on April 5, 1973, in another nighttime launch. After travelling for 606 days and covering about 600 million miles (1 billion kilometers), the craft shot by Jupiter three times closer than Pioneer 10. Because it was penetrating so deeply into the intense radiation belts of Jupiter, with the danger that energetic particles there could damage electronic equipment, scientists decided that the spacecraft should be sent past Jupiter on a different trajectory from that of its predecessor. Thus, Pioneer 11 approached the giant world from beneath and climbed at enormous speed, passing very rapidly south to north through the radiation belts that concentrate near Jupiter's equatorial plane. This trajectory had other dividends: Not only did it protect the spacecraft from

serious radiation damage, but also it allowed unprecedented views of the north and south poles of the giant planet. Additionally, the fly-by had been chosen so that the gravity and motion of Jupiter would fling the spacecraft across the Solar System for a subsequent rendezvous with Saturn.

The path after Jupiter was somewhat unusual for a spacecraft, in that it climbed high above the plane of Earth's orbit around the Sun. The path also brought Pioneer back toward the Sun crossing Jupiter's orbit for the second time on June 10, 1977. Then, kicked outward by the giant planet's powerful gravitational field, it headed outward along a trajectory that allowed it to make some new observations of the magnetic field of the Sun, as described in an earlier chapter. Arrival at Saturn was scheduled for September 1979. Originally, Pioneer 11 was to fly inside the innermost ring and only 2,485 miles (4,000 kilometers) above the planet's cloud tops, but scientists finally decided to send the spacecraft outside the rings to check the environment there for a much larger Voyager spacecraft scheduled to explore Saturn a few years later. Pioneer 11 will head out of the Solar System in a nearly opposite direction to Pioneer 10.

SCIENTIFIC RESULTS OF THE PIONEER ODYSSEY

Although the Pioneer instruments could not probe beneath the clouds to the core of the planet, precise tracking of both spacecraft on their Jovian odyssey provided details of the gravitational field of the giant planet and some of its satellites. Analysis of the tracking data showed that Jupiter is perfectly symmetrical, as though turned on a giant lathe. It has no unusual features in its gravitational field. The conclusion is that the planet is essentially a great ball of liquid, spinning rapidly on its axis and distorted as expected by this spin. Both probes accurately measured the planet's polar diameter at 84,209 miles (135,492 kilometers) and its equatorial diameter at 88,727 miles (147,761 kilometers). While the equatorial diameter is very close to that determined from the Earth, the polar diameter is a little more than 1,240 miles (1,995 kilometers) less than measured from Earth.

John D. Anderson, of the Jet Propulsion Laboratory, and William B. Hubbard, of the University of Arizona, developed a post-Pioneer model

(a)

The Pioneer spacecraft is seen (above) under test before being mounted on an Atlas-Centaur rocket for launch to Jupiter. Below, the main components of the spacecraft are indicated on this drawing, which also identifies the various scientific experiments carried on-board. (NASA)

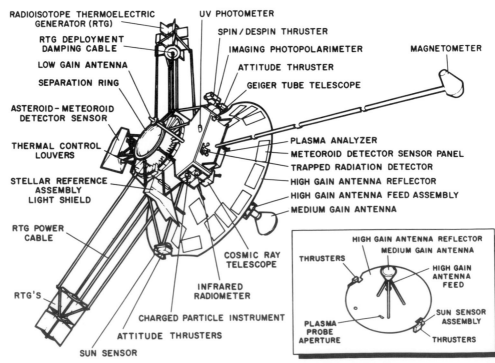

RADIOISOTOPE THERMOELECTRIC GENERATOR (RTG)

UV PHOTOMETER

RTG DEPLOYMENT DAMPING CABLE

SPIN / DESPIN THRUSTER

LOW GAIN ANTENNA

IMAGING PHOTOPOLARIMETER

SEPARATION RING

ATTITUDE THRUSTER

ASTEROID-METEOROID DETECTOR SENSOR

GEIGER TUBE TELESCOPE

MAGNETOMETER

THERMAL CONTROL LOUVERS

PLASMA ANALYZER

METEOROID DETECTOR SENSOR PANEL

TRAPPED RADIATION DETECTOR

HIGH GAIN ANTENNA REFLECTOR

STELLAR REFERENCE ASSEMBLY LIGHT SHIELD

HIGH GAIN ANTENNA FEED ASSEMBLY

MEDIUM GAIN ANTENNA

RTG POWER CABLE

COSMIC RAY TELESCOPE

RTG'S

INFRARED RADIOMETER

CHARGED PARTICLE INSTRUMENT

ATTITUDE THRUSTERS

SUN SENSOR

HIGH GAIN ANTENNA REFLECTOR

MEDIUM GAIN ANTENNA

THRUSTERS

HIGH GAIN ANTENNA FEED

PLASMA PROBE APERTURE

SUN SENSOR ASSEMBLY

THRUSTERS

(b)

of Jupiter. At the center of the planet, they suspect, there is a small core of iron and rocks, where the temperature is about 53,500° F (30,000° K), about six times as hot as the bright photosphere of the Sun. At the core of Jupiter the pressure is about 100 million times that of Earth's atmosphere at sea level. The mass of the core is thought to be about ten to twenty times that of Earth.

Above the core to about 28,600 miles (46,000 kilometers) from the center is a thick shell of liquid metallic hydrogen mixed with helium, believed to contain about twenty atoms of hydrogen for every atom of helium. How the helium is mixed with the hydrogen is not certain. In fact, the helium may conceivably be separated as a layer between the core and the metallic hydrogen. Temperatures in the metallic hydrogen shell range from 17,500° to 21,000° F (10,000° to 12,000° K), and the pressure is thought to be about 3 million times Earth's sea-level atmospheric pressure; i.e., about 22,000 tons per square inch.

Above the shell of metallic hydrogen, another shell extends to about 43,500 miles (70,000 kilometers) above the core. This shell also consists mostly of hydrogen, but now in a molecular state of two hydrogen atoms bound together.

The gravitational measurements were not precise enough to reveal positively that the planet does have a solid core. It seems logical, however, to assume that other substances in addition to hydrogen and helium were present in the primordial nebula from which the planet formed and that these probably sank to the center of the lighter hydrogen and helium that make up the bulk of Jupiter. But once scientists had shown that the planet was almost entirely liquid, their next question was how the planet generated its heat. We recall that Jupiter emits more than twice as much energy as it receives from the Sun, which the Pioneer spacecrafts confirmed. However, a liquid is virtually incompressible, so that it seems unlikely that Jupiter could be radiating heat into space because it is contracting. Rather, it seems that the planet is cooling and the contraction results from the heat loss. Thus the excess heat from Jupiter is most likely primordial heat and is not being renewed.

While metallic hydrogen conducts electricity very well, it is not such a good conductor of heat. So the interior of Jupiter must seethe in cells of metallic hydrogen like water coming to a boil in a saucepan. This seething process carries heat toward the surface. Deep within the interior of Jupiter, enormous loops of electric current are believed to flow. Through a dynamo process acting within the electrically conductive liquid metallic hydrogen, a powerful magnetic field is created, one that was measured by the two Pioneers. This field is tilted some 11° to Jupiter's axis of rotation (about twice the present tilt of Earth's field). The field is not a simple field like that of Earth but corresponds, rather, to the combined field of several different magnetic systems. Thus we believe that while Earth's field is generated deep within the planet, probably within the core, Jupiter's field is generated closer to the planet's surface, at least in part.

Jupiter's field is not only more than ten times stronger than Earth's field, but is also more complex. About four fifths of the Jovian field strength is accounted for by a simple north-south dipolar (two-pole) field analogous to that of a bar magnet. The remaining fifth is due to far more complicated four- and eight-pole components, which reflect the turbulent nature of the electrical currents flowing within the giant world.

At the base of the atmosphere of Jupiter, where the liquid hydrogen gradually merges into the atmospheric hydrogen, the temperature is probably about 3,100° F (2,000° K), which might be referred to as Jupiter's "surface" temperature.

A major discovery by the two Pioneers was that the temperature is fairly constant everywhere around the planet at a given level in the atmosphere. The temperature remains the same, day and night; and more surprisingly, from the equator to the poles. The day and night temperature can be easily explained on the basis of the rapid rotation and cloud movements, but the equator was expected to be warmer than the poles, because it receives more radiation from the Sun. The explanation seems to lie in the way heat is transferred from deep within the planet to its visible surface. This is a convective process, like heating water in a pan on a stove. The convection is most active where the greatest temperature difference occurs, which would normally be at the cold poles. So in what logically should be the cold polar regions, the material of the planet bubbles more rapidly toward the surface, thus carrying more heat to these regions and keeping their temperature up. It is just as though the heat flow within the planet is governed by thermostats that automatically cause more heat to be sent to those surface regions where the temperature would tend to fall because of insufficient solar heat.

From the "surface" to a pressure level of about 4.5 terrestrial atmospheres, the Jovian atmosphere is believed to be a uniformly heated mixture of 99 per cent hydrogen and helium. Above this level, water vapor and other substances condense out of the atmosphere. The lowest clouds are probably water droplets; above them, clouds of water-ice crystals; higher still there are crystals of ammonia. Traces of other substances lend color to the clouds.

At a pressure of about four and a half times Earth's sea-level atmosphere, the lowest clouds form a transition between the clear, lower atmosphere and a region of weather that reaches upward some 45 miles (70 kilometers). When the pressure falls to about two atmospheres, sulfur compounds, hydrogen and ammonia polysulfides, perhaps even organic molecules, form clouds that are gray or grayish brown.

Higher still in the Jovian atmosphere are white ammonia clouds that condense where the pressure is about three quarters that of Earth's atmosphere at sea level. The temperature of these clouds, which are at the top of the majority of cloud formations, is about −190° F (150° K). Above the ammonia clouds is the region of the Jovian stratosphere, a transparent shell containing hydrogen, helium, ammonia, and methane. It seems

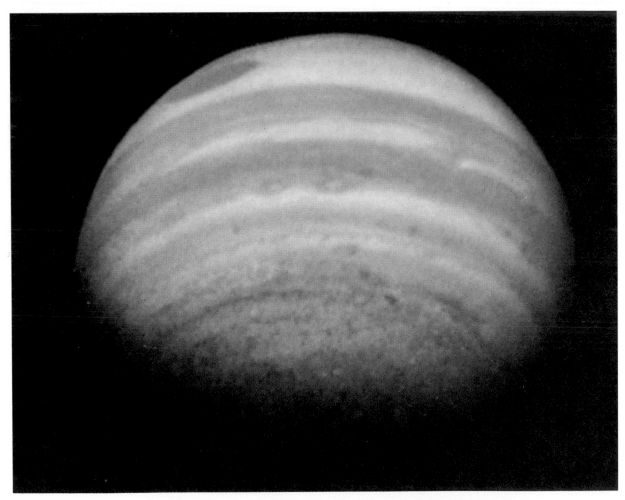

A view of Jupiter impossible to obtain from Earth. The color picture, of the north polar region of Jupiter, shows the Great Red Spot. The blue or grayish areas might be blue skies, as seen on Earth, caused by scattering of sunlight by particles in the atmosphere. (NASA)

to be thicker over the polar regions than over equatorial regions.

At 8.7 miles (14 kilometers) above the ammonia clouds, the temperature measured by the Pioneers was 145° F (418° K) and the pressure was 0.3 that of Earth's atmosphere at sea level. Above that level, temperature, but not pressure, rose with increasing altitude. A haze of aerosols —possibly hydrocarbons—was discovered at high altitudes in the stratosphere.

The Pioneer probes discovered that the clouds making up the generally light-colored zones are higher and colder than those in the darker belts. Though ammonia ice crystals predominate in both these banded features, those clouds in the belts are mainly reddish brown and gray and hence less reflective of light than the lighter zones. Both zones and belts change in size, distribution, and color intensity, especially at higher latitudes. Above 50° latitude the banded structure disappears and a mottled pattern takes its place. The polar regions are thus characterized by circular features which suggest considerable turbulence.

What seems to be happening on Jupiter is that warm, moist equatorial- and temperature-region air rises within high-pressure zones and spills over into the lower-pressure belts. In the belts, the masses of atmosphere fall to lower levels and become drier. Some of the air spreads into the adjacent polar belt, some into the adjacent equatorial belt. Jupiter's rapid rotation deflects the flow of poleward-moving air toward the east because it is moving faster than the regions of air into which it is moving and of equator-moving air toward the west because it is moving more slowly than the regions into which it is moving. The result is the creation of the horizontally oriented bands across the planet. While weather systems on Earth develop under these same effects into swirling cyclones and anticyclones, the rotation of Jupiter is so much faster that the cyclones and anticyclones are stretched around the planet into belts and zones.

Here on Earth, atmospheric flow is driven mainly by heat energy received from the Sun, which is concentrated in equatorial regions. The flow of air masses is from the equator toward the poles, swirled by the rotation of the Earth. The flow on Jupiter is different; it is driven by heat from within the planet. Moreover, Earth's weather patterns are distorted by interaction of

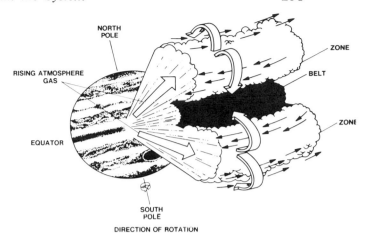

This drawing illustrates how the belts and zones form in the atmosphere of Jupiter. Rising masses of gas spill over north and south and form belts on either side of the zones. The zones are rising masses of gas, the belts descending masses of gas. (NASA)

the air flows with the land masses. On Jupiter there are no such solid objects beneath the atmosphere to affect the atmospheric circulation patterns.

But there is one major similarity between the two weather systems: both are affected greatly by the presence of water vapor. As columns of atmosphere containing water vapor rise on Jupiter, as on Earth, water vapor condenses and releases latent heat, which pushes the column of gas still higher. The rising column sucks in more moist air at its base, and the process is accelerated further. As rising columns in a given zone grow, they tend to join one with another, ultimately producing the planet encircling bands.

Atmospheric features of Jupiter, particularly the Great Red Spot, tend to last much longer than comparable features on Earth. This is because there is virtually no friction between the weather layer and the hydrogen atmosphere beneath it. The energy in the weather system is not dissipated, as it is on Earth, by interaction with land masses and mountains. Still another factor is that a hydrogen-rich atmosphere holds on to its energy. It takes some twenty times as long to remove heat from the Jovian atmosphere as it does from the nitrogen-and-oxygen atmosphere of Earth. Moreover, the Jovian atmosphere receives its heat more slowly even when the internal heat is taken into account. The sum of the internal heat flow and the heat from the Sun at Jupiter is still only about one twentieth that received by Earth from the Sun.

The result is that despite winds that can reach hundreds of miles per hour, over-all atmospheric flow on Jupiter is relatively modest compared to its rotation speed of 22,000 miles (35,500 kilometers) per hour, and its enormous circumference of 271,500 miles (436,800 kilometers).

The Pioneers obtained some new information

(a)

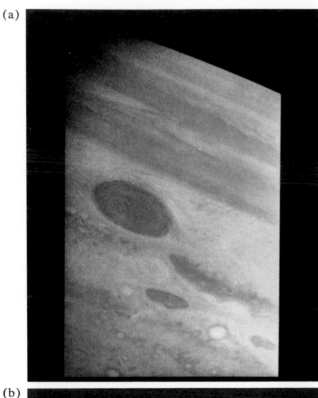

Pioneer II obtained the best picture of the Great Red Spot from a distance of 338,000 miles (545,000 kilometers), in December 1974. (a) The color picture shows a wealth of detail within the spot and its obvious circulatory nature. Note how the spot causes peaks projecting into the zone on either side of it. (b) and (c) are the same picture taken in red and blue light respectively. The blue picture seems to show a counterclockwise spiral within the spot. There also appears to be a narrow jet stream of brown material flowing southwestward about and to the left of the spot, and there is a series of white cloud puffs strung out along the boundary between the South Tropical Zone (which contains the spot) and the brown belt north of it. The white oval below the right end of the spot is one of three such ovals that are about 120° apart on Jupiter and have persisted for about thirty years. The circular feature in the center of the oval suggests that it, too, is rotating. (NASA)

(b)

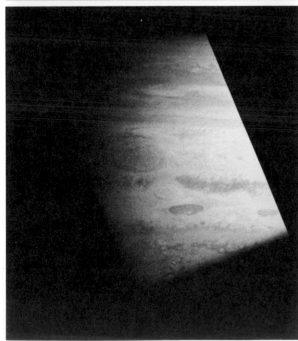

(c)

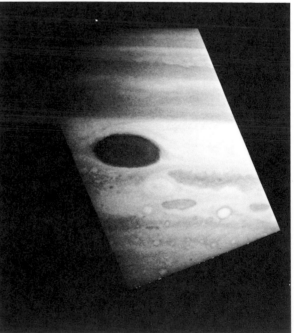

Other red spots have been observed on Jupiter, but they do not last long. This one, photographed by Pioneer 10, had disappeared when Pioneer 11 visited Jupiter, a year later. Note, however, that its shape is very similar to that of the Great Red Spot, even to the pointed ends. (NASA)

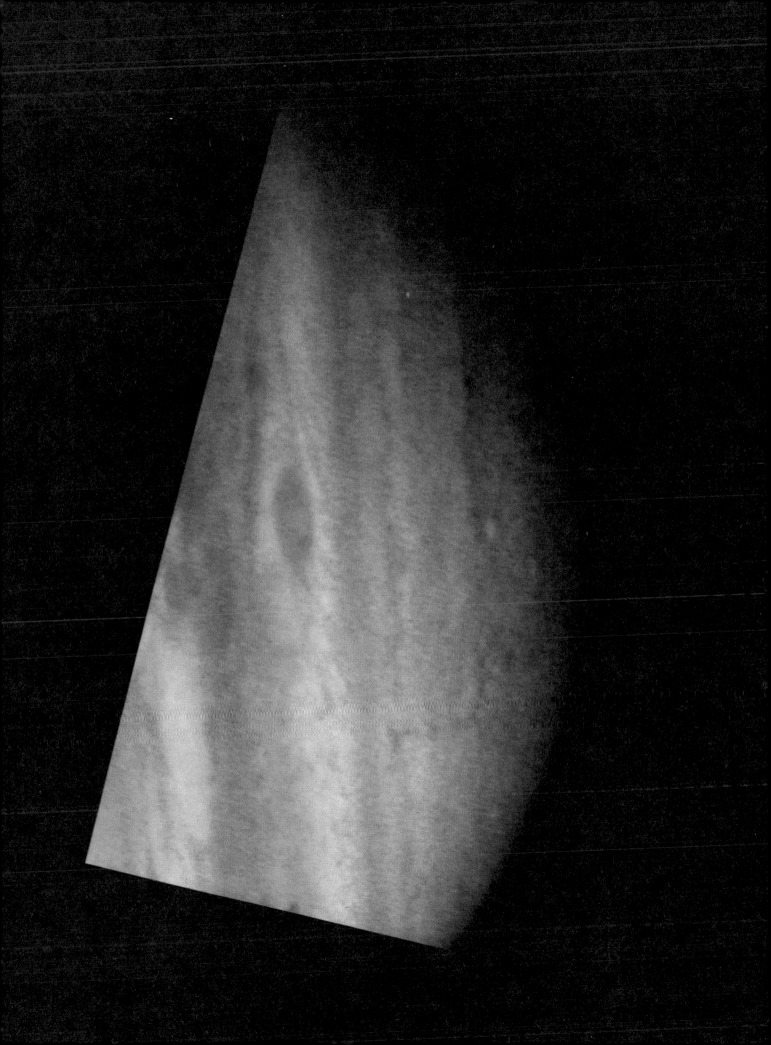

about the Great Red Spot. First, the liquid nature of Jupiter ruled out its being a column of gas extending above a feature on a solid surface. Since it splits a zone, and zones are rising air masses, the Great Red Spot appears to be explained by a rising column of warm air, or is at least maintained by one. The spot might be likened to an enormous hurricane rotating in a counterclockwise direction.

Pioneer photography disclosed a huge red cloud overlying the spot approximately five miles (eight kilometers) above the general level of the uppermost clouds. At this altitude it is cool enough for small amounts of phosphorous to condense, thereby creating the vivid red color of the spot.

How long will the spot last? It seems to be a free-wheeling phenomenon that does not dissipate much energy. So it could last indefinitely. By contrast, much smaller red spots do not last long. Pioneer 10 showed one of these in its pictures taken in 1973, but the spot had disappeared the following year, when Pioneer 11 flew by the planet.

Space biologists have from time to time suggested that the Great Red Spot may be a caldron in which organic matter is being brewed. How true such a speculation might be is anyone's guess today. But one thing does seem clear: the elements needed for life are present in the Jovian atmosphere and there are regions where conditions of temperature, pressure, and humidity are suitable for life to develop.

What are the elements?

First, there are the raw materials, such as ammonia, water, methane, sulfur, and water vapor; all were detected by Pioneer instrumentation. Second, abundant energy is available at Jupiter from the Sun and from the interior of the planet. Third, countless liquid and solid particles are suspended in the Jovian atmosphere, many of which could conceivably provide protective sites for the development of complex organic molecules and perhaps even micro-organisms. Fourth, temperature and pressure conditions for water to exist in a liquid form are appropriate within a wide band of the atmosphere.

But the real question is whether the liquid-water zone has been stable over long enough periods of time for life-precursor organic molecules to have developed and later to have evolved into life itself. Cyril Ponnamperuma has pointed out that the region in which organic evolution could have taken place, and may still be taking place, is a thousand times larger than the comparable region on our Earth. Such a vast expanse could encompass myriad microenvironments with a better chance even than on Earth for nature to work its miracle of creating living things. In fact, it has been speculated that the red color of the Great Red Spot may result from photosynthetic organisms there.

The outer regions of the Jovian atmosphere contain electrically charged particles that form an ionosphere produced by incoming solar radiation. The outer atmosphere of Jupiter above the level at which the pressure is one thousandth that of Earth's atmosphere at sea level extends for 1,860 miles (3,000 kilometers), over five times the height expected. Five different layers were found in the atmosphere where electrons were concentrated, and there was some evidence that two other layers might also be present.

The ionosphere of Jupiter also proved to be five times hotter than expected. Scientists believed that the heating arises from energetic particles precipitating from the radiation belts and sharing their energy with the upper atmosphere, in addition to the incoming solar ultraviolet radiation. Also, energy-carrying waves might be propagated upward from Jupiter into the higher atmosphere. Another possibility is that the strong gravitational field of Jupiter may be pulling in many more meteorites than a small planet such as Earth does. These meteorites would also add energy to the upper atmosphere as they burned up during passage through it.

JUPITER'S MIGHTY MAGNETOSPHERE

Some of the most significant findings of the two Pioneer spacecraft were in connection with the magnetic field of Jupiter and its interaction with the solar wind. In fact, the first major event in the odyssey to Jupiter was Pioneer 10's crossing the bow shock on November 26, 1973. This signified that the spacecraft had left interplanetary space and had entered the environment of the Jovian system, because the bow shock is formed where the magnetic field of Jupiter stops the solar wind from flowing uninterruptedly toward the planet. The crossing of the bow shock took place at a distance of 4,772,000 miles (7,678,000 kilometers) from Jupiter, much farther out than had been expected. Later the instruments aboard Pioneer detected other bow crossings, which meant that the bow shock of Jupiter did not remain at a constant distance from the planet but pulsed in and out as it was buffeted by the solar wind.

Once deep in the magnetosphere, the region where the magnetic field of Jupiter predominates over the fields carried by the solar wind, the space probe began to detect extremely intense radiation—charged particles such as electrons and protons moving at very high velocities. This radiation was most energetic as the probe crossed the magnetic equator of Jupiter, a little over one hour before making its closest approach to the surface of the planet.

The radiation belts of Jupiter were found to be incredibly intense—up to ten thousand times more so than Earth's counterparts. Even more astonishing was the discovery that the total energy of the particles in the Jovian belts is millions of times greater than the total energy of particles in the Earth's radiation belts.

The magnetosphere of Jupiter is maintained by the planet's powerful magnetic field, which is about seventeen thousand times stronger than that of Earth. Radiation is trapped in much the same way that it is trapped in Earth's magnetosphere. But there is a major difference: While the particles in Earth's radiation belts are thought to originate from the solar wind, those in the radiation belts of Jupiter may originate mainly from the planet itself but are accelerated by the effects of the solar wind flowing past the planet. Strong centrifugal forces created by the spin of the planet throw off particles continuously from the top of Jupiter's ionosphere. They flow outward along magnetic field lines in the form of a flat, 435,000-mile (700,000-kilometer) thick equatorial current sheet or ring called a magnetodisc. Because the particles are electrically charged, their outward flow gives rise to an electric current, which in turn generates a magnetic field. This adds to the general field of Jupiter and tends to expand the magnetosphere beyond what would be expected from the planetary field alone. But by the time this outward flow of charged particles reaches the boundaries of the magnetosphere its effects are weakened and the solar wind starts to play a dominant role. During particularly strong gusts of the solar wind the plasma of charged particles is forced back toward Jupiter, sometimes at speeds of up to 20,000 miles

per hour (32,000 kilometers per hour). On such occasions, the great, bloated magnetosphere suddenly collapses to an eighth of its earlier diameter.

This squeezing of the great magnetic bag of charged particles has some unusual effects. High-energy electrons are squirted through leaks in the magnetic bag at such high velocities that they permeate the Solar System. (Some were even detected at Mercury by another spacecraft.) Because the magnetic field of Jupiter is tilted relative to the rotational axis of the planet, the magnetosphere is lopsided. The field is actually not only tilted, but its center is 435 miles (700 kilometers) north of the center of Jupiter and slightly more than 4,350 miles (7,000 kilometers) outward from the axis of rotation. This lopsidedness causes the magnetic bag to receive a particularly hard squeeze once each Jovian day. The electrons are squirted about the Solar System in bursts every ten hours. They have been recorded for many years by instruments carried by Earth satellites but were not confirmed as originating from Jupiter until the Pioneer missions provided the data. Thus, except for the Sun, Jupiter seems to be the only important source of high-energy particles in the Solar System.

A big question was how the particles of the radiation belts could acquire such enormous energies. What seems to be happening is that instead of following magnetic field lines from one pole to another as particles do in Earth's magnetosphere, the Jovian particles recirculate between an inner and outer magnetosphere and pick up additional energy each time around until they are either shot out into space from the leaky bag or are swept up by one of the big, Galilean satellites, three of which plow through the radiation belts as they orbit the giant planet. Amalthea also sweeps up some of the particles, and others fall back into the ionosphere. If it were not for these effects the energy in the radiation belts would increase almost indefinitely and no spacecraft would be able to approach safely into Jupiter's magnetosphere.

Though it may be leaky and change its contours continuously as it reacts to gusts of the solar wind, the Jovian magnetosphere nevertheless has an identifiable structure, which scientists refer to as the inner, middle, and outer regions. The inner magnetosphere is doughnut-shaped, with the planet in the hole of the doughnut. It includes the multishelled inner radiation belts and the satellites Amalthea, Io, Europa, and

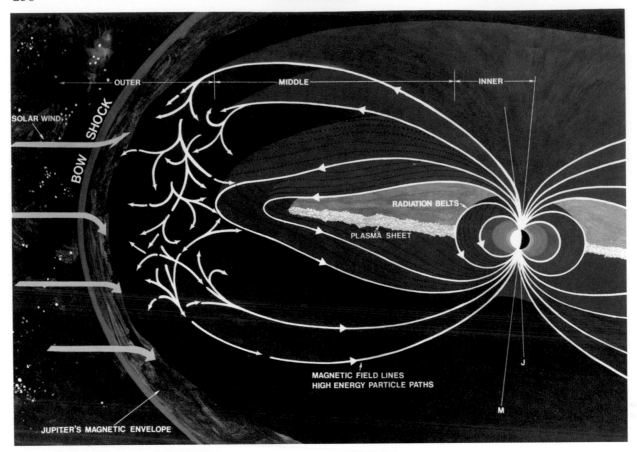

Some of the complexity of the huge magnetosphere of Jupiter is shown in this drawing, in which the inner, middle, and outer regions of the magnetosphere are identified. Particles appear to be recirculated many times between inner and outer regions as they acquire enormous energies. (NASA)

Ganymede. This region extends to a distance of about 0.9 million miles (1.5 million kilometers) from the planet.

The middle magnetosphere contains the magnetodisc current sheet. It extends from the border of the inner region out to some 2.67 million miles (4.3 million kilometers) from the planet. This disc flops around each day as the particles streaming outward start off along the planet's equator and attempt to align themselves along the magnetic equator.

The outer magnetosphere is shaped by the solar wind changes. It is an extremely turbulent region where energetic particles and magnetic

fields behave erratically. From it protons and electrons escape from the Jovian system into interplanetary space.

The Jovian environment's most dangerous radiation was measured in the inner magnetosphere. Pioneer 10, traveling close to the center of the region's inner belts, received almost one hundred times the radiation dosage that would kill people. In fact, the intensity was so great that about 1 billion electrons struck each square centimeter of the spacecraft's hull every second. And if this were not enough, that same square centimeter was also bombarded by about 6 million protons each second. Ninety per cent of the electrons had energies ranging from 3 to 30 million electron volts.*

Pioneer 11, which approached along the different trajectory described earlier, sustained about the same electron bombardment as Pioneer 10 but about twenty times more proton bombardment. However, the spacecraft's path took it through the radiation zone more quickly, with the result that the total radiation was reduced and the spacecraft survived the ordeal.

JUPITER AS A RADIO TRANSMITTER

The Pioneer missions confirmed all the radio emissions from Jupiter and added some more information about them. The motions of molecules in the Jovian atmosphere produce the thermal (centimeter) radio waves. The decimeter waves are produced in the radiation belts by the gyrations of electrons along magnetic field lines. But the decimetric radio waves are somewhat more complex. They appear to be related to flashes of lightning in the Jovian atmosphere, flashes that release energy equivalent to that of several large hydrogen bombs. Pioneer confirmed Io's involvement with these decimetric radio bursts. One of the spacecraft passed behind Io as seen from Earth, with the result that the radio signals to Earth from the spacecraft had to pass through Io's atmosphere. The way in which the characteristics of the received signals changed enabled radio astronomers to determine the nature of Io's ionosphere. They found that it is extensive and conductive enough to allow the linkage of magnetic lines of force to the ionosphere of Jupiter. Thus, an electrical current could flow from Jupiter's ionosphere along a magnetic field line of Jupiter to one side of Io's ionosphere. It could then flow through Io's ionosphere to the other side of the satellite and then back down another flux line to the ionosphere of Jupiter, thus completing a full circuit.

As Io orbits through the Jovian magnetic field, an electrical potential of 400,000 volts is generated across the satellite. This drives the current

---

* An electron volt is a measure of the energy carried by the particle. Analogous to defining the energy of impact of a weight falling to the floor from a certain height, an electron volt represents the energy acquired by an electron that falls through a potential difference of one volt.

flow, which is probably about 10 trillion watts. The ionosphere of Io is made conductive by the presence of electrically charged atoms of sodium, which have been derived from salts on the satellite's surface being bombarded by high-energy particles from the Jovian radiation belts.

THE BIG SATELLITES

The Pioneer spacecraft in passing through the Jovian system made relatively close approaches to several of the Galilean satellites. All the satellites were photographed to better detail than is possible from the Earth. The effects of the gravity fields of the satellites on the spacecraft enabled us to obtain better estimates of masses and densities. The effects of the satellites on the magnetosphere of Jupiter were also detected and measured.

Coupled with better observations from Earth and the application of new techniques of observation using complex electronic devices to interpret the light reflected from the satellites, the spacecraft information gave us a new interest in these outer worlds of the Solar System, which rival the inner planets in size.

The four satellites of Jupiter observed by the two Pioneers seem to be composed mainly of water and rocky materials. Sufficient radioactive thorium, uranium, and potassium may have been present on the satellites to heat them after their formation. Such heating may have led to considerable changes in their internal structure, particularly a differentiation of their materials in which lighter ones floated toward the surface and heavier ones sank to form a denser core.

In general, the hemispheres facing the direction of orbital motion, the leading hemispheres, differ somewhat in color and composition from the trailing hemispheres. All the Galilean satellites are locked to Jupiter so that they continually face one hemisphere to the giant planet, just as our Moon does to Earth. The difference between leading and trailing hemisphere probably results from a collision between the leading hemisphere and the energetic particles of the radiation belts and with micrometeoroids which appear to be concentrated into the system by Jupiter's gravity. While Io, Europa, and Ganymede are so deeply immersed in the radiation belts, interesting worlds that they are, they may be difficult to explore except by machines. Callisto, in contrast,

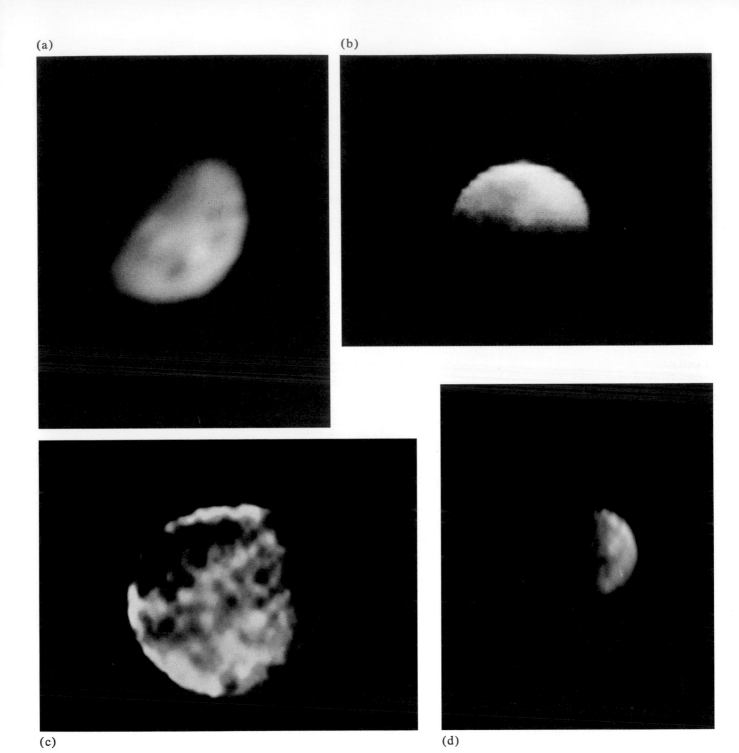

(a) (b)

(c) (d)

These four pictures are the best photographs obtained of the big satellites of Jupiter.
(a) This Pioneer 11 image is a view from over the north pole of Io. It shows how the
orange coloration of the polar regions contrasts with the whitish equatorial regions.
(b) Europa, about the size of our Moon, is shown with its north pole at about the tip
of the half-moon shape in this Pioneer 10 view of the satellite. The northern hemisphere has dark areas, the southern hemisphere lighter areas. The bright areas are
known to be covered with ice or frost. The dark areas may be rocky or an unfrosted
solid ice. (c) The biggest satellite, Ganymede shows features that are craterlike and
very large dark circular areas that may be big impact basins like those on the Moon.
Ganymede is as big as the planet Mercury. Radar evidence suggests that the surface of
Ganymede is a mixture of ice and rock. (d) The outermost of the Galilean satellites,
Callisto is about the size of Mercury too. The very low density of this satellite suggests
that it consists mainly of water with an icy crust. This picture was taken by Pioneer 11.
(NASA)

may be safe enough for astronauts to explore its fascinating surface.

Pioneer confirmed that Io and Europa are denser than Ganymede and Callisto. All four moons are frozen hard; their surfaces temperatures were confirmed as being between —234° and —225° F (125° and 130° K). The four worlds compare as follows:

THE GALILEAN SATELLITES

Satellite	Average distance from Jupiter		Diameter		Density water=1	Mass Moon=1
	miles	km	miles	km		
Io	261,860	421,400	2,262	3,640	3.52	1.22
Europa	416,650	670,500	1,895	3,050	3.28	0.67
Ganymede	664,280	1,069,000	3,275	5,270	1.95	2.02
Callisto	1,168,980	1,381,200	3,107	5,000	1.63	1.47

The outer satellites are less dense because Jupiter was much hotter early in its history than it is now. Lighter elements would be prevented from condensing on the inner satellites, so they consist mainly of rocks. The outer satellites contain much more water and other lightweight substances. This confirms speculations voiced before the Pioneers made their journeys.

The spacecraft proved that Io has an atmosphere. Although it has a density of only one twenty-thousandth that of the Earth, it extends some 70 miles (115 kilometers) above the satellite's surface. The ionosphere reaches even higher, especially on the day side, where it is excited by solar radiation; ionized gases were detected as high as 450 miles (700 kilometers). The gas mixture making up the inosphere contains sodium, hydrogen, and perhaps nitrogen. Astronomers have also discovered that Io is surrounded by a cloud of sodium vapor (which has even now been photographed from Earth) that extends 10,000 miles (16,000 kilometers) above its surface. And the satellite gives rise to a torus of hydrogen that extends along its orbit.

Almost as dense and rocky as Io, Europa was photographed once only by Pioneer 10 and exhibited dark features in the northern hemisphere and a bright, light-colored region in the southern. We think that billions of years ago radioactive heating warmed Europa sufficiently to liquefy its water content and force it toward the surface, where it then froze into a layer some thirty miles (fifty kilometers) thick. Since then, meteorite bombardment has shattered some of the rocky covering and exposed some of the ice beneath it. The dark features seen on the surface of the satellite might be large basins of the type known to exist on the Moon, but we can still only speculate about what causes the colors on the Galilean satellites.

Water is the main ingredient of Ganymede. In fact, the satellite may be basically a great globe of water almost as big as the planet Mercury. The deep ocean is thought to be covered by an ice crust impregnated with rocks. There may also be a muddy core deep within the satellite. The Pioneer pictures showed circular features on Ganymede and dark and light spots. Radar measurements from Earth suggest that there are large numbers of rocks embedded in an ice crust. But how features could remain on such an icy world without deforming is unknown. Ganymede appears to have a thin atmosphere, whose makeup is unknown although methane and ammonia seem likely constituents.

The Galilean satellite with the lowest density, Callisto, is probably also mainly water.

EVOLUTION OF THE JOVIAN SYSTEM

Studies of the Jovian satellites are important for cosmogonists to understand how the Jovian system evolved during its 4.5-billion-year history. They also shed light on possibly analogous situations in the other giant planets, Saturn, Uranus, and Neptune, as well as the Solar System as a whole.

The Solar System, as we know, contains three

major kinds of bodies: the Sun, the hard-surfaced terrestrial-type worlds (including satellites and asteroids), and the four giant planets. Pluto seems more like a satellite than a planet in its own right. The terrestrial planets are concentrated in a fairly narrow band around the Sun. The outer, giant planets are very much farther away and separated by the asteroid belt from the inner planets. It may be, therefore, that the division of the planets into these two separate groups represents the result of a different process of formation. The terrestrial planets may have formed in a high-density shell of material surrounding the early Sun, while the Jovian planets may represent abortive attempts to form stars that would have been companions to the Sun. Multiple-star systems are quite common in the universe.

Thus Jupiter and the other giant planets could be stars that failed. Support for this thought comes from the fact that these planets are all low-density worlds composed of essentially the same mix of materials that make up the Sun.

Some astronomers feel that the term failed stars is misleading, for it implies that if conditions had been slightly different the giant planets would have become self-luminous as their internal nuclear fires ignited. In fact, none of them possess anywhere nearly enough mass to start a nuclear reaction. The largest of the giant planets, Jupiter, possesses only 0.1 per cent the mass of the Sun. It would have to have at least thirty times as much mass as it does for it to become a star. Even so, astronomers feel that clues important to a full understanding of the origin and evolution of the Solar System can be revealed only by a penetrating study of the huge world.

Pioneers 10 and 11, and supporting ground-based research, bore out this conviction. Measurements of Jupiter's internal heat and of the decreasing density of the Galilean moons with increasing distance from Jupiter has permitted an acceptable scenario for the formation of the system. About 4.5 billion years ago, at the same time as Earth was being formed, an enormous flat, disc-shaped cloud of gas and dust began to contract. It was far out from the Sun, in the region where Jupiter orbits today. It was a cold cloud, only about 55° above absolute zero temperature. Within the space of a few million years the diameter of the cloud was only 600 million miles (1 billion kilometers) and shrinking rapidly. Within another million years it was only 400

million miles (650 million kilometers) across. As it shrank, its temperature rose until it reached 3,100° F (2,000° K), and the hydrogen gas of the cloud began to ionize as the atoms were stripped of their electrons. The proto-Jupiter flared with energy and then collapsed within a few months to a diameter of 400,000 miles (640,000 kilometers). The planet was now producing considerable amounts of heat, which it radiated into space as it glowed redly, converting gravitational energy into radiation.

The planet cooled rapidly, and within a few hundred thousand years it no longer glowed but, rather, had to shine by reflected light from the Sun. Jupiter has continued to cool over the billions of years to the present.

The rocky core of the giant world probably condensed within the cloud at an early stage as grains of silicates and metals accreted. And the Galilean satellites probably formed from similar heavier grains in the original cloud of gas and dust that formed the planet. But the satellites formed at differing distances from the redly glowing planet, so that they were affected in various ways. Those closest to Jupiter were too hot for much water and other lightweight, volatile materials to be incorporated into their substance. They formed as rocky worlds. Those farther out were much cooler and could collect the volatile substances.

WHERE DO WE GO FROM HERE?

Although we can now reconstruct the general history and evolution of the Jovian system, we still lack a vast amount of information to confirm what we learned from the Pioneer missions.

To follow these exploratory missions, in the late spring and summer of 1978, NASA dispatched two larger, Mariner-class spacecraft to the outer planets. Called Voyagers, they were sent along trajectories taking them first to Jupiter, where they will observe the planet, its environment, and satellites. Then the probes will proceed to Saturn to perform similar investigations and to examine the magnificent ring system as well. One of the spacecraft is expected to pass close to Uranus after its Saturnian visit.

In 1983, another mission to Jupiter is planned by the U.S. space agency. Known as Galileo, the spacecraft being developed will consist of an orbiter and a probe. The former will survey the

(a)

A future mission to Jupiter will consist of an orbiting spacecraft from which a probe will plunge into the Jovian atmosphere. (a) An artist's conception of the orbiter making a close approach to one of the Jovian satellites. (b) A red-hot nose cone separates from the probe portion of the spacecraft as it descends by parachute deep into the Jovian atmosphere to sample its constituents. The Jupiter orbiter probe, named Galileo, is to be launched in 1982 from Earth orbit after being carried there by a space shuttle. It will be the first interplanetary mission to be launched by this new, recoverable launch vehicle. (Jet Propulsion Laboratory)

(b)

giant world over many months, much as Mariner 9 and the Viking orbiters did at Mars. And the latter will penetrate into the Jovian atmosphere to make direct measurements en route before plunging to destruction. During descent, the Galileo probe will relay its observations via radio link to the Galileo orbiter, above.

Still later, in the 1990s, other automated spacecraft may be sent to survey the four large Galilean satellites and perhaps to land on one or more of them. When we realize that the surface areas of those moons are comparable to those of Mercury, Mars, and our own Moon combined, we gain a fresh insight into the magnitude of the task before us.

The exploration of the Solar System has only just begun.

11
DISTANT GIANTS

Before the invention of the telescope, the most lavishly equipped observatory in the world was situated on a small island, called Hven, in The Sound, north of Copenhagen. There, at the observatory of Uranienborg, Frederick II, king of Denmark, had sponsored Tycho Brahe in the most precise astronomical observations yet undertaken. For over twenty years ending in 1597, Tycho meticulously recorded the apparent positions of the planets among the "fixed" stars. Uranienborg was, indeed, just about the last observatory for naked-eye observations of the heavens.

While Tycho still held to the old belief of an Earth-centered universe, he took all his observations with him to Prague when invited there by Emperor Rudolf II in 1599. There he met a young German, Johannes Kepler, who became his assistant. Two years later, in 1601, Tycho died, but his assistant, Kepler, carried on his work of trying to compile accurate tables for predicting the motions of the planets. But Kepler broke from the traditional thinking of Tycho and accepted the concept of the planets moving in ellipses around the Sun. The result was that Kepler was able to define and publish his three famous laws and thereby remove the last obstacles to general acceptance of the Copernican system. These laws relied heavily upon the precise observations made by Tycho, and among the planets he had observed was Saturn.

Saturn was one of the planets known to the ancients. Its wanderings among the stars were described by Babylonian astronomers as long ago as

650 B.C. Because it moved more slowly than any of the other planets, it was regarded as the most distant of them.

Saturn moves along a slightly eccentric elliptical orbit around the Sun, taking 29.46 Earth years to complete one revolution. The planet is so far away—an average of 886.7 million miles (1.427 billion kilometers)—that it takes its light nearly eighty minutes to reach us.

The planet rotates quickly on its axis; its day is only 10 hours 14 minutes long. The centrifugal effect from this swift rotation flattens the planet, much as it does Jupiter, to the extent that the polar diameter is almost 10 per cent less than the equatorial diameter of 74,500 miles (119,900 kilometers).

While Jupiter has a very low density compared with the Earth, Saturn is even more bizarre. If we could find an ocean large enough to hold the huge world, Saturn would float in it. The material of the planet has a density of only 70 per cent that of water; while the planet has a volume 755 times that of Earth, it has only just over 95 Earth masses. So, in spite of its huge size, the surface gravity of Saturn is only 1.2 times that of Earth, which means that if there were a solid surface to stand on, a person on Saturn would weigh somewhat more than on Earth.

When pale, yellowish Saturn approaches its closest to Earth, at opposition, it is a conspicuous sight in the night sky, but not nearly so bright as Jupiter. Nevertheless only two stars shine brighter than Saturn: Sirius and Canopus. One reason Sat-

Earth and Saturn to scale. (NASA)

urn is so bright despite its great distance is that it reflects three quarters of the sunlight falling upon it. In addition, Saturn has a brilliant ring system that also reflects much light, especially when seen in a position that tilts the rings toward Earth.

THE MAGNIFICENT RING SYSTEM

Saturn's rings have fascinated people ever since

their discovery, accredited to Galileo, soon after the invention of the telescope.

The aperture of Galileo's telescope was small and its ability to magnify the images of planets rudimentary: it could increase their apparent diameters only about thirty-two times. Moreover, the optical quality of the lenses was far from good. Thus it is not surprising that Galileo was confused by his discovery of the rings. He announced his observation in a letter to the Secre-

These two views of Saturn show how the ring system appears to change as viewed from Earth because of the inclination of the system to the orbit of Saturn. (Hale Observatories)

tary of State to the Grand Duke of Tuscany. The letter, dated July 30, 1610, said: "The fact is that the planet Saturn is not one alone, but is composed of three [companion stars] which almost touch one another and never move nor change with respect to one another. They are arranged in a line parallel to the zodiac, and the middle one is about three times the size of the lateral ones."

The aging Galileo could scarcely believe his eyes when he looked at the planet Saturn again a couple of years later. Perhaps the Church was right and he was imagining things in the heavens, for he was astounded to find that the two attendants of Saturn had vanished. On December 4, 1612, he wrote: "What is to be said concerning so strange a metamorphosis? Are the two lesser stars consumed after the manner of solar spots? Have they vanished or suddenly fled? Has Saturn, perhaps, devoured his own children?"

But he still held on to the reality of his earlier observation and predicted that the companion stars would be seen again at the next opposition of Saturn. What we know now, and Galileo did not and could not know, was that on December 28, 1612, the ring system of Saturn had been so positioned as to be edge on toward the Earth and the Sun. And since the rings are extremely thin, they were invisible in a small telescope, such as Galileo had, for several weeks before and after that date. Galileo never recorded his observations as indicating a ring around Saturn, so why he predicted they would become visible again is an unsolved mystery.

After Galileo's untimely death, other observers turned their telescopes on Saturn. Pierre Gassendi, a Frenchman, published several drawings made between 1633 and 1656 that raised more questions than they gave answers. Finally, a young Dutchman, Christiaan Huygens, found the truth. In his celebrated treatise *Systema Saturnium,* published in 1659, he wrote: *"Annulo cingitur tenui, plano, nusquam cohaerente, ad eclipticam inclinato."* This translates into: "It is surrounded by a thin, flat ring, nowhere touching, [and] inclined to the ecliptic." The inclination of the ring plane to the plane of the ecliptic is important, because it accounts for the rings sometimes being seen edge on and at other times tilted toward or away from Earth. The rings are actually inclined slightly more than 20° to the ecliptic plane.

Huygens, who with his brother had built a greatly improved telescope, also discovered a satellite of Saturn, Titan, which orbited the ringed planet once each sixteen days.

The next major contribution to our understanding of the Saturnian system was made by that same Cassini who had so diligently studied Jupiter. Invited to Paris from Italy by Louis XIV to build and direct the Sun King's new observatory, the Italian astronomer discovered (in 1671) Saturn's second satellite, Iapetus. A year later he discovered a third, Rhea.

For reasons that are not clear, it was another four years before he announced an important discovery that tied his name permanently to studies of Saturn's rings. He stated that the breadth of the ring was divided into two parts by a dark line, now known as Cassini's Division. This finding should have cast grave doubts on a theory, widely accepted after Huygens' interpretation of Saturn's strange appendages, of the rings being solid, rigid, and opaque. But the belief that this was so prevailed throughout the eighteenth century and well into the nineteenth, even though in 1785 the world-famous French mathematician Pierre Simon Marquis de Laplace had clearly demonstrated that even two solid rings separated by Cassini's Division was impossible. Their material could not remain stable. If the centerline of each ring moved at local orbital speed, the inner parts would be dragged outward, the outer parts inward, and the ring would crumble into pieces. But it was difficult to kill the "modern" myths. The ring of Saturn had been described as circling the jewel of the planet with a ring of solid gold, and such exotic fantasies were too entrenched to dispel them with the logic of mathematics in an age of romanticism.

But progress toward the truth could not be halted forever. In 1848, a young French mathematics professor, Édouard Roche, further developed Laplace's theories. He calculated that within a circle of radius 2.44 times Saturn's any plastic object such as a moon held together by its own gravitational field must inevitably fall apart. The minute differences in orbital speeds of the particles resulting from the breakup would scatter these particles along the orbital path. Subsequent collisions would also spread them into a ring. Actually, all of Saturn's rings are within what is now referred to as Roche's limit, while the moons of Saturn are, of course, outside it. But nobody paid much attention to Roche's calculations when they were published.

A few years went by and the scene shifted to the University of Cambridge, in England. The university invited contributions to a competition in honor of the astronomer John Couch Adams. The subject for the 1855 Adams Prize Essay was an analysis of Saturn's ring system. Two years later, the winner was a brilliant young mathematician whose name would be immortalized by his defining a set of equations that would ultimately revolutionize the art of communications. His name was James Clerk Maxwell. In his thesis he concluded: "The only system of rings which can exist is one composed of an indefinite number of unconnected particles, revolving round the planet with different velocities according to their respective distances." Maxwell's explanation of the true nature of the rings, based on his careful mathematical analysis of the dynamics of particles revolving around a planet, was verified before the turn of the century through a careful study of the rings' spectrum. Because particles moving toward or away from an observer shift spectral lines to the blue or red end, respectively, of the spectrum depending upon the speed of movement, observations of the spectrum of the rings could be interpreted to show that the innermost parts of the rings move faster than the outermost. The rings could not be solid. And the observation was reinforced by another type of observation: Sometimes a star passes behind the rings as seen from Earth and is occulted by them. But the rings do not completely obscure the light from bright stars, again showing that the rings are not solid.

As studies continued, it became apparent that Saturn has not one but a system of rings. The easily visible outermost parts have a diameter of about 171,000 miles (275,000 kilometers). But the rings cannot be more than a few miles thick. There are three distinct rings that are fairly easy to observe and two extremely faint rings. The outer bright ring is called the A ring. It is about 10,500 miles (16,900 kilometers) wide and is separated by the 1,860-mile (3,000-kilometer)-wide Cassini's Division from a brighter inner, or B, ring. This ring is about 16,800 miles (26,900 kilometers) wide. Inside this ring is a very faint ring called the Crêpe ring or C ring. It is about 10,000 miles (16,000 kilometers) wide and is separated, by an elusive gap, from the B ring.

This gap, which is about 600 miles (1,000 kilometers) wide, is called Encke's Division, after its discoverer.

There are two very faint rings: One of them is between the Crêpe ring and the surface of the planet. On the inner edge it approaches very close to the atmosphere of Saturn, but its outer edge appears to be separated from the Crêpe ring by a 1,250-mile (2,000-kilometer) gap which has been named Guérin's Division after the French astronomer who discovered it. This very faint inner ring is called the D ring. Outside the visible rings there is also some material that forms an extremely faint outer ring which has been named the E ring. How far out this material extends is unknown, but it could present a hazard to spacecraft flying by Saturn.

The sizes of the particles making up the rings may be as small as grains or as large as city blocks, though most range from a few centimeters to a meter or two in diameter. By analyzing the light from the rings, especially at wavelengths beyond those of visible light, astronomers have decided that the ring particles may consist of rocky cores surrounded by layers of ice. Collisions among the particles have probably acted and still act to reduce their average size. Smaller particles may also spiral inward and fall to the planet.

The first radar contact with Saturn was made during the winter of 1972–73 with the 210-foot (65-meter)-diameter Goldstone antenna, operated by the Jet Propulsion Laboratory. The beam was sent out at 400 kilowatts. It took just over an hour to travel the distance to Saturn. Some of the radio energy was reflected by the planet and its ring system, and just over another hour later it arrived back at Earth. The signal was now so faint that it required extremely sensitive receiving devices to detect it. But weak though they were, the radio signals carried important information. Their characteristics indicated that they must have bounced off large chunks of material in the rings, bodies about the size of a compact automobile. It did not appear from the radar signals that the rings could consist of tiny ice crystals, as some astronomers had speculated. Nor could they be merely dust or gas. The echoes indicated particles with rough, jagged surfaces, of at least 3-foot (1-meter) diameter and probably larger. But no radar echoes came from Saturn at this time; it was evidently a very poor reflector compared with the ring system.

The total amount of matter in the rings is unknown. Astronomers can place an upper limit to the matter in the ring system because it does not perturb the motions of the innermost satellites. The total amount of matter within the rings cannot accordingly be much more than that in our Moon. Probably it is somewhat less.

As for the origin of the rings, that is still a matter of much controversy. Many astronomers look on the rings as debris from a satellite that either broke up or was never formed, invoking theories similar to those for the asteroids. Other speculations are that the rings resulted from the breakup of an asteroid that approached too close to Saturn, or they represent condensations from the original nebula of insufficient mass to form a satellite. The seeming uniqueness of the ring system was also a theoretical problem that did not present an easy solution. Why did Saturn, of all the planets, have a ring system? The answer came only very recently. Saturn is not unique: Uranus, too, possesses a system of rings, but whereas those of the former consist of bright particles, the rings of Uranus consist of dark particles, which reflect hardly any light. But an answer to the question of why planets have ring systems may have to wait until spacecraft fly by Saturn and Uranus, in the coming years.

SATURN'S INTERIOR

If we were able to lower a probe into the atmosphere of Saturn (as we may be able to do one day in the not too distant future), we would find that it is much akin to the atmosphere of Jupiter. It is predominantly a mixture of hydrogen and helium with relatively small amounts of ammonia, ethane, acetylene, ethylene, water, and perhaps hydrogen sulfide and hydrogen cyanide.

Below the outer atmosphere there is most probably a deep ocean of liquid hydrogen, at the bottom of which the hydrogen takes on its metallic state, just as it does in Jupiter.

In 1938, Rupert Wildt calculated that there should be a core to Saturn formed by the separation of the original gaseous mixture into its various components under the pull of gravity, somewhat in the same way that an iron core formed in the Earth. He estimated that Saturn's core of metals and rocks is about six times as dense as water and that it is surrounded by a thick layer of ice with a density of about 1.5 times that of water. The ice contains water, frozen carbon dioxide, and other lightweight substances. The outer layers of the ice shell would be frozen hydrogen which gradually becomes liquefied at greater distances from the core. This model was widely accepted for a number of years. Then, in 1951, W. H. Ramsey, with some of the newly gained knowledge of atomic structures, pointed out serious flaws in Wildt's model.

First, the model would be consistent with Saturn's exceptionally low over-all density only if the central core possessed an unrealistically low density and if the amount of hydrogen within Saturn were very different from the proportion of this gas in the other big planets.

Second, there was no allowance for the fact that the density of hydrogen more than doubles during the transition to the metallic phase, a transition that clearly had to be expected within Saturn because pressures would be so high. The Wildt model had to be abandoned.

Today, we are still uncertain about the interior of Saturn. However, many astronomers accept a core of heavy metals and silicates of about ten Earth masses. Close to the core there may be a shell of hydrogen mixed with heavier elements, then a massive shell mainly of hydrogen, stratified in various states according to temperature and pressure—first metallic hydrogen, then liquid hydrogen, then the atmosphere of gaseous hydrogen.

The question of whether or not Saturn possesses a magnetic field has not been resolved, though many astronomers contend that one is likely; the planet spins rapidly and it probably has a shell of metallic hydrogen circulating under the influence of internal heat. Weak radio emissions were detected from Saturn in 1957, but they were not of a frequency associated with radiation belts that would prove the existence of a magnetic field. But it may be that Saturnian radiation belts are not very intense, because of the presence of the ring system.

MANY SATELLITES

Like Jupiter, Saturn has a retinue of satellites, at least ten. Cassini, who discovered Iapetus and Rhea, later discovered two more: Tethys and Dione. A century later, William Herschel discov-

(a)

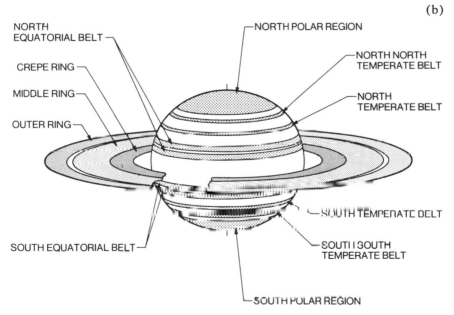

(b)

NORTH
EQUATORIAL BELT

CREPE RING

MIDDLE RING

OUTER RING

NORTH POLAR REGION

NORTH NORTH
TEMPERATE BELT

NORTH
TEMPERATE BELT

SOUTH TEMPERATE BELT

SOUTH SOUTH
TEMPERATE BELT

SOUTH EQUATORIAL BELT

SOUTH POLAR REGION

(a) *Saturn photographed in ultraviolet (UV), blue (B), and infrared (IR) light in August and September 1969. The image marked M was photographed in the wavelength of an absorption line produced by the gas methane. The dark regions on it signify an abundant methane atmosphere, while the light regions indicate little or no methane present. All pictures are reproduced with south at the top. (University of Arizona) (b) The drawing indicates the banded structure of Saturn as recorded by astronomers. This drawing is, however, reproduced with north at its top. (Jet Propulsion Laboratory)*

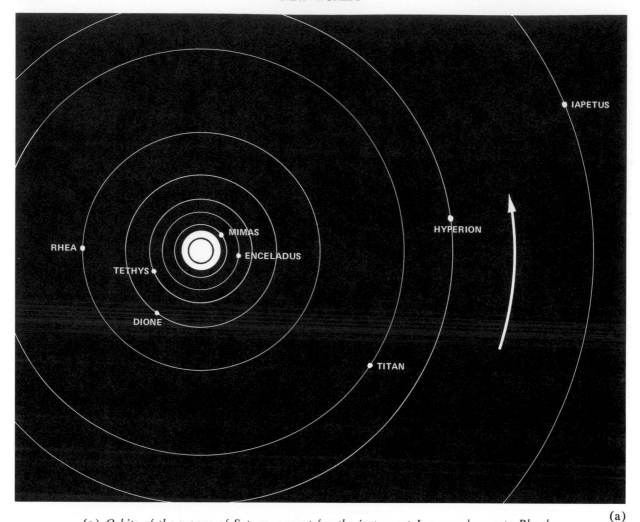

(a) *Orbits of the moons of Saturn, except for the innermost Janus and remote Phoebe, which cannot be included on this scale.* (b) *Photograph of Saturn and, left to right, its satellites Titan, Rhea, Enceladus, Mimas, Tethys, and Dione. (Gerard P. Kuiper)*

(b)

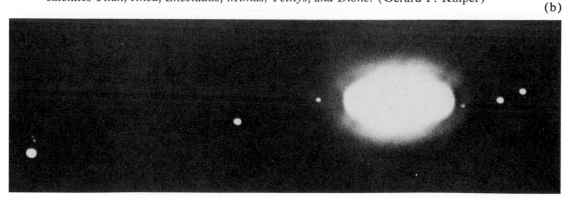

ered Mimas and Enceladus. Two more, Hyperion and Phoebe, were first observed in the second half of the nineteenth century, and the tenth moon, Janus, was not found until 1966.

The Saturnian system differs markedly from that of Jupiter. The ringed planet has only one large satellite, Titan, compared with Jupiter's four. There are two medium-sized moons, but the rest are very small bodies, just slightly larger than Jupiter's outer satellites. Titan and the inner Saturnian moons follow orbits that are prograde and almost circular. Phoebe follows an eccentric retrograde orbit. In general it also seems that the density of the satellites increases with distance from Saturn. Except for one: Iapetus.

This interesting body exhibits unusual changes in brightness, noticed by Cassini over three hundred years ago. The satellite seems to face one hemisphere toward Saturn. This results in the satellite presenting one face to Earth when it is on one side of the planet as seen from Earth, and another face when on the other side of the planet. When Iapetus is east of Saturn as viewed from Earth it presents its leading edge to us. When west of Saturn it shows its trailing hemisphere. The leading hemisphere is one sixth the brightness of the trailing hemisphere, meaning either that the surface is very different or the satellite is an unusual shape so that it presents different areas of surface to us, like a sausage viewed broadside or end on. Only a mission to Saturn is likely to clear up this centuries-old mystery.

Titan is an intriguing object. Larger than the planet Mercury, it may turn out to be the largest moon in the Solar System. Titan's great distance from us makes measurement of its diameter tricky. On March 29, 1974, the dark limb of our Moon occulted the Saturnian system. This provided an excellent opportunity to measure the diameter of Titan by timing how long it took for the edge of the Moon to cover the satellite. J. Veverka, J. Elliot, and J. Goguen, of Cornell University's Laboratory for Planetary Studies, seized the opportunity. Their measurements showed a diameter of 3,604 miles (5,799 kilometers) for Titan, making it larger than Ganymede. But Titan has an atmosphere, and a very thick one at that. So if this atmosphere is about 90 miles (145 kilometers) thick, the surface diameter of Titan is about 3,418 miles (5,500 kilometers).

The tiny, unblurred image at the intersection of the indicator marks is Phoebe, photographed on June 9, 1960, at the U. S. Navy's Flagstaff Station, in Arizona. (U. S. Navy)

Were Titan closer to Earth it would be even more fascinating to study. But, as it never comes closer than some 746 million miles (1.2 billion kilometers), even large telescopes cannot resolve details on its surface and we are led to considerable speculation. The atmosphere of Titan is much denser than that of Mars and may even rival the pressure of our own atmosphere. Unfortunately, we are very uncertain about the composition of the atmosphere. There are indications that it consists of methane, ammonia, hydrogen, and helium. There may also be nitrogen, but we cannot detect it from Earth. The puzzle is how a relatively small body such as Titan can hold an atmosphere of hydrogen. There have been speculations that the atmosphere is rich in hydrocarbons and that Titan may be an abode of living things.

An object of Titan's reflective properties (albedo) at its distance from the Sun would be expected to have a surface temperature of about $-297°$ F ($90°$ K) to balance incoming solar radiation with radiation reflected. At such a temperature most of the heat radiated from Titan into space would be at a wavelength of 20–30 microns (millionths of a meter), which is infrared. But scientists find that Titan emits mostly at a substantially shorter wavelength, of 12 microns, and this would correspond to a surface temperature of $-234°$ F ($125°$ K). The question then arises as to where Titan is getting its heat. It is not likely to have an internal source to account for this much higher temperature. Scientists have concluded that the answer may lie in an opaque atmosphere that traps incoming solar radiation and leads to a greenhouse effect similar to that on Venus.

Fortunately, three space probes are headed for close encounters with the mysterious planet, so during the decade of the '80s we may have some answers. In the meantime the satellite system of Saturn remains as intriguing as that of Jupiter and somewhat more difficult to explore.

SATURN-BOUND SPACECRAFT

As mentioned in the previous chapter, after its successful encounter with Jupiter, Pioneer 11 was aimed to meet with Saturn in the late 1970s. Even before its encounter with Jupiter, the option of flying to Saturn was kept open by a course cor-

rection to alter the trajectory past Jupiter so that the spacecraft could be directed to Saturn later. Slightly more than a year after the Jupiter encounter, Pioneer's trajectory was again changed slightly, followed by another correction, in 1976, that aimed the spacecraft to pass within 60,000 miles (100,000 kilometers) of Saturn.

Much soul-searching then took place as to exactly how Pioneer should fly by Saturn. Should it be sent far out, so as to avoid any possibility of colliding with ring particles and be sure of making a close approach to Titan? Should it be sent between the Crêpe ring and the planet, as perhaps our only opportunity for decades to make such a close pass? Scientists interested in finding out as much as possible about the magnetic field of Saturn favored the closest possible passage. The bigger Voyager spacecraft would also not be risked in flying through the rings, so Pioneer 11 gave a unique opportunity.

The decision was finally made late in 1977: Be cautious. Not, really, to safeguard Pioneer, but to safeguard the much larger investment in the two Voyager spacecraft that were following it to Saturn. It was decided to send Pioneer along the path chosen for the first Voyager, to see if that path was a safe one. Pioneer 11 was directed onto a trajectory that would carry it out beyond the rings.

The spacecraft will still be able to gather a wealth of scientific information about the rings and the planet and perhaps some of its satellites. It will perform essentially the same experiments as those at Jupiter: close-up photography of the clouds and of the rings, and investigations of the charged-particle environment of Saturn and the interaction of the planet with the solar wind. Radio tracking of the spacecraft will also refine our understanding of masses and sizes of the objects within the Saturnian system, and occultations by the rings, the planet, and some of its satellites will provide information about atmospheres and ionospheres and concentrations of particles.

Two Voyager spacecraft are also on their way to Saturn, carrying high-resolution television cameras that will enable them to take detailed pictures of the moons of Saturn and of the ring system and the planet's cloud structures. Not yet scheduled but possible is a spacecraft similar to the Jovian Galileo probe, which would perform similar functions at Saturn.

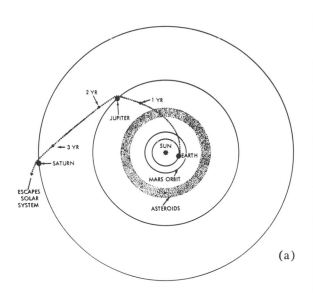

(a)

(b)

Two large spacecraft follow a small Pioneer spacecraft to Saturn. (a) This drawing shows the path of the Voyager spacecraft from Jupiter to Saturn. (b) is an artist's impression of this fantastic journey into the outer Solar System. (c) Taking the measure of Saturn's rings. Painting shows how Voyager spacecraft will fly behind the rings of Saturn and use cameras and radios to measure how sunlight is affected as it shines between the ring particles. (NASA)

(c)

THE KOWAL OBJECT

In 1977, Charles Kowal, of the California Institute of Technology, was examining photographs of star fields that he had obtained on October 18 and 19 with the 48-inch (1.22-meter) Schmidt telescope at Palomar Mountain. He discovered a faint object on the plates that had moved slightly over the two nights. The new object did not resemble anything previously seen in the Solar System. It was not a comet. It was too small to be a planet. It was also far out in the Solar System, near the orbit of Uranus, and appeared to be following a planetary type of orbit. From its pinpoint size and brightness, the object was estimated to be about 200 to 300 miles (125 to 185 kilometers) in diameter, which made it more like an asteroid than anything else.

As further observations were made, the orbit of the object could be estimated more precisely. It appears to move like a planet, orbiting the Sun between Saturn and Uranus in a period of about seventy years in a nearly circular path and quite close to the plane of Earth's orbit around the Sun.

Astronomers began to question if the Kowal Object was only one of a number of such bodies, another asteroid zone in the outer Solar System. But the object could also be a satellite of one of the outer planets that escaped from its primary, as Pluto is thought to have been a satellite of Neptune at one time.

However, in the course of a few more months the orbit was shown to be highly eccentric, taking the object from the orbit of Uranus to within that of Saturn. Thus it looked more and more like an escaped satellite. Its discoverer commented that the object would one day encounter either Uranus or Saturn or would be thrown by the gravity of one of those planets into some other part of the Solar System. In the meantime, Kowal named it Chiron, after the mythological centaur (half horse and half man).

URANUS, ANOTHER RINGED WORLD

Beyond Saturn is remote Uranus, about which we know even less. The planet is so far away that even at its closest to the Earth it is scarcely visible without a telescope.

Since antiquity, some magic has been attached to the number seven, so it was widely taken for granted that the Sun and the Moon, plus the five visible planets Mercury, Venus, Mars, Jupiter, and Saturn completed the total number of heavenly bodies that moved among the fixed stars. Thus, it came as a shock to many traditionalists when word came from England that on March 13, 1781, a German-born organist in Bath by the name of William Herschel had discovered another planet. He had been testing a new home-built telescope, wandering his gaze among the stars of the constellation Taurus, when he accidentally saw Uranus, appearing as a star that was not on the charts. He watched it move and knew that it was a planet. He named it after George III, *Georgium Sidus*. But astronomers did not like the departure from tradition. So Johann E. Bode suggested another name, from mythology, and the planet has ever since been known as Uranus.

Astronomers could backtrack on the movement of Uranus among the stars. They discovered that it had been seen several times before Herschel's discovery, but no one had recognized it as a planet. The English Astronomer Royal, John Flamsteed, had seen Uranus a number of times from 1690 into the early 1700s. Pierre-Charles Lemonnier, of France, had noted it in his records at least eight times between 1768 and 1769. But who had thought that there might be another planet? So these worthy astronomers had just added to their star charts and forgotten about it.

The early observations were nevertheless extremely valuable, because they enabled the mathematicians to calculate the orbit of Uranus more quickly than if they had had to wait for observations to pile up after the discovery. So an accurate orbit quickly became available. Uranus travels around the Sun once every 84.015 years at an average distance of 1.783 billion miles (2.87 billion kilometers), or 19.2 times Earth's distance from the Sun. The best measurement of the diameter of Uranus was obtained in 1970 from a high-altitude balloon that carried a telescope above much of the Earth's atmosphere. This diameter is about 32,190 miles (51,800 kilometers). The mass of Uranus is believed to be about 14.6 times that of the Earth, and its density is about 1.21 times that of water.

We can get an idea of the planet's immense distance from us when we compare the Sun as seen from Earth and from Uranus. The Sun appears from Uranus as a blindingly brilliant star, but it

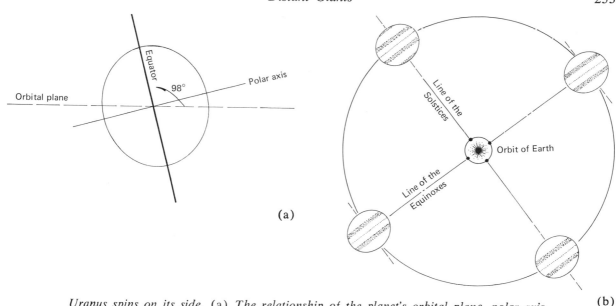

Uranus spins on its side. (a) *The relationship of the planet's orbital plane, polar axis, and equator.* (b) *The resultant changes in orientation of Uranus with respect to the Earth.*

shows hardly any disc. Despite this, the light from it is still brighter than that from a full moon on Earth.

Astronomers believe that the clouds of Uranus reflect light very efficiently, about 90 per cent of the sunlight falling on them being reflected back into space. But our knowledge about what this means in terms of atmospheric temperature and composition is extremely scant. There is some indication, however, from radio waves emitted by Uranus, that the planet has a higher temperature than would be expected from solar heating alone. Like Jupiter and Saturn, it may possess an internal heat source. One explanation is that the planet has a haze of methane and deep, dense ammonia clouds below the methane layer. The methane might be warmed by solar ultraviolet radiation. But all such theories are highly speculative as yet. Some temperatures have been established for the outer regions of the Uranian atmosphere, placing it at around —360° F (55° K).

The planet appears bluish green when seen through a telescope, with a suggestion of very faint whitish belts along its equator. Otherwise the atmosphere seems to be without markings. As to the interior of the planet, we can only speculate that it may be similar to those of Saturn and Jupiter but with a greater proportion of ices of

water and ammonia. Uranus spins rapidly on its axis in 10 hours 48 minutes, and it does this on its side. The axis of Uranus is very nearly parallel to the plane of the planet's orbit around the Sun. As a result, the north pole points sunward for half the planet's year and then outward into the darkness of space for another half year. So Uranus does not experience the day and night cycle of a normal day, but, rather, long, seasonal days and nights that last for forty-two Earth years each. There are forty-two years of night, then forty-two years of day, during which the sun appears to perform a ten-hour circle in the sky, the diameter of which depends upon how far you may be from the pole. For two short periods each year, however, there are normal, five-hour days and nights, when the axis of Uranus is aligned along the planet's orbit.

Uranus has five known satellites. They, too, follow unusual orbits, because they travel around Uranus in approximately the plane of the planet's equator. As seen from Earth, they sometimes orbit the planet in wide circles, at other times, twenty-one years later, they move up and down, above and below the planet as seen from Earth. And, of course, they exhibit motions between these two extremes when Uranus is at other parts of its orbit around the Sun.

(a)

(a) *These three pictures show Uranus and its satellites as photographed with the 82-inch telescope of McDonald Observatory; the changing positions can be seen as they orbit the planet.* (NASA and University of Arizona) (b) *The five satellites identified; Ariel, Umbriel, Titania, Oberon, and Miranda.*

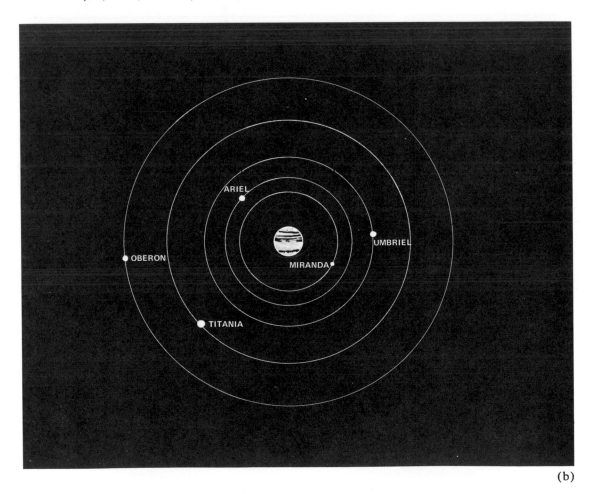

(b)

The moons Oberon and Titania were discovered in 1787 by William Herschel, six years after he had first seen Uranus. Two more were discovered, by William Lassell in 1851, and were named Ariel and Umbriel. The fifth moon, Miranda, was not discovered until comparatively recently, by Gerard Kuiper in 1948.

But the really big surprise about Uranus did

not come until 1977. At that time an unusual astronomical event occurred: the planet passed between Earth and a fairly bright star. The disc of Uranus occulted the star. Astronomers prepared batteries of highly sophisticated instruments to observe this occultation, for it would provide an accurate measurement of the size of Uranus and more information about the planet's atmosphere. They used highly sensitive light-measuring devices attached to big telescopes. These devices would show how the light from the star changed as it was occulted and again when it emerged from behind the planet.

James L. Elliot, of Cornell Observatory, traveled to Australia to observe the occultation. So did Robert Millis, of Lowell Observatory. But while Millis observed the occultation from the ground at Perth Observatory, Elliot made use of a new technique for precise observation, the use of a high-flying NASA aircraft equipped with telescopes. The aircraft carried the astronomer's instruments high above the distorting layers of Earth's lower atmosphere. Everything went well and the equipment worked perfectly. But both astronomers were surprised to find that there was a diminution of the star's light before it encountered Uranus. Even more intriguing, Elliot found that this absorption of starlight occurred again after the occultation as Uranus moved away from the star.

At first it was thought that Uranus had some other small satellites that had caused the light loss. Then, as observations from other astronomers poured in and powerful digital computers were set to work to analyze the light curves, it became apparent that Uranus must also have a ring system. And the system appeared to consist of several rings. There had been speculation that the outermost planets might have rings of debris circling them, but apart from the faint belts sometimes seen across the equator of Uranus, no evidence of such rings had been obtained from Earth. Theory suggested that the big giants might have rings of debris from their formation and that this debris would consist of ices. At Jupiter, the temperature was high enough for the ices to vaporize, but from Saturn outward they could persist. However, the rings of Uranus are extremely dark, and it is difficult to imagine that they are ice, as those of Saturn seem to be. Scientists hope that one of the Voyager spacecraft may be able to travel beyond Saturn to Uranus and throw some light on these mysterious appendages to the planet that spins on its side.

Although there have been many suggestions for NASA to send a mission to Uranus, none has become an actual project. The hope now is that the Voyager spacecraft—although not designed for an eight-year mission—will fill the bill. One of them could be sent to Uranus if the spacecraft fulfills its mission at Jupiter and Saturn. It would reach the ringed planet in 1985, and if all its equipment is still working, could perform there the same types of experiments and measurements made at Jupiter and Saturn. The performance of the spacecraft is not in itself enough. Money has to be allocated to keep the tracking networks following the spacecraft, to keep the teams of scientists available when the spacecraft reaches Uranus and to monitor its progress all through its long voyage. In the past, obtaining money to support long-term space projects has been extremely difficult—witness the difficulties in seeking support for adequate preservation of samples and pictures from the Apollo expeditions to the Moon and for the maintenance of scientific equipments left there by the astronauts, which, although they continued to function perfectly, had to be shut down.

THE MATHEMATICIANS' PLANET, NEPTUNE

Between 1800 and 1820, Alexis Bauvard busied himself preparing tables dealing with the motions of the then-known three outer planets: Jupiter, Saturn, and Uranus. In so doing, this French mathematician took into account the mutual perturbations of these bodies, based on the calculations of Laplace. But no matter how hard Bauvard tried, he could not make the calculated positions of Uranus agree with the observed positions.

For over a decade everyone was mystified and no one came up with a solution. Then, in 1834, the Reverend Thomas J. Hussey, followed three years later by Friedrich B. Nicolai, proposed that the differences between the calculated and observed positions of Uranus might be caused by another planet, even farther from the Sun. In 1840 a German astronomer, Friedrich W. Bessel, agreed that a still unknown planet might be tugging at Uranus and disturbing its orbit around the Sun.

The next year, John C. Adams, a student at Cambridge University, in England, calculated where such a planet would have to be to account for the observed differences. Although he sent his results to Sir George B. Airy, the Astronomer Royal, no action was taken.

Meanwhile, the French astronomer Urbain Jean Joseph Leverrier, unaware of Adams' work, started to investigate the problem. In 1846 he communicated his findings to the French Academy of Sciences, and fortunately he also mailed a copy of his data to Johann Gottfried Galle, of the Berlin Observatory. That observatory had excellent star charts, and Galle seized the opportunity to start a search, on September 23, 1846. Within the first hour he had found a star that did not appear on the charts. It was close to the position predicted by Leverrier. The missing planet had been found. It was named Neptune.

Today, while we accept as normal calculations of such great precision as to allow two spacecraft to rendezvous in Earth orbit or around the Moon, and to aim a spacecraft to pass through the gaps in the rings of Saturn, almost a billion miles from Earth, we should not forget that Leverrier had to work with very simple tools. He had no high-speed digital computers—only a paper and pencil, an inquiring intellect, and enormous patience. His prediction of Neptune's position in the sky will always rank among the triumphs of mathematics and logical reasoning. Ironically, Adams, too, had predicted the correct co-ordinates for locating Neptune, but the officious English astronomers had ignored the student from Cambridge.

Neptune, moving along its orbit at a leisurely 3.4 miles per second (5.47 kilometers per second), requires 165 years to make one complete revolution of the Sun. Its average distance is 2.8 billion miles (4.5 billion kilometers). So, although, like Uranus, it reflects much of the sunlight falling upon it, the planet cannot be seen without a telescope. In a telescope it appears as a very small, bluish green disc marred by a few very faint, irregularly shaped spots.

Neptune rotates in 15 hours 48 minutes on an

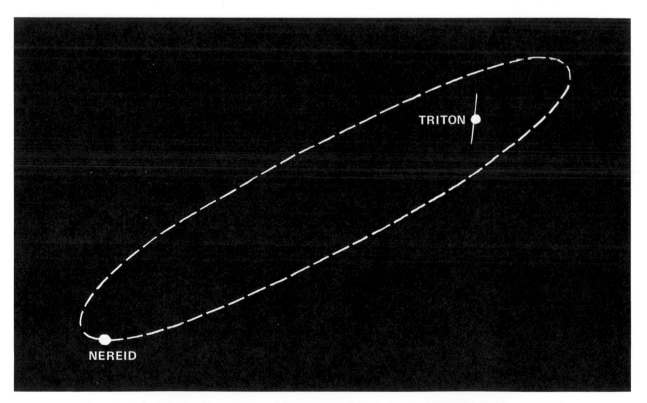

Drawing identifying the satellites of Neptune and their orbits. Note the highly elliptical orbit of Nereid, compared with the close-in circular orbit of the massive Triton, seen edge on.

axis more normally inclined than that of Uranus, a tilt of about 29° to the plane of its orbit.

Physical characteristics of the planet are difficult to define at such an enormous distance, but its diameter is believed to be about 30,760 miles (49,490 kilometers) and its mass about seventeen times that of Earth. The density of Neptune is about 1.67 times that of water. The diameter was measured by a stellar occultation in 1968.

As would be expected from its greater distance from the Sun, Neptune is colder than Uranus, but not much. Weak microwave radiation from Neptune suggests a temperature of about —99° F (200° K) within the lower atmosphere and hundreds of degrees colder in the upper atmosphere.

The atmosphere of Neptune, similar to that of Uranus, is believed to be clear of clouds. It probably contains more methane and less ammonia than the atmosphere of Uranus. But the composition of the atmosphere and the internal structure of this distant world can only be speculated upon. Probably there are large quantities of hydrogen and helium forming the bulk of the planet, together with ices of water and ammonia.

As the 1970 decade drew to a close, astronomers had found only two moons of Neptune. There may be others. William Lassell discovered the larger of the two, Triton, which circles the planet in a close retrograde orbit with a period of 5 days 21 hours. Brighter than any of the satellites of Uranus, Triton may have a diameter of 3,700 miles (6,000 kilometers), which would make it larger than Mercury and Pluto. The moon's high reflectivity has led scientists to speculate that it possesses an atmosphere of methane.

Neptune's other satellite, Nereid, was discovered by Gerard Kuiper in 1949. It is a very small body, of perhaps 300 miles (500 kilometers) diameter, that travels once every 360 days along a very elongated orbit, its distance from Neptune varying between 0.87 and 6.0 million miles (1.4 and 9.7 million kilometers). Unlike Triton, Nereid follows a retrograde orbit.

Just when spacecraft will be sent to Neptune is uncertain, but we can be sure that it will not be until after completion of the current missions to Jupiter and Saturn with a possible extension to include Uranus. There has been some speculation that one of the Voyager spacecraft may be sent on to Neptune, but there are no definite plans for such a mission. In 1986, however, Uranus and Neptune will be so aligned in space that a spacecraft could be sent to visit both worlds in a single mission. But funding does not look promising. Initial reconnaissance of Neptune may not begin until the 1990s. When it does, mankind's reach will stretch to what was for a long time considered the outer limit of the Solar System. But we have now known for over half a century that another world beckons our questioning minds.

12

PLUTO, THE LONELY SENTINEL

Like Neptune, that world is Pluto. It was discovered not by accident but as the result of a determined scientific effort. Shortly after Neptune had been discovered, astronomers noticed that its orbit was perturbed. Moreover, they found that the presence of Neptune did not account in full for the deviations of Uranus from its calculated position. They suspected that another, still undiscovered planet was responsible for these effects.

In 1880, D. P. Todd began looking for this planet X, whose diameter he felt would be nearly 50,000 miles (80,000 kilometers). His efforts were in vain. Early in the twentieth century Percival Lowell tackled the problem. Working with other astronomers at the Lowell Observatory, in Arizona, he commenced a detailed survey that lasted until his death, in 1916. Work temporarily halted, then was resumed, in April 1929, under the guidance of Clyde W. Tombaugh. With improved telescopic equipment at his command he found the planet on February 18, 1930. It was named Pluto, and as a tribute to Percival Lowell the symbol ♇ was given to the new planet. Ironically, the calculations that led to the predictions of Pluto's existence now appear to be incorrect. The planet's small mass is insufficient to have had as much effect on the orbital paths of Neptune and Uranus as Lowell and others believed.

As a consequence, the search continued for other planets of the outer Solar System. Lowell Observatory astronomers carefully inspected the regions where planets might be expected. During the period from 1930 through 1945 about 90 million stars and 30,000 other objects were examined to make sure that they were not planets. The search was spearheaded by Tombaugh himself, who used a blink microscope technique. He photographed the same area of sky about two or three nights apart and then carefully compared the negatives for a trace of anything moving between the two pictures.

Tombaugh explained:

The switching of views from one plate to the other of the comparison pair was accomplished by a pair of crescent-shaped shutters just in front of the half-silvered surfaces of a special prism which diverted the images in the field from either plate to a common eyepiece. As the one pair of shutter blades closed, the other pair opened. The brief pause in the reciprocation motion permitted a brief exclusive view of one plate only. In this operation the light intensity in the field remains constant, and the viewer is spared the annoying effect of flickering. From experience, I found that about three alternate views per second permitted the greatest efficiency in speed and thoroughness. If alternate views were accomplished by turning lights alternately on and off, the afterglow would have slowed the work of examination by a factor of four or five. The search I made would then have required about 50 years, instead of 14.

A total of 338 pairs of 14×17-inch (35.56×43.2-centimeter) plates were examined. Tombaugh found four thousand asteroids on them but nothing that could be identified as a distant planet. He stated that the search was done with

(a)

(b)

(a) *Small sections of the plates used in the discovery of Pluto are reproduced here. The images of the planet are marked by arrows. Upon examination of these plates, Clyde W. Tombaugh made the discovery in a part of the sky where Percival Lowell had predicted that the planet might be found. The plates were taken on January 29 (left) and January 23 (right), 1930, and the planet was discovered on February 18, 1930.* (Lowell Observatory)

such thoroughness that it was most unlikely that any planetary object with a visual brightness exceeding magnitude 16 had been missed.

But the search has continued, though sporadically. Joseph L. Brady looked at another possible way of locating a distant planet by its perturbations on the orbit of a comet. He picked Halley's comet and checked records of it going back seventeen hundred years. But Tombaugh cautions that comets are not reliable, since they emit explosive jets of gas that act as rocket engines to push them from their paths. Between 1968 and 1972 Brady nevertheless worked with computer models of a hypothetical ten-planet Solar System and showed that a planet three times as massive as Saturn and orbiting the Sun at a distance of 5.9 billion miles (9.5 billion kilometers) would account for the changes in the comet's orbit. Brady estimated that such a planet would circle the Sun in a period of 464 years on an orbit inclined at an angle of 120° to the plane of Earth's orbit. Such a planet would be brighter than Pluto. But it has never been seen.

Whether or not other planets exist is still far from settled. The discovery of Chiron shows that there are surprises waiting for us in the outer Solar System.

As far as we can now ascertain, Pluto is the lonely sentinel at the outer limits of our Solar System. It is a rather small and comparatively dense body without an atmosphere. It is thus quite different from the gaseous giants and more closely resembles their satellites. Pluto follows a peculiar orbit, which crosses the orbit of Neptune but at a relatively high inclination. There is a strange combination of the motions of the two planets around the Sun whereby, seen from Neptune, Pluto appears to make great loops around that planet as though gravitationally attached to the bigger, gaseous planet. The orbit of Pluto is more inclined and more eccentric than that of any other planet: 17.2° and 0.205, respectively.

Pluto takes 248 Earth years to travel completely around the Sun. Its eccentric orbit swings it to 2.7 billion miles (4.35 billion kilometers) from the Sun at perihelion and carries it out to 4.6 billion miles (7.4 billion kilometers) at aphelion. Thus, Pluto travels some 1.8 billion miles (2.9 billion kilometers) beyond the orbit of Neptune.

On December 11, 1978, Pluto crossed the orbit of Neptune, thereby losing, for twenty-one years,

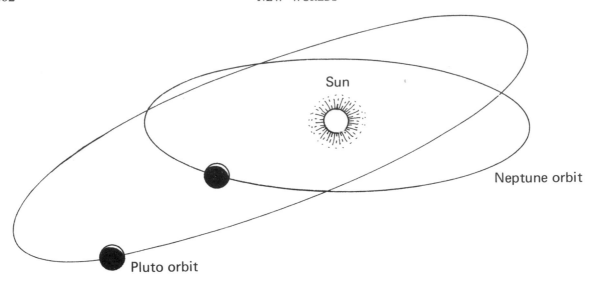

The orbit of Pluto is compared with that of Neptune to show the high inclination of Pluto's orbit.

the position of most distant member of the Solar System. Its perihelion passage occurs on September 12, 1989, and it reaches aphelion again in 2113.

Astronomers are not certain of the length of Pluto's day. Observations of the way in which the planet varies in brightness in a regular fashion suggest that we might be seeing surface features brought into view by the planet's rotation. Such observations can be interpreted as a rotation period of 153 hours, nearly six and a half Earth days.

There have been other changes in the observed brightness of Pluto, which might be explained if the spin axis of Pluto lies close to the orbital plane of the planet, similar to conditions for Uranus. Then, it is reasoned, if Pluto's pole were facing Earth at the time of the planet's discovery, the intervening years would have gradually presented a different aspect of the planet to Earth so that we would have seen new areas of its surface. This, in turn, might have caused the observed changes in brightness. Astronomers Leif Anderson and John D. Fix, of Indiana and Iowa universities respectively, who suggested this explanation, also suggest that polar regions of Pluto reflect more light than the equatorial regions, that is, they have a higher albedo.

Because Pluto is so far away from Earth, astronomers have great difficulty in determining its true size. But during the last decade or so their estimates have ranged from 3,000 to 4,225 miles

(5,000 to 6,800 kilometers). An observation of the occulation of a star by Pluto in 1966 set the upper limit. Estimates of the mass of Pluto range from 0.11 to 0.17 per cent of that of Earth, and density (if former value is correct) about 88 per cent that of the Earth, namely, about 4.86 times that of water. Such a dense planet seems odd after the four gaseous giants of the outer Solar System. Again, this type of density is more akin to that of the satellites of the outer planets.

In mid-March 1976, three astronomers, using the 150-inch (4-meter) -aperture telescope at the Kitt Peak National Observatory, in Arizona, discovered a strange characteristic in the light coming from Pluto. Dale Cruikshank, David Morrison, and Carl Pilcher, of the University of Hawaii, observed Pluto through two very-narrow-band filters, one that could admit the wavelengths of light reflected from water ice and the other from methane ice. The astronomers found that the light of Pluto came through the filters as if it were reflected from methane ice.

This is the first time that solid methane has been found in observations of planetary objects of the Solar System. Since it reflects sunlight strongly, methane on the planet's surface may have caused astronomers to overestimate the size of Pluto. It may, indeed, be a smaller but brighter object. Based on these new results, some astronomers have revised their estimates of the diameter of Pluto down to a figure of 2,500 miles (4,000

The huge radio telescope in the hills inland from Arecibo, Puerto Rico, is about one thousand feet (three hundred meters) in diameter. As mentioned in earlier chapters, it was used to help explore the Solar System by bouncing radar echoes from the planets. It has also been used several times to beam experimental messages from mankind to the stars. This telescope could indeed communicate with an identical telescope to the limits of our Milky Way Galaxy if we are prepared to wait for many thousands of years for an answer. (National Radio Astronomy and Ionospheric Center, Arecibo Observatory)

kilometers) and perhaps less. This would make Pluto only a little larger than our Moon and further strengthen the idea that it may be an escaped satellite of Neptune.

Pluto's slow rotational period also adds to this concept for the origin of the outermost planet, since all the independent planets except those locked to slower motions by their proximity to the Sun rotate much faster. One speculation is that Pluto was originally a satellite of Neptune together with Nereid and perhaps others. There was an encounter with another body, now Triton,

Control console of the great radio telescope at the National Radio Astronomy and Ionospheric Center, Arecibo Observatory, Puerto Rico.

which flung Nereid into its present, elliptical orbit, hurtled Pluto into its peculiar orbit which still keeps it attached to Neptune, and pulled Triton into its retrograde orbit. But Pluto's origin still remains unexplained, because we do not have sufficient supporting facts.

Because Pluto is so far away from the Sun, whatever atmosphere the planet might otherwise possess must lie frozen on its surface. The discovery of methane ice seems to confirm that this is so. It is also evident that the temperature of Pluto's surface cannot rise much above —369° F (50° K), the freezing point of methane.

At times, though, conditions may permit argon, neon, and methane vapors to form a slight haze, for a considerable variation in surface temperature should occur between aphelion and perihelion.

Seen through our most powerful telescopes, Pluto is slightly yellowish and its surface seems to be rough on a small scale because of the way that light is polarized by it. Much conjecture surrounds the nature of Pluto's internal structure. If, as most astronomers believe, the planet originated at the edge of the Solar System, it should contain principally water, ammonia, and methane, for these substances would be the last to condense from the cool, outer regions of the solar nebula. It is possible that Pluto possesses a core of water ice overlain by methane and ammonia layers topped by a frozen methane crust.

Until very recently, no moon had ever been observed to orbit the outermost planet. If Pluto were, indeed, a one-time satellite itself, none would be expected. But if it is a planet in its own right, then it clearly differed in still another way from the gaseous giants Jupiter, Saturn, Uranus, and Neptune in not possessing a family of its own. However, James Christy of the U. S. Naval Observatory observed a slight bulge in the image of Pluto on several photographic plates taken at Flagstaff, Arizona, in April and May 1978. Checking earlier plates taken from 1965 to 1970, he found the same oddity. Tentatively, Christy and coworker Robert Harrington feel there is, indeed, a satellite of Pluto some 15,000 to 20,000 kilometers out.

When probes are dispatched to Pluto, late in this century or early in the next, man will have examined—at least remotely—all the major worlds of the Solar System. But his thrust into the universe will have only just begun. For beyond the sentinel of the Sun's empire beckon the stars.

Already, we are more than casually speculating about life among the stars, of strange yet sentient beings that may inhabit planets orbiting distant suns. Films like *2001: A Space Odyssey, Star Wars,* and *Close Encounters of the Third Kind* are box-office hits, and the flow of fictional and nonfictional books on life in the universe is endless. More concretely, attempts have been made to capture signals from hypothetical extrasolar (beyond the Solar System) communities, and gigantic arrays of antennas have been proposed to continue the quest for our equals or our masters in the universe.

And the quest will continue. "For man there is no rest and no ending. He must go on—conquest after conquest. This little planet and its winds and ways, and all the laws of mind and matter that restrain him. Then the planets about him, and at last across the immensity to the stars. And when he has conquered all the deeps of space and all the mysteries of time—still he will be but beginning."

H. G. Wells, from the film
Things to Come

There are plans to set up huge radio listening posts on Earth to see if we can detect the whispers of civilizations on planets circling other stars. Such a program to search for evidence of extra-solar intelligence has started already in the Soviet Union and is in the formative stages in the United States. Unfortunately, funding is still minuscule and enthusiasm for joining a cosmic community does not seem great among those who hold the nation's purse strings.

Wernher von Braun, father of America's space program, was director of the George C. Marshall Space Flight Center, in Huntsville, Alabama. He was most recently vice-president of Fairchild Industries, in Germantown, Maryland, and president of the National Space Institute, a nonprofit educational organization in Washington, D.C. Along with Frederick Ordway, he wrote the widely acclaimed *History of Rocketry and Space Travel* and *The Rockets' Red Glare,* an illustrated history of rocketry published in 1976 by Anchor Press. Dr. von Braun died in 1977.

Frederick I. Ordway III started his career in rocketry at America's pioneering rocket-engine firm Reaction Motors, Inc., and later joined the Von Braun team at the Army Ballistic Missile Agency and NASA's Marshall Center. More recently, he became a professor at the University of Alabama in Huntsville, School of Graduate Programs and Research. He is now with the U. S. Department of Energy.

INDEX